Fisheries Management

Fisheries Management
A Manual for Still-Water Coarse Fisheries

Ash Girdler CBiol, FIBiol, MIFM, FIFM, CEnv
Ian Wellby FIFM, CEnv
Robin Welcomme PhD, HonFSE, FIFM, CEnv

WILEY-BLACKWELL
A John Wiley & Sons, Ltd., Publication

This edition first published 2010

Blackwell Publishing was acquired by John Wiley & Sons in February 2007. Blackwell's publishing programme has been merged with Wiley's global Scientific, Technical, and Medical business to form Wiley-Blackwell.

Registered office
John Wiley & Sons Ltd, The Atrium, Southern Gate, Chichester, West Sussex, PO19 8SQ, United Kingdom

Editorial offices
9600 Garsington Road, Oxford, OX4 2DQ, United Kingdom
2121 State Avenue, Ames, Iowa 50014-8300, USA

For details of our global editorial offices, for customer services and for information about how to apply for permission to reuse the copyright material in this book please see our website at www.wiley.com/wiley-blackwell.

Library of Congress Cataloging-in-Publication Data
Girdler, Ashley.
Fisheries management : a manual for still-water coarse fisheries / Ashley Girdler, Ian Wellby, Robin Welcomme. – 1st ed.
p. cm.
Includes index.
ISBN 978-1-4051-3332-6 (hardback : alk. paper) 1. Fishery management. I. Wellby, Ian. II. Welcomme, R. L. III. Title.
SH328.G69 2010
639.3'12–dc22
2009038733

A catalogue record for this book is available from the British Library.

Set in 10 on 12.5 pt Sabon by Toppan Best-set Premedia Limited
Printed and bound in Singapore by Fabulous Printers Pte Ltd

1 2010

Contents

Acknowledgements

We express our gratitude to Dr Chris Williams for his work on the fish health chapter and other help he has offered. We also thank Nigel Hewlett for his advice during the preparation of this book. Many thanks too to Jenny Wellby for the help and support she has offered and her contribution in editing the various drafts. We also thank the staff of the A.G.A. Group, who have contributed to this book by taking the time to extend their efforts during the working day to provide the expansion of our image library, which has ensured that the text is populated with up-to-date graphics and images.

Part I
Ecology of Still-Water Fisheries

Before looking at the management of a still-water fishery, it is important to understand how lake and pond ecosystems function in the absence of human interference. Still-waters are fascinating and complex aquatic environments that are very different from the terrestrial environment in which we live. It is important to acknowledge and understand these differences so that we fully understand the possible consequences of any management decisions.

Although we describe our still-waters as freshwater lakes, suggesting that the water is in some way pure water, in reality it is never pure. Even rain is not pure water as it contains significant amounts of dissolved gases and other chemicals. Water has the ability to dissolve a vast number of natural and synthetic elements and compounds. It is often referred to as the 'universal solvent'. As water falls from the sky and runs through or across the land as surface run-off, it picks up a myriad of chemicals as dissolved substances. It may also collect further materials, including chemicals that are insoluble but are suspended in the water. This veritable soup supports a whole range of algae and invertebrates, which support the fish that are of greatest interest to the manager.

Animals that live in water are very different to animals that live on land. Although this is obvious, the difference is rarely fully appreciated. To understand how to manage a fishery we must understand the sorts of conditions the fish live in: their environmental conditions.

Part I of this book looks at still-waters from the basic principles of how they work, considering some of the conditions prevalent in still-waters and how these fluctuate over time and space, and describes the various animals that live in them.

1 Introduction

Ecologically, still-waters are described as standing 'open' water, a term that covers wetlands including fens, marshes and temporary flashes as well as inland water bodies such as lakes, ponds, meres and reservoirs. It is estimated that there are 3344 km^2 of still-waters in the UK, of which 95% are less than 1 hectare (1 hectare (ha) is equal to 10 000 m^2) in area, which means that there are at least 400 000 lakes, ponds and dams in the UK. It is also estimated that the number of still-waters has decreased by 35–75% over the past 100 years, although numbers are considered to have increased in recent years.

The management of still-waters as ecologically important systems is therefore of high priority for the UK's biodiversity and, as fisheries are often the primary use of such waters, there will be a requirement for fishery management skills that deal with the suitability of the environment to support fish. It is thought that as many as 90% of the nation's still-waters may be in private ownership and thus removed from the direct influence of conservation bodies, regulators and freshwater scientists, further emphasising the role of fishery managers. In addition, UK managers often manage fisheries on continental Europe, mainly for UK fishermen.

There has been a growing trend for the recreational fisheries in England to be based in such still-waters, in particular those for coarse fish. This has created a need for the systematic management of many ponds, dams, gravel pits and lakes to maintain good populations of fish and good-quality fisheries. At the same time, the management of such fisheries is becoming increasingly complex. It is no longer sufficient to keep a few fish in a pond, as increasing regulation and demands for high-quality fishing experiences by the public force the manager to stricter policies. Underlying this is the fact that still-water pond management is a business in many instances and must satisfy customers to make a profit by providing good-quality fishing; the manager can only achieve this by providing a suitable environment and good-quality fish.

Fisheries management, in common with most other activities that involve a combination of problem solving and improvement that result in physical solutions, requires that well-developed planning is a necessary part of the process. Fishery managers are no different from us all in as much as we all plan. We all plan to lesser or greater degrees: we plan what we are going to eat, for example, and where and when we are going to buy the food we have decided to eat. However, the way we approach fisheries management planning often needs to be more prescriptive. Each fishery will ideally have a written management plan that will have been developed for it; certainly where the fishery management involves more than a solitary individual a plan should be developed and put in place as a priority. It is important to ensure that the aims and objectives are clear and understood, that the actions required to achieve them are well considered and planned, and that the tools to measure or monitor the management actions are in place at the onset of the plan and not developed as an afterthought.

Of course, the degree of complexity of the plan can vary greatly across a broad spectrum. A fishery can vary from a small one acre pond used exclusively for angling as the only single interest to large water bodies such as the Norfolk Broads that will have many users. Most of these users will have as an overriding priority their own single interest. They will often include botanists, conservationists, ornithologists, sailors and walkers, among many more.

This book is, therefore, intended to provide the information that the vast number of enthusiastic but amateur fishery managers who are responsible for maintaining and enhancing our still-water heritage need to increase their effectiveness. It is more a compendium than a novel and is thus not intended for reading from cover to cover but to address, in a series of chapters, the various aspects of ecology and management of still-water fisheries. It is written in three parts.

The first part describes the basic ecology of still-waters and catalogues the various species that are significant to the fishery manager. Here it should be noted that most personnel associated with still-water fisheries should have a strong naturalist bent and be aware of the various living things that inhabit their premises. This part refers mainly to the UK although the fauna and flora are very similar to those of the rest of northern Europe.

The second part deals successively with the details of actual management, providing information on construction, assessment, and monitoring and control methodologies. The methods and procedures set out here are drawn from the north temperate region, but similar approaches would be appropriate to recreational fisheries worldwide.

The third part touches on the complex area of legislation and the social context of still-water fisheries. We have been restricted to considering the legal context of England and Wales because that of Scotland is in a state of flux at the time of writing. The institutional section is of more general application, however.

Examples have been provided of some of the procedures and methods described. These are to be found in a series of appendices.

It has been decided not to include references within the text of this book, in order not to break up the flow of words for the user. Instead, a further reading list has been included for those who wish to explore particular issues in greater depth.

2 Ecology of Lakes

2.1 Water

Water is the basic medium for all fisheries. It falls on the earth as precipitation (rain- and snowfall) and moves to the sea through streams and rivers. It can be temporarily stored in ponds, lakes and reservoirs that form the environment for still-water fisheries (Fig. 2.1).

2.1.1 The structure of water

The fact that water will dissolve a vast range of substances is one of the many properties that make it such a vital compound for all life on earth. Its chemical structure is the important factor that allows water to act in the way it does.

A water molecule is made up of two hydrogen atoms and one oxygen atom. In the molecule these ball-like atoms come together in a V-shaped structure (Fig. 2.2a). This structure means that the shared electrons of the atoms are unevenly distributed in its outer layer, which creates slight positive and negative charges on the ends of the molecule.

Because all the molecules in water are charged particles, each hydrogen atom can form a weak attachment with the oxygen atom of another water molecule. At the same time, the oxygen can form a weak attachment with two hydrogen atoms of another water molecule (Fig. 2.2b). These weak bonds give a molecule of water the potential to link simultaneously

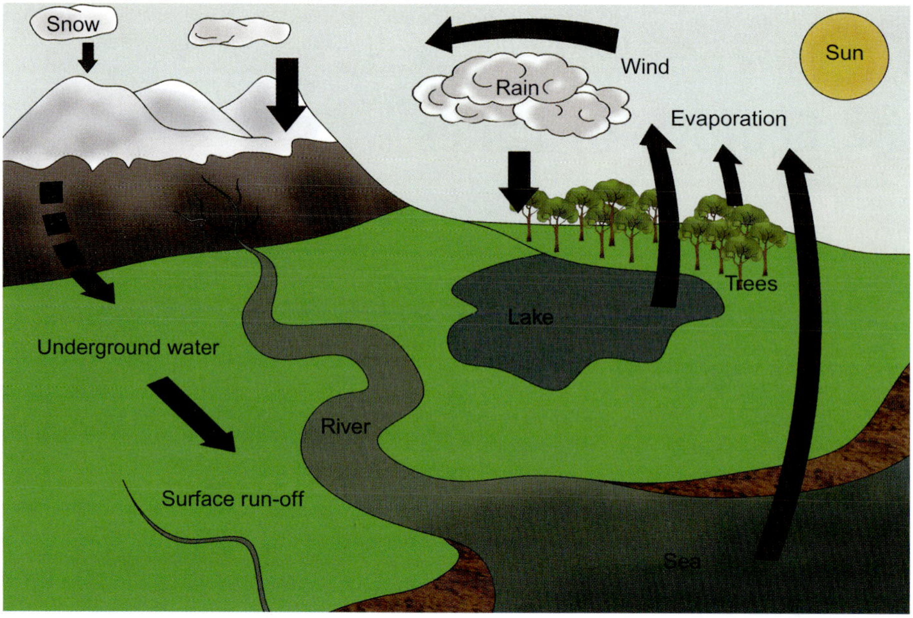

Figure 2.1 The water cycle.

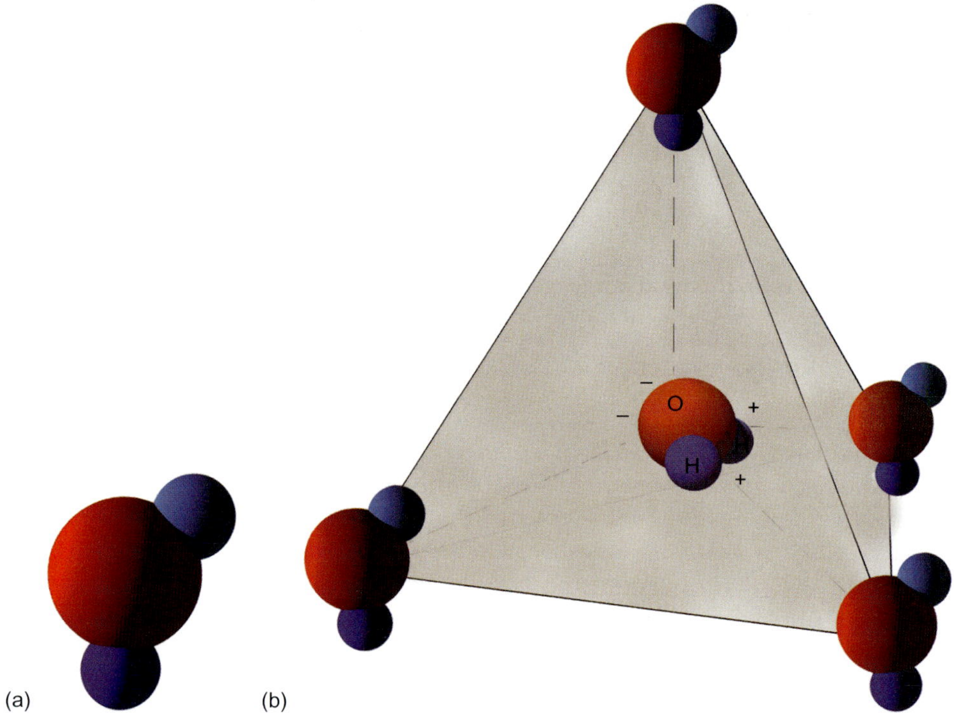

Figure 2.2 (a) The structure of a water molecule. (b) The hydrogen bond.

with hydrogen bonds to four other water molecules. It is the polarity of water and its ability to form hydrogen bonds that have made it essential as the 'medium of life'. We need to understand these properties to appreciate fully the environmental conditions that fish live in and to enable the fishery manager to improve these when required.

2.1.2 Aspects of the chemical nature of water

We encounter water every day of our lives and to a large extent take it for granted. However, as well as its importance as a universal solvent, many chemical and biochemical processes can only take place in water. It differs from all other liquids in this respect.

Conductivity and total dissolved solids

Pure water is impossible to find in the natural world. Indeed, what is pure water? Pure water is water that has no other substance dissolved in it. For commercial and scientific reasons, water can be made pure to varying degrees by an industrial process called distilling, giving us distilled water. Water does not naturally exist in this form as there is always something dissolved in it. For example, seawater has about 37 grams (g) of various salts dissolved in every litre, and freshwater has around 0.5 g of dissolved salts.

The concentration of salts in water is measured by its conductivity, which is the ease with which any sample of water conducts an electric current. It is measured in siemens per metre (S/m) in the UK and Europe, and microohms per centimetre (mmho/cm) in the USA and sometimes in the UK literature. Typical values range from 5.5×10^{-6} S/m for ultra-pure water, to 0.005–0.05 S/m for normal surface freshwaters, and 5 S/m for seawater. The other measure, total dissolved solids (TDS), is a measure of the total ions in solution and is reasonably related to conductivity by the relationship:

$$\text{TDS (in milligrams per litre: mg/l)} = 0.5 \times \text{conductivity (dS/m or mmho/cm) or}$$
$$= 0.5 \times 1000 \times \text{conductivity } (\mu\text{S/cm}).$$

Hard and soft water

In the absence of human inputs, the chemical content of substances dissolved in water depends on the nature of the rock in which the still-water is located and the geology of its drainage basin. The main dissolved substances are calcium and magnesium carbonates; their concentration determines the hardness of the water. In general, still-waters situated in areas of hard rock, such as granite, have very soft water, whereas those situated in chalk or limestone areas have hard to very hard water. Hardness is evaluated against a simple scale (Table 2.1) and several instruments and measuring kits are available to allow it to be easily measured.

The chemical content of the water is strongly linked to its potential to support life; soft water ponds tend to be less rich in organisms, harder water tends to be richer and support larger biomasses and a greater diversity of species. The hardness of the water directly affects its 'buffering capacity', which is its ability to resist the effects of acids or alkaline substances that would change its pH: the harder the water the greater its resistance to pH fluctuations. In some soft-water ponds and lakes, exposed to rainfall contaminated with sulphur-rich industrial discharges, the water has become so acidic that no living things can survive. In

Table 2.1 Chart for evaluating water hardness

Calcium + magnesium (mg/l)	Classification
Less than 17.1	Soft
17.1–60	Slightly hard
60–120	Moderately hard
120–180	Hard
Over 180	Very hard

such cases the water has to be treated with a calcium-based substance such as lime or gypsum to neutralise the acid to return it to an acceptable pH level.

2.1.3 Aspects of the physical nature of water

How water responds to temperature changes

How water behaves as it warms and cools is important. First, it takes a lot of energy to warm water, which then is relatively efficient at retaining the heat compared with the atmosphere. This means that in nature water is a very stable environment. For example, in temperate regions the air temperature may be 10°C at 2 a.m. and 27°C at 2 p.m. So any land animals have to tolerate a 17°C change in temperature in a 24-hour period. A reasonable-sized still-water in the same area may be 18°C at 2 a.m. and 19°C at 2 p.m. In deep areas it will be cooler and even more stable. This stability is of great benefit to cold-blooded (poikilothermic) animals such as fish because their body temperature is linked to that of the water. This fact can be important and should be remembered when transferring fish from one water body to another, as fish find it difficult to tolerate sudden changes in water temperature; this is known as thermal shock.

As most liquids cool, the molecules within them become less active and the distance between them is reduced. As a result, the liquid becomes ever denser as the temperature decreases and eventually freezes to a solid. We know that a solid should be denser than a liquid and therefore the frozen solid sinks through the liquid or forms on the bottom. This works for all liquids except for water. As you cool water down towards freezing, it behaves in exactly the way you would expect, with the cooler, more dense water sinking to the bottom. This occurs until the water is at approximately 4°C (actually 3.94°C), at which point further cooling makes the water less dense. This means that water cooled from this point on actually starts to rise up. When it freezes at 0°C it is still less dense than the slightly warmer water below so the ice forms at the surface; the solid water floats on the liquid as ice. It is the only molecule to behave in this way.

The reason water behaves like this is all to do with the electrical charges in the water molecule. As the molecules cool and become less active they start to form into a lattice stabilised by the attraction of the positive and negative charges in them. This lattice framework is very stable and creates spaces between the molecules. Therefore it is less dense than the slightly less stable water at 4.1°C.

On any still-water, as the temperature falls to freezing point, frozen water forms as ice on the surface and denser water at 4°C sinks to the bottom. If the cold weather continues then the ice at the surface may get thicker, as more water freezes, but it also forms an insulating layer. This insulates the rest of the water and therefore makes sure that at the bottom of the water body, deeper than around 1.25 m (4 feet), there is always a layer of water at 4°C or higher. This is where most of the fish will be found during this time and they thereby obtain a varying degree of relief from freezing water.

The behaviour of water as it changes in temperature has further profound effects on life in still-waters.

Stratification of water

This may begin during spring, as the water temperature is slowly increased by the energy from the sun and the surface layers get warmed. The warmth is not lost appreciably over night and the temperature remains stable. The lower layers, cut off from the energy of the sun, warm more slowly, relying on mixing from the surface layer above. In deeper still-water lakes (commonly considered to be over 10 m) these two layers become separate as the temperature difference increases and the mixing of the two layers declines. This is known as thermal stratification (Fig. 2.3).

The depth at which true thermal stratification actually occurs will vary considerably depending on several factors. The most important is the clarity of the water. The clearer the water the deeper the sun's energy can penetrate. The area that separates the warmer upper layer (epilimnion) from the cooler lower layer (hypolimnion) is called the thermocline.

In temperate regions, thermal stratification varies through the seasons. Once a thermocline has formed in the summer it is generally fairly stable. During the summer the hypolimnion becomes isolated and cold, with limited dissolved oxygen. The two layers are separated by a significant difference in temperature. However, as autumn approaches the warm upper layer declines in temperature. As the two layers get closer in temperature then some mixing will take place and eventually the thermocline will break down as the whole water body settles at around the same temperature, usually about 10°C. It is possible for the thermocline to break down very quickly during a period of stormy weather when there is a rapid fall in ambient temperature and a strong wind. This will result in the sudden mixing of the warmer, oxygenated epilimnion with the colder, poorly oxygenated hypolimnion. This can result in

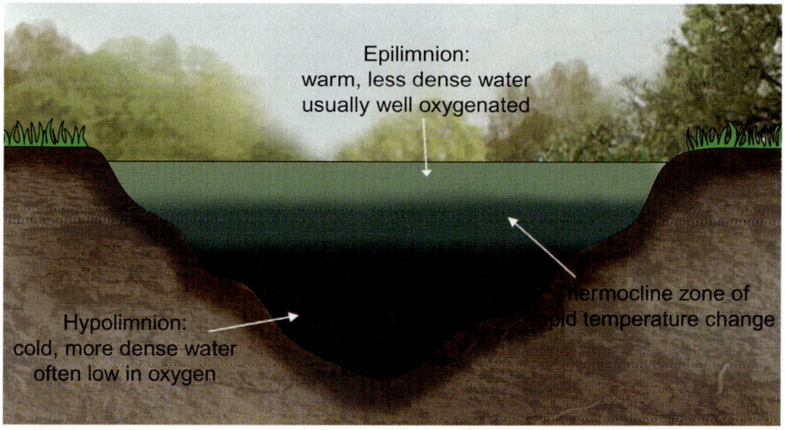

Figure 2.3 Stratification of a water body.

a sudden drop in water quality and resultant stress and even occasional mortalities in fish. This is known as the autumn turnover.

It should be noted that a complete and lasting thermocline only develops in relatively deep water bodies.

The ecological importance of temperature

Water temperature is a very important factor as it regulates many of the chemical and biological processes that occur in water. Temperature affects the complex interactions of all the factors that form the basis of how we define water quality. The degree to which this occurs varies both daily and throughout the year.

Being able to control water temperature, especially in larger water bodies, is at best impractical but it is important to understand the changes that temperature exerts as an environmental parameter. An example of this can be seen from studies of optimum temperature ranges for the transport of live fish (see Chapter 11). These suggest that the generally applicable temperatures, needed to provide a conducive environment, are 6–8°C for cold-water fishes and 10–12°C for warm-water fishes in summer, 3–5°C for cold-water fishes and 5–6°C for warm-water fishes in spring and autumn, and 1–2°C for all in winter.

2.2 Trophic status of still-waters

The nature of the bedrock, together with the soils and vegetation within the drainage basin, regulates the natural inorganic and organic nutrient content of lake or pond water. These factors combined are used to classify still-waters into the following types.

- Oligotrophic: usually soft water lakes with few nutrients, capable of supporting only low biomass and a small number of species. Representative fish are lake whitefish (coregonids), charr and trout.
- Mesotrophic: moderately soft to slightly hard water with a medium amount of nutrients, capable of supporting a good and diverse population of organisms. Representative fish species include trout, charr, perch and rudd.
- Eutrophic: soft to hard water with a high degree of nutrients that supports a specialised fauna capable of resisting the lowered dissolved oxygen conditions. Representative fish include roach, bream and pike.
- Hypertrophic: soft to hard water highly charged with nutrients to the point that only highly specialised and resistant communities can survive. Representative species include the common carp.
- Dystrophic: a specialised category of waters, usually associated with moorlands, swamps or certain types of wet forest, usually with soft, acid water and heavily stained to a tea colour with tannins and other vegetative substances, with a very low capacity to support life.

Inputs of nutrients from, for example, human activities can move the trophic status of still-waters from a lower trophic category (one with fewer nutrients) to a higher one, a process known as eutrophication.

2.3 The physical nature of still-waters

The nutrient status of still-waters described earlier in this chapter has the most profound effect on the ecology of the system. The physical characteristics play an equally important role, although sometimes this is difficult to quantify.

The main zones of a typical still-water are shown in Figure 2.4.

2.3.1 Origin of lakes

The size and depth of a still-water is often suggested by describing it either as a pond or a lake. It may or may not be useful to consider the difference between a pond and a lake from a scientific or practical perspective. Here are some definitions for you to consider.

- A pond is any man-made body of water where light is found in the entire body of water.
- A lake is any body of water that has a profundal zone (a layer below the limit of effective light penetration for organisms).
- A lake forms a thermocline, a pond does not.
- A lake has at least one windswept shore.

A third category, reservoirs, groups man-made water bodies of various sizes that are usually impounded behind a dam.

Lakes are often classified as to whether they are 'on line', in other words connected directly into the watershed by an inflowing and outflowing watercourse, or 'off line' in which they are isolated from the main surface flow patterns and are connected mainly through

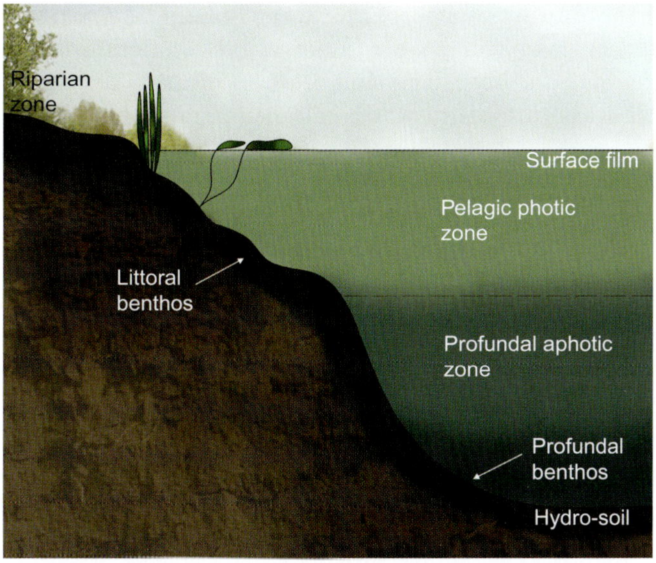

Figure 2.4 The main zones of a still-water.

groundwater inflow and seepage. Clearly, lakes that are 'on line' can be affected by, and have more serious effects on, the water basin environment.

As you can see, there is no one description that is completely acceptable, so in this book, ponds, lakes and reservoirs are grouped as still-waters.

The physical characteristics of a still-water depend much on its origins. Many lakes and ponds in the UK are natural, usually arising from glacial activity or are ancient river features such as oxbows or fluvial scour. Glacial lakes are commonly long and narrow and are often very deep with little riparian development. Old river features are usually shallower with a somewhat better shoreline development.

Most lakes in England and Wales, however, are manufactured, being created either by extraction of minerals or by damming.

Extraction lakes include peat workings, which are usually shallow with extensive shoreline development, such as the Norfolk Broads, and mines such as gravel pits and quarries that are usually deep and steep sided with a limited riparian zone. Recently, however, there has been a tendency for gravel pits and quarries to be landscaped for wildlife and fish, with shallows, islands and bays.

Dams range from ancient millponds and village duck ponds to reservoirs created for power generation, water supply or flood alleviation. Reservoirs differ from other lakes in that their deepest point is adjacent to the dam wall and the upper part of the reservoir is usually much shallower. Reservoirs often have side arms and convoluted shorelines. They also suffer from drawdown; when the water level is lowered as it is used for power generation or drinking.

2.3.2 *Size and depth*

The size and depth of a lake are important factors and are often linked, namely the larger a still-water the deeper it may be. The main effect appears to be in the stability of the system and the rate of change of the water parameters, particularly temperature. Generally a large, deep lake will have a more stable temperature regime than a small, shallow lake. Furthermore a deep lake will almost certainly develop a full thermocline, which will mean all areas of the still-water below a certain depth will have a stable, cool temperature.

2.3.3 *Marginal zone and the bank profile underwater*

The marginal zone is the area where the land meets the water and is the most important part of the environment for many resident species. The marginal and emergent plants that surround the water are vital to the health of the ecosystem. The larger this zone, therefore, the more of this habitat is available and generally the healthier and richer in life the water body will be. This is often represented by a shoreline/area ratio, where the length of the shoreline is divided by the area of the lake to indicate the degree of development of the riparian zone. Thus, a circular lake without an island and a steep bank will have a low shoreline/area ratio and will have the worst possible marginal habitat. By contrast, a highly indented shoreline, typical of reservoirs and dams, has a high shoreline/area ratio, which indicates a long riparian zone relative to the size of the lake.

2.3.4 *Surrounding landscape*

The water quality and the chemical profile of a still-water are often dictated by the water that runs into it from the surrounding catchment, so the surrounding landscape has an important impact. A still-water in a lowland valley on rich clay or loamy soil will have a very different profile from a highland still-water on granite bedrock. As has been discussed previously in this chapter, the surrounding landscape has a dominant effect on the nutrient status of the water and will also affect other aspects such as the pH and temperature regimes that a still-water will be subjected to.

In simple terms, the richest and most productive lakes are likely to be low lying, with relatively shallow depths over 80% of the area with only 20–30% being below 3 m deep. The banks should fall away at a shallow angle and the bottom should be undulating with some areas being very shallow. The surrounding landscape should be low lying and on a clay or loam soil.

2.4 Nutrient and chemical cycles

There are many chemicals that are important but two key nutrients for a plant are nitrogen and carbon. Without these chemicals, plant life could not thrive. Still-waters are often considered to be largely self-contained units, with little input from beyond the water's edge. Of course, most water bodies do have some sort of inflow, even if it is only run-off from surrounding fields or groundwater. This inflow will bring in nutrients. However, most nutrients are continuously recycled within the aquatic system. So some nutrients come in with the inflow of water and some are lost to the sediment in the form of organic silt, but most are recycled in the various nutrient cycles.

2.4.1 *Nitrogen cycle and the breakdown of ammonia*

Nitrogen makes up about 78% of the air we breathe and, as an essential component of proteins, it is found in the cells of all living things. Inorganic nitrogen may exist in its free state as a gas, or as the compounds nitrite, nitrate or ammonia. Organic nitrogen is found in proteins and other compounds. Nitrogen is an essential nutrient for plants but they cannot use free nitrogen in the air; most can only use nitrate compounds whereas some can use ammonium, although this is toxic to most organisms. Ammonia is produced as a breakdown product of the metabolism of proteins, so it is produced by all animals. This is then broken down, or mineralised, by decomposers such as bacteria and fungi into ammonium compounds.

These facts mean that nitrogen and its compounds are cycled continuously by plants absorbing nitrates, animals eating the plants and releasing ammonia, and bacteria breaking down ammonia to nitrates. This natural process is known as the 'nitrogen cycle'. A diagrammatic description of this is shown in Figure 2.5.

The nitrogen cycle is immensely important as it provides plants with a vital nutrient and breaks down the compound ammonia, which would otherwise build up to toxic levels, particularly in the aquatic environment.

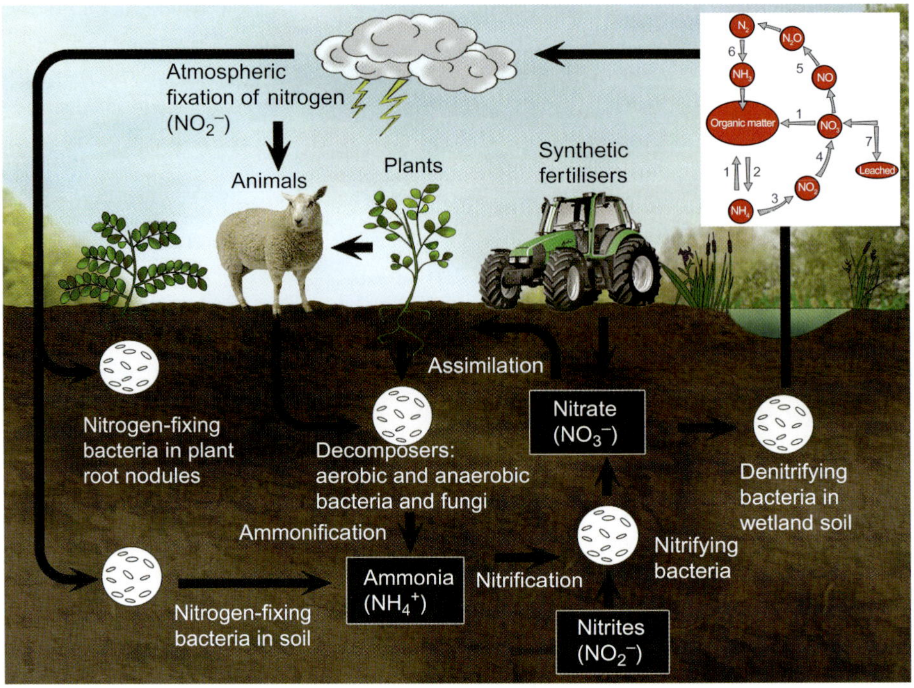

1. Uptake of NH or NO by organisms
2. Release of NH by decomposition
3,4. Microbial oxidation of NH (yields energy in aerobic conditions)
5. Denitrification (NO respiration) by microbes in anaerobic conditions (NO is used instead of O as the terminal electron acceptor during decomposition of organic matter)
6. Nitrogen fixation
7. Nitrate leaching from soil

Figure 2.5 The nitrogen cycle.

At any one time a large proportion of the total fixed nitrogen will be locked up in the biomass, or in the dead remains, of organisms (shown collectively as 'organic matter'). For plants, the only nitrogen available to support new growth will be that supplied by nitrogen fixation from the atmosphere (pathway 6 in Fig. 2.5) or by the breakdown of ammonium to nitrate through the decomposition of organic matter (pathway 2 in Fig. 2.5). Many of the important stages in this cycle are mediated by specialised groups of microorganisms.

Nitrification

One of the most important parts of the nitrogen cycle is known as nitrification, the breakdown of ammonium to nitrate. Ammonium is actually very toxic and although some plants can use it, generally it is toxic both to plants and animals. However, a type of autotrophic bacterium (belonging to the genus *Nitrosomonas*) can chemically alter the ammonium into a much less toxic compound, nitrite (NO_2^-). The bacterium does this as part of its normal metabolism to produce energy. Further modification of nitrite by another type of bacterium (belonging to the genus *Nitrobacter*) further converts the nitrite to nitrate (NO_3^-), again as part of its normal metabolism to produce energy. Both of these processes involve chemical oxidation and are known collectively as nitrification.

Nitrogen in the form of nitrate can easily be used by plants for growth and is non-toxic to aquatic life. Under normal natural circumstances, nitrification occurs rapidly and it would be very rare for levels of ammonia to build up to a significant level before they are broken down in the process described above.

Almost all of the nitrogen found in any ecosystem originally came from the atmosphere. In still-waters most will enter through rain or run-off from the surroundings.

Ammonia

Ammonia is toxic to fish and aquatic organisms, even in very low concentrations; when un-ionised levels reach 0.06 mg/l, fish may suffer gill damage. When un-ionised ammonia levels reach 0.2 mg/l, sensitive fish like trout and salmon begin to die; as levels near 2.0 mg/l, even ammonia-tolerant fish like carp begin to die. Ammonia levels greater than approximately 0.1 mg/l usually indicate some degree of pollution.

The danger ammonia poses for fish depends on the water's temperature and pH, along with the dissolved oxygen (DO) and carbon dioxide (CO_2) levels. The higher the pH and the warmer the temperature, the more toxic the ammonia; also, ammonia is much more toxic to fish and aquatic life when water contains very little DO and CO_2.

Nitrite

Nitrite is generally relatively short-lived because it is quickly converted to nitrates by bacteria. High nitrite levels in freshwater produce a serious condition known as brown blood disease in fish. In this disease, nitrites react directly with haemoglobin to produce methaemoglobin, which destroys the ability of blood cells to transport oxygen. Nitrite levels greater than 0.60 mg/l are toxic to fish, although in some conditions nitrite can be toxic to fish at levels above 0.03 mg/l.

Nitrate

Nitrate is the stable compound that is the end result of nitrification. It is a solid that is usually dissolved and found in solution in the aquatic environment. Nitrates form a fundamental nutrient for plants and so stimulate the growth of plankton and water plants that provide food for fish. Additional nitrates can find their way into the aquatic environment because they are a major ingredient of farm fertiliser; when it rains, varying amounts may wash from farmland into nearby waterways. Nitrates may also enter waterways from leaking septic tanks and cesspools, manure from farm livestock, animal wastes (including fish and birds) and many other sources.

Although nitrate can be toxic to aquatic life at levels of over 100 mg/l, it is not likely to be found at these levels in natural or semi-natural freshwater fishery environments because it will be used by the plant life.

Elevated levels of ammonia, nitrite and nitrate can often indicate that the water has been polluted or that the natural nitrogen cycle of the aquatic system has broken down.

2.4.2 The carbon cycle

All life is based on the element carbon and it is the major chemical constituent of organic matter (Fig. 2.6). Most of the carbon coming into a still-water system comes in the form of the atmospheric gas carbon dioxide (CO_2). Plants are autotrophic and have specialised

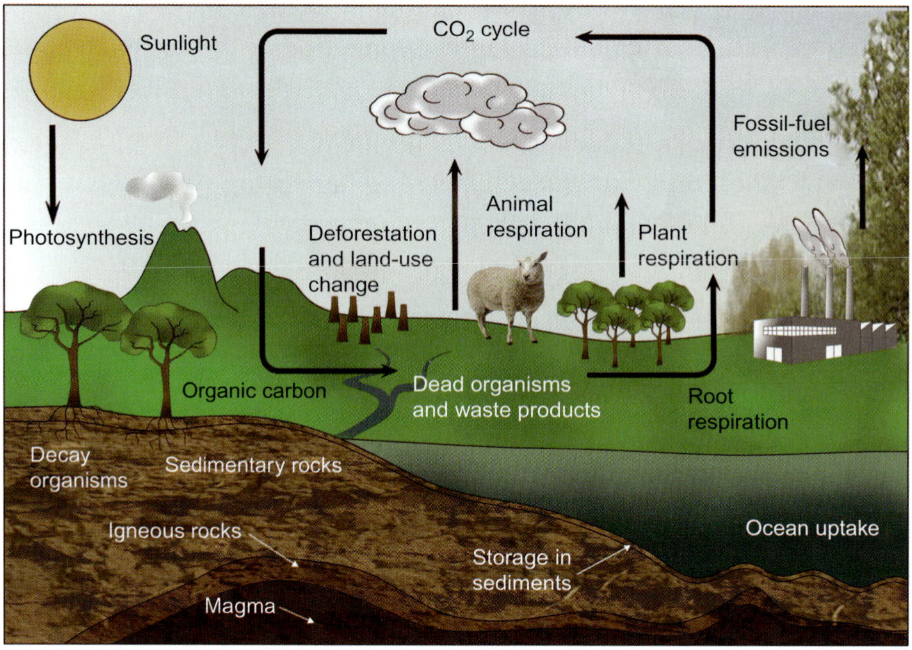

Figure 2.6 The carbon cycle.

mechanisms that allow for absorption of CO_2 and, with the addition of water and sunlight, these organisms use the process of photosynthesis to convert the CO_2 chemically to carbon-based sugar molecules (Fig. 2.6). These molecules can then be chemically modified by these organisms, through the metabolic addition of other elements, to produce more complex compounds like amino acids, proteins and cellulose. This organic matter, produced in the plants, is passed to heterotrophic animals through consumption.

Carbon is released back into the water as CO_2 gas by the process of respiration. Respiration takes place in plants and animals and involves the breakdown of carbon-based organic molecules into CO_2, water and various waste compounds. Importantly in still-waters, organic-rich detritus (sediments) contains a host of organisms whose primary ecological role is the decomposition of organic matter into its various, abiotic, components. All these decomposers use respiration for their energy production.

Carbon dioxide can also enter waters by simple diffusion. Here it can be converted into carbonate (CO_3^{2-}) or bicarbonate (HCO_3^-) ions or calcium carbonate ($CaCO_3$). Carbonate and bicarbonate ions also enter the water from erosion of rocks such as limestone or chalk. These forms of carbon can be used to produce shells and other body parts of invertebrates.

Photosynthesis and respiration not only allow for the cycling of carbon in the still-water ecosystem but also have important consequences for the amount of oxygen in the water, as we will see later in this chapter.

Further nutrients such as potassium and magnesium cycle through an ecosystem in similar ways to these two important examples. These cycles are vital for the maintenance of a healthy balance in a still-water. They also lead onto the next section, which is the flow of energy (food) from the autotrophic plants, which store energy from sunlight, all the way through to the top predator in a system. Before moving on to the flow of energy, we need to consider a couple of other key components of water that are vital to the life process.

2.4.3 *Dissolved oxygen*

Dissolved oxygen (DO) is oxygen that is dissolved in water. It gets there by diffusion from the surrounding air, aeration of water that has tumbled over falls and rapids, and as a by-product of photosynthesis.

Fish and aquatic animals cannot produce oxygen from water (H_2O) or other oxygen-containing compounds. Only green plants and some bacteria can do this through photosynthesis and similar processes according to the following formula, with the addition of sunlight energy:

Carbon dioxide + water $\rightarrow$ sugar + oxygen

Box 2.1 Measuring dissolved oxygen

Dissolved oxygen concentrations are most often measured in parts per million (p.p.m.) or milligrams of gas per litre of water (mg/l; 1 mg/l is equivalent to 1 p.p.m.). The most important measure of dissolved oxygen for fish and other aquatic organisms is the percentage saturation, which is calculated as the percentage of dissolved oxygen concentration relative to that when completely saturated at the temperature of the measurement depth. As temperature increases, the concentration at 100% saturation decreases (see table), although the altitude of the lake, the barometric pressure and the salinity of the water also affect this but to a lesser extent.

Values for dissolved oxygen saturation at a range of temperatures at different ranges of salinity are as follows.

Temperature (°C)	Salinity (parts per thousand)		
	0	5	10
0	14.6	14.11	13.64
2	13.81	13.36	12.91
4	13.09	12.67	12.25
6	12.44	12.04	11.65
8	11.83	11.46	11.09
10	11.28	10.92	10.58
12	10.77	10.43	10.11
14	10.29	9.98	9.68
16	9.86	9.56	9.28
18	9.45	9.17	8.90
20	9.08	8.81	8.56
22	8.73	8.48	8.23
24	8.4	8.16	7.93
26	8.09	7.87	7.65
28	7.81	7.59	7.38
30	7.54	7.33	7.14

Water's capacity to absorb, or dissolve, oxygen into it is limited. When the point is reached where no more oxygen can be absorbed by the water, the water is termed 'saturated'. As water temperature increases, so the water's capacity to hold dissolved oxygen decreases; so as the water temperature rises there may not be enough oxygen in it for the organisms that rely upon it to respire normally (see Box 2.1). When there are too many bacteria and/or aquatic animals in a defined enclosed area of water, they may overpopulate it, using the dissolved oxygen quicker than the natural processes can replenish it, and thus exceeding the water's carrying capacity.

Oxygen levels can also be indirectly reduced through over-fertilisation of water-plants by run-off from farm fields containing phosphates and nitrates (the ingredients in fertilisers). Under these eutrophic conditions, the numbers and size of water plants can increase significantly. If the weather then becomes cloudy for several days, respiring plants will use much of the available DO. When these plants die they become food for bacteria as they decompose, which in turn multiply and use large amounts of oxygen. Indeed there is also the relationship between the organic waste material that feeds the bacteria and the bacteria that breakdown the organic material, so it is taken back into the system as energy.

How much DO an aquatic organism needs depends upon the species, its physical state, the water temperature, pollutants present, and more. Consequently, it is impossible to predict accurately the minimum DO levels for specific fish and aquatic animals. For example, because fish are poikilotherimic (cold-blooded) animals, they use more oxygen at higher temperatures when their metabolic rate increases; so although at 5°C (41°F) trout use about 50–60 mg of oxygen per hour, at 25°C (77°F) they may need five or six times that amount.

Numerous scientific studies suggest that 4–5 mg/l (or 4–5 parts per million (p.p.m.)) of DO is the minimum amount that will support a large, diverse fish population. The DO level in good fishing waters generally averages about 9.0 mg/l. However, DO concentrations fluctuate over a 24-hour cycle (a diurnal rhythm), so critical DO levels should be measured at dawn when levels will be at their lowest.

Daily dissolved oxygen fluctuations and dawn lag

Just as oxygen may be added to the water by photosynthesis, so it is removed from water by all aerobic plants and animals to drive their energy use in the process of respiration.

$$\text{Sugar} + \text{oxygen} \rightarrow \text{carbon dioxide} + \text{water} + \text{energy}$$

This process is continuous for all plants and animals during both night and day.

The combined effects of photosynthesis and respiration, and the way natural diffusion from the air works, mean that the amount of dissolved oxygen in a water body can vary quite dramatically not only in the medium term but also daily. It is very important to understand these daily fluctuations in DO as they probably represent one of the most common reasons for stressed fish and associated mortalities.

In the presence of an established plant community, the daytime photosynthetic production of oxygen will put oxygen into the water more rapidly than it is removed by respiration. At night, however, all the aerobic communities including the plant community will continue to respire and therefore remove oxygen from the water, resulting in a decline in the DO level. This will result in the lowest DO concentrations being at dawn just before sunrise and the

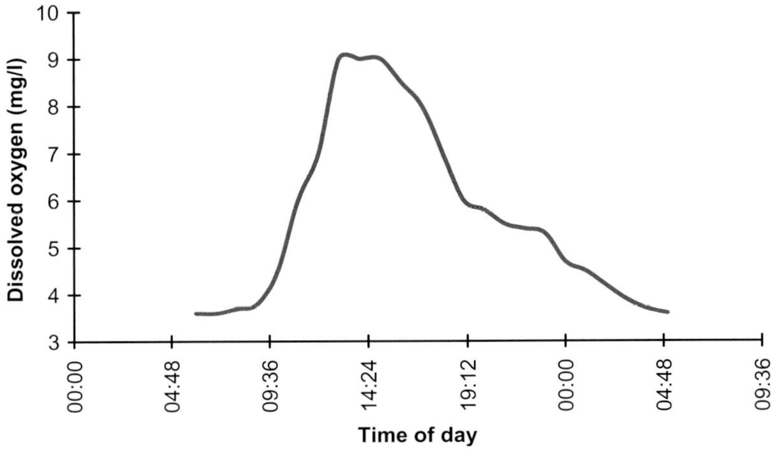

Figure 2.7 Diurnal fluctuation of dissolved oxygen.

start of conditions suitable for photosynthesis. This is known as the dawn lag. The daily changes in DO are represented in Figure 2.7.

The scale of these daily changes in DO will vary through the year, but will commonly stay within the acceptable range for the whole animal community. However, problems can occur when climatic or environmental conditions create circumstances whereby the overnight oxygen demand, needed for respiration, outstrips the amount of oxygen added by photosynthesis the day before and that diffusing in from the air. Under these circumstances, DO levels first thing in the morning fall below what is required by the lake animals, particularly fish, and stress and mortalities can occur within the fish community.

2.4.4 Phosphorus

The element phosphorus is necessary for plant and animal growth. Phosphates (chemical compounds containing phosphorus) enter waterways from human and animal wastes (the human body releases about half a kilogram of phosphorus per year), phosphate-rich rocks, wastes from laundries, cleaning and industrial processes, and farm fertilisers. In the case of enclosed still-waters, varying amounts of phosphates may still enter with the run-off from nearby farm soils when it rains.

Phosphates stimulate the growth of plankton and water plants that provide food for fish. This may increase the fish population and improve the quality of life for all inhabitants of a still-water. If too much phosphate is present, however, algae and water plants can grow strongly, becoming invasive and creating a nuisance by choking the water body and using up large amounts of oxygen as they respire.

The phosphorus cycle is said to be 'imperfect' because not all phosphates are recycled. Some simply drain off into lakes and oceans and become lost in sediments. Phosphate loss is not serious because new phosphates continually enter the environment from other sources.

Although phosphate is unlikely ever to be at levels that will be toxic to freshwater fish, it does have a large part to play in the eutrophication of our freshwaters along with nitrate. The effect of varying levels of phosphorus and phosphate is shown in Table 2.2.

Table 2.2 Effects of varying levels of phosphorus/phosphate

Amount of total phosphate/phosphorus	Effects
0.01–0.03 mg/l	Amount of phosphate/phosphorus in most uncontaminated lakes
0.025 mg/l	Accelerates the eutrophication process in lakes
0.1 mg/l	Recommended maximum for rivers and streams

2.4.5 *pH*

The balance of positive hydrogen ions (H^+) and negative hydroxide ions (OH^-) in water determines how acidic or basic (alkaline) the water is. When analysts measure pH, they are determining the relative concentration (expressed in exponential or 'power' form) of hydrogen ions; the term 'pH' comes from the power of Hydrogen. The logarithmic pH scale ranges from 0 (a high concentration of positive hydrogen ions; strongly acidic) to 14 (a high concentration of negative hydroxide ions; strongly basic). In pure water, the concentration of positive hydrogen ions is in equilibrium with the concentration of negative hydroxide ions, and the pH measures exactly 7.

In a lake or pond, the water's pH is affected by its age and the chemicals discharged by communities and industries. Most lakes are basic or alkaline when they are first formed and become more acidic with time because of the build-up of organic materials. As organic substances decay, carbon dioxide (CO_2) forms and combines with water to produce a weak acid, called carbonic acid. This is the same chemical that is in carbonated soft drinks. The formation of large amounts of carbonic acid will lower the water's pH. Conversely, the removal of large amounts of CO_2 during photosynthesis will raise the pH. So, in addition to the long-term trend for still-waters to become more acidic, the activities of the flora and fauna that result in the daily fluctuations in DO also cause daily fluctuations in CO_2, which further result in daily fluctuations in carbonic acid and hence pH (Fig. 2.8).

The effect that pH has on fish is slightly complicated by the synergistic effect that it may have on other compounds. Synergy is the process whereby two or more substances combine and produce effects greater than their sum. Thus 2 + 2 = 4 (mathematically), but synergisti-

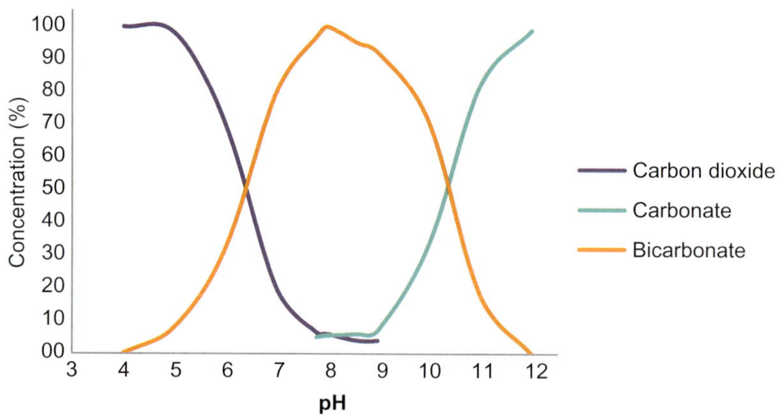

Figure 2.8 Relationships between carbon dioxide, carbonate, bicarbonate and pH.

cally 2 + 2 = much more than 4! Synergy is a mathematical impossibility, but it is a chemical reality. For example, when acid waters (waters with low pH values) come into contact with certain chemicals and metals, they often make them more toxic than normal. As an example, fish that usually withstand pH values as low as pH 4.8 will die at pH 5.5 if the water contains 0.9 mg/l of iron. Mix an acid water environment with small amounts of aluminium, lead or mercury, and you have a similar problem: the effect of the combination far exceeds the usual impact of the individual substances.

There has also been a historical tendency for soft-water lakes in areas of hard bedrocks to become more acidic due to the effects of acid rain. This is because energy generation by burning fossil fuels releases large quantities of sulphur into the atmosphere, which is then dissolved into solution by the rain to form weak sulphuric acid. Excessive acidity (low pH levels) is traditionally combated by liming lakes (see section 23.2.2).

2.5 Energy movements and trophic levels

The basis of most biological activity is primary production. This is the production of organic material (plants) from nutrient chemicals (CO_2 and water) and sunlight through the process of photosynthesis. This organic material is used by all the other organisms within the aquatic ecosystem. It is useful to visualise the flow of food (and therefore energy) through the ecosystem in the form of a chain, known as a food chain. Each organism forms a link within the chain that feeds on and therefore derives energy from the preceding one. This is in turn eaten and provides energy for the following organisms in the chain. These different levels are known as trophic levels (Fig. 2.9a).

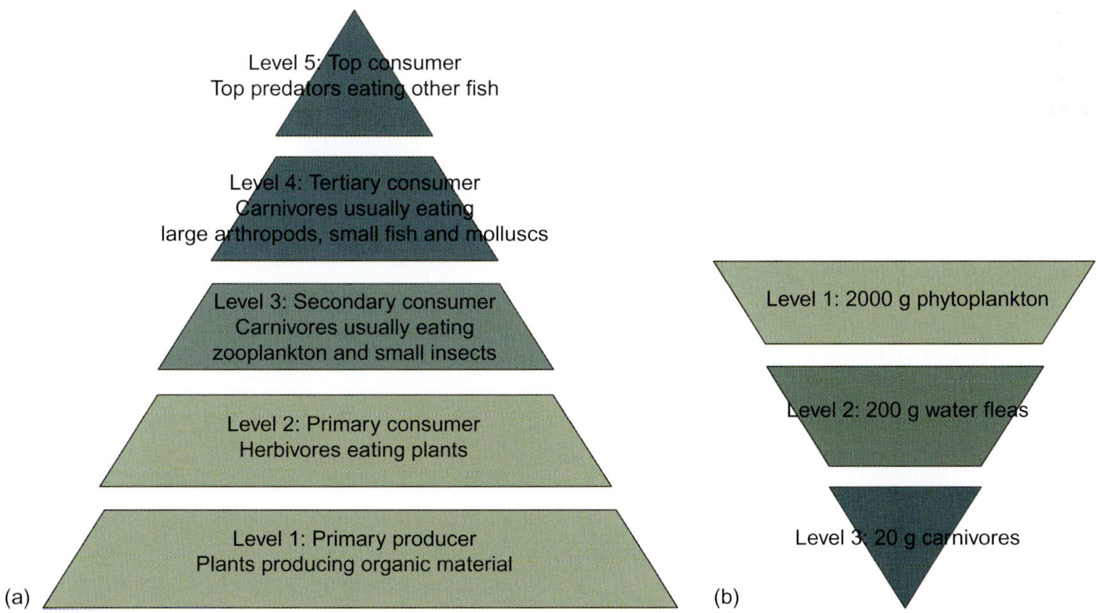

Figure 2.9a (a) Trophic levels. (b) Pyramid of biomass.

It is very important to note that energy transfer from one level to the next is not 100% efficient and that there is a substantial loss of energy at each stage. In fact, it is rare for any more than 20% of energy to transfer to the next level. This is due to heat loss, respiration and loss as waste products (excreta).

It is possible to have food chains with as few as three links, and it is very unusual to have food chains with more than six links. It is possible for the same animal to be in different parts of a food chain and to be in completely different food chains. So, from an ecological point of view, they have limited value in the study of the way an ecosystem works; however, they do show the movement of energy very well. A more visual way of looking at this energy movement is using the pyramid of biomass (Fig. 2.9b).

The fact that only a small amount of energy is taken from one level of the food chain to another means that each successive stage is made up of fewer organisms. For example, 1000 phytoplanktonic algae will feed 100 zooplankton (water fleas), which in turn will feed one roach. This is shown in the pyramid of biomass.

This diagram graphically illustrates the fact that the larger carnivores at the top of a food chain require a great deal of support from the animals lower in the food chain, and without these large food supplies the top levels would exist in much lower numbers.

Both the food chain and pyramid of biomass allow us to understand the movement of food (energy) through a system and are important in our understanding that to obtain the desired numbers of higher organisms, any management measures must protect the lower organism (food supply) that they rely on. However, this is a very simple way of looking at the overall communities within a still-water. No animal exists only in one food chain with one source of food and one predator. The relationships are much more complex than this. These relationships are better visualised using a food web (Fig. 2.10). This is similar to a food chain but all possible relationships are included. As a result, they can become very complicated but they are important in understanding the variety of predator–prey relationships and the way that each organism relies on so many other organisms.

2.5.1 Population and community organisation and dynamics

By looking at food webs, we start to appreciate the complexity of the interrelationship between organisms in an ecosystem like a still-water. Still-waters are also very varied habitats with many different components in a seemingly simple system. They change dramatically with each season and over the period of a day. For the management of any still-water, a clear understanding of these relationships is vital for success. Much of the rest of this book is concerned with managing these relationships to benefit the target species of a management plan.

The primary productivity of a lake is built on the ability of plants to grow. There are many factors that affect this ability. In perfect conditions plants, particularly algae, can grow and reproduce extraordinarily quickly. However, in most natural systems there are usually several factors that limit this growth. These are known as limiting factors and are usually to do with water temperature, sunlight (intensity and day length), nutrient availability and suitable habitat. In most coarse fisheries, conditions are good for primary production in the spring and summer, with plenty of sunlight and warm, shallow waters, which have plenty of nutrients.

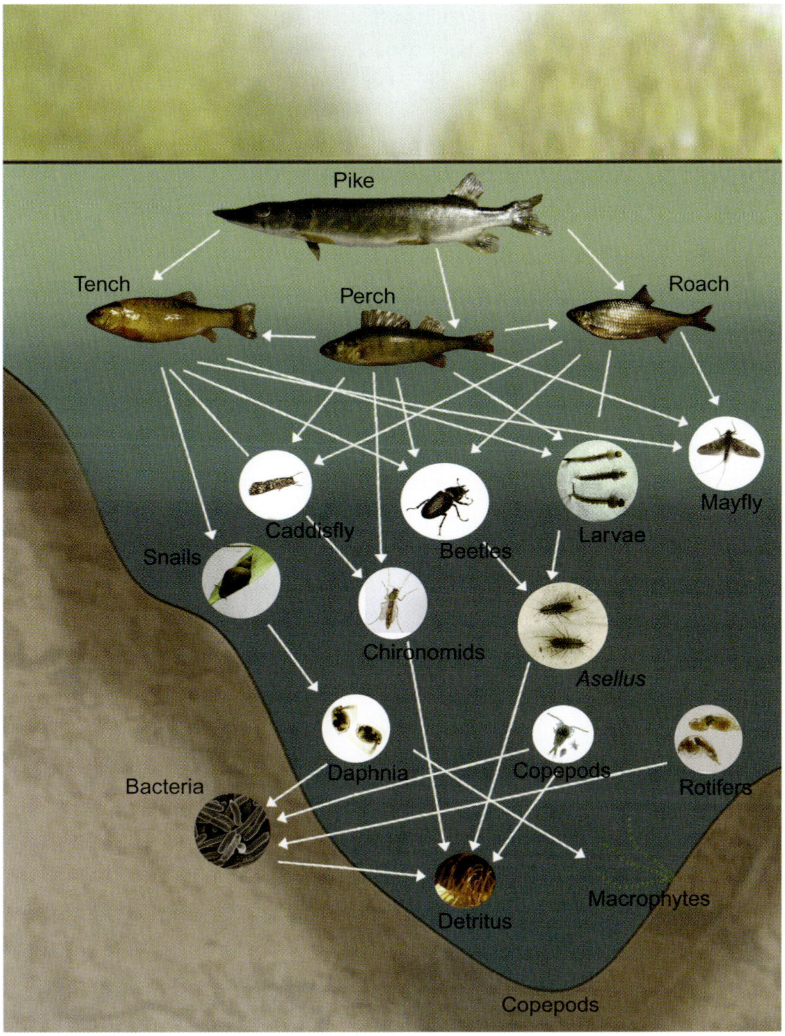

Figure 2.10 A food web.

2.5.2 Biomass

Biomass is the weight of any given group of organisms present in the system at any one time. It may be applied only to one species at a time or to the whole community present in the lake or pond. Biomass may build up over a period of time to a sustained level, characteristic of any particular still-water, known as the carrying capacity. It is sometimes the job of a still-water manager to alter the conditions in the still-water to increase, or decrease, its carrying capacity.

2.5.3 Productivity

Productivity, in the ecological sense, is the total amount of biomass achieved by any organism in any given period. The biomass includes the weight of any reproductive products and any dead organisms. It is, therefore, expressed as a rate: so many grams of organism per unit of time. Thus it is indicative of the speed with which the carrying capacity can be reached. Again, management can intervene to increase or decrease the productivity.

2.5.4 Predator–prey relationships

For the purposes of management, predator–prey relationships are calculated from the amount of predators (biomass) present in a system relative to the amount of their prey. They therefore give numerical values to the various stages in the pyramid in Figure 2.9, Predator–prey relationships will vary with season, the availability of nutrients and several factors that affect the primary production of the system. Simple predator–prey relationships can be seen in the way that an abundance of primary productivity drives a sharp increase in the herbivore population as there is plenty of food available. The abundance of herbivores can then lead to a similar increase in predators, because the predator–prey relationships show that with an abundance of prey so there will be a promotion of survival in the predators. Increased productivity is passed through the food web all the way to the top, and the fishery is blessed with an abundance of fish at the top of the food web. Too many top predators, however, will reduce the prey food available and thus reduce the number of predators the system can support, keeping the system in balance by natural fluctuations.

3 Bacteria

Bacteria are often considered the simplest of the life forms. They are single-celled organisms that sometimes form colonies. They are very small and all are microscopic, the largest being 25 micrometres (μm; 1 μm is equal to one-thousandth of a millimetre). They are prokaryotic, meaning they have little structure to their internal organelles, such as the lack of a defined nucleus (Fig. 3.1). They come in many shapes and forms, and some are motile.

Bacteria are often associated with disease, and these pathogenic bacteria will be dealt with in the section on fish health. However, bacteria also serve an absolutely vital role in the decomposition of all organic material, helping to release the nutrients stored in dead material; as described in the previous chapter, they also serve a vital role in the nitrogen cycle.

3.1 Nitrifying bacteria

Nitrifying bacteria are found in most waters of moderate pH, but are not active in highly acidic water (pH < 5). They are usually found as mixed-species communities (termed 'consortia') because some of them, for example *Nitrosomonas* species, are specialised to convert ammonium to nitrite (NO_2^-); whereas others, for example *Nitrobacter* species, convert nitrite to nitrate (NO_3^-). In fact, the accumulation of nitrite inhibits *Nitrosomonas*, so it

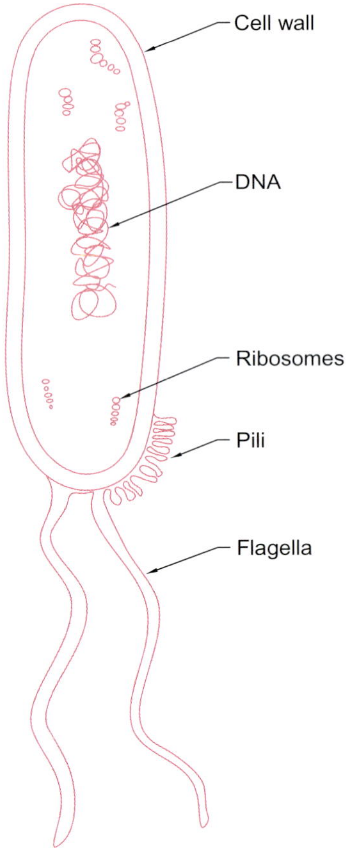

Figure 3.1 A prokaryote cell.

depends on *Nitrobacter* to convert this to nitrate, whereas *Nitrobacter* depends on *Nitrosomonas* to generate nitrite.

Nitrifying bacteria have some important environmental consequences. Because they are commonly occurring, most of the ammonium in oxygenated natural waters is readily converted to nitrate. However, the process of nitrification has some undesirable consequences. The ammonium ion (NH_4^+) has a positive charge and so is readily adsorbed onto negatively charged clay colloids and organic matter. In contrast, the negatively charged nitrate ion (NO_3^-) is not held on soil particles and so can be washed down the soil profile in the process of leaching (arrow marked 7 in Fig. 2.5). In this way, valuable nitrogen can be lost from the soil, reducing the soil fertility. The nitrates can then accumulate in groundwater and other waters, adding to their eutrophic nature.

3.2 Cyanobacteria (blue-green algae)

Traditionally the cyanobacteria have been included among the algae, referred to as the cyanophytes or blue-green algae. Taxonomically they are now placed with the prokaryotes (bacteria). However, because of their general behaviour and the considerable impacts they have upon aquatic ecology, it should suit the fishery manager to consider them alongside other algae.

The cyanobacteria were one of the first groups of living things to appear in the fossil record, dating back some 3800 million years ago to the Precambrian period, when they may have played a major role in creating Earth's oxygen atmosphere.

3.2.1 Blue-green algae and the need to ensure public health

Some species of blue-green algae can emit toxins that can be damaging to fish stock and are potentially a health hazard to humans and animals. The risk from blue-green algae, although not being minimised, should be put into context. There have been no reports of a human death attributed to blue-green algae, but there have been several cases where algal toxins have been implicated in illness, including rashes in those who have a reaction to water containing blue-green algae on their skin. Pets, livestock and wildlife would appear to be far more susceptible to the toxins. It is also an undeniable fact that reported occurrences of blue-green algae go hand in hand with increases in eutrophication.

There can be no doubt that the safety of people using a still-water as an amenity must be paramount. The public should be alerted to the possible dangers if a blue-green algal bloom is present and particularly if a scum has formed; if toxins are released it is generally as the blue-green algae cells die, forming a scum if cell numbers are high enough. Not all species produce toxins and, even if a bloom is of a potentially toxin-producing species, it may not do so under all conditions. So although the use of the word 'dangers' may seem a little dramatic, it is advisable to notify users of a possible risk.

4 Plants

4.1 Introduction

Plants form the base of the food chain and therefore play a major role in controlling the overall productivity and function of freshwaters. They also provide structure to the lake or pond and create a diversity of habitats that are essential to many animals including fish. It is important for those managing fisheries and freshwater ecosystems generally to understand the environmental factors that control the distribution and abundance of plants. It can be simply stated that, generally, the more plants in your fishery the greater its ability to support other animals and the fish living within it.

Plants help create the natural balance that should be the aim of all fishery managers. Fish that live in environmental conditions that are as near optimal for them as possible will be less stressed than fish that live in conditions that are suboptimal, and plants are the first and most important step in creating this balance.

Plants have several important functions in the ecology of a still-water; these include:

- the provision of habitat;
- serving as a substrate for spawning;
- acting as a food source;
- water filtration and purification;

Box 4.1 Photosynthesis

Plants, cyanobacteria and some bacteria are able to convert water, carbon dioxide and light into sugar, water and oxygen in the presence of chlorophyll *a* through a process called photosynthesis. Photosynthesis can be divided into two stages. The first is light dependent and needs the energy collected from light to produce adenosine triphosphate (ATP), which is the energy unit of the cell. The light reactions of photosynthesis are associated with or occur in membranes. The second stage is light independent and occurs in the aqueous phase (i.e. not in membranes). Here the energy that has been captured in ATP is used (along with other factors) to create sugar. According to the formula

$$6CO_2 + 12H_2O \rightarrow C_6H_{12}O_6\,(\text{glucose}) + 6H_2O + 6O_2$$

plants use light in the violet to blue range, as well as the red–orange to red range. Light in the green range is reflected. This accounts for why plants are green.

In most cyanobacteria the light reactions occur on membranes that are arranged in sheets of lamellae next to the periplasmic membrane. The dark reactions generally occur in the centre of the cell. Eukaryotic algal cells have special organelles called chloroplasts, which contain alternating layers of lipoprotein membranes and aqueous phases.

The more general conversion of carbon dioxide into organic carbon compounds by photosynthesis is called carbon fixation.

- the production of oxygen by photosynthesis;
- regulation of the nitrogen cycle.

Plants also have a function in protecting banks against erosion caused by wind and wave action as well as human and animal impact such as that produced by simply walking along a bank. Marginal and emergent aquatic plants form the all-important intermediate zone or interface between the aquatic and terrestrial environments.

It is important to appreciate that some species of plant can become a nuisance owing to their invasive nature or to other undesirable impacts such as those caused by algal blooms. Before considering what plants are beneficial and what purpose they serve, it is essential for the fishery manager to have an understanding of the types of aquatic and semi-aquatic plant.

Two main groups of plants are significant in freshwaters: algae and macrophytes. The two groups form separate but linked components of these ecosystems as they have very different requirements to flourish and equally have very different functions and impacts within a system.

The capacity to create their own food through photosynthesis is the single most important characteristic of plants and distinguishes them from other forms of life (see Box 4.1).

4.2 Algae

Algae are unicellular or multicellular rootless plants that never differentiate into stems or leaves. They grow in proportion to the amount of available nutrients, oxygen concentrations and sunlight. They can affect water quality by both increasing and decreasing the amount of oxygen dissolved in the water. They are an essential form of food for fish and small aquatic animals. Algae are found in most habitats on earth, though most occur in freshwater or marine environments. There are thought to be around 5000 species of algae that inhabit freshwaters in the UK.

4.2.1 *Morphology*

Algae come in a variety of shapes and in varied colours due to their different photosynthetic pigments. Algae can be unicellular and microscopic or colonial, forming plate-like colonies, thread-like filaments, net-like tubes or hollow balls. Many planktonic algae species bear horns, ridges or wings to increase their surface area to volume ratio, which not only increases their ability to obtain scarce nutrients from the environment, but also protects them from herbivores and slows the rate with which they sink in water.

Sizes of algae vary hugely. Diatoms may range in size from 2 μm to several millimetres, although there are only a few species larger than 200 μm; whereas colonial brown algae, in which some species form branched filaments or foliose plants, may be many metres long. Some algae, like diatoms, are encased in a siliceous cell wall, which takes the form of a box, and come in many unique and beautiful forms.

4.2.2 *Reproduction*

Algae reproduce either sexually or asexually. They use various methods that are fairly group specific, but some features are shared. When reproducing asexually, most groups develop spores within a parent cell, which are released into the environment. Each of these spores then develops into a single algal cell, or they divide by mitosis into numerous cells. For example, reproduction in the cryptomonads is asexual and primarily by longitudinal cell division with the cell dividing in either a free-swimming or non-motile form. Sexual reproduction is not rare but occurs less often than asexual reproduction. Gametes are produced in different algal cells and released into the environment. Male gametes usually attach themselves to female gametes and their genetic material is absorbed through the cell coat of the female. In freshwater species, this is followed by the formation of a cyst in which meiosis takes place. Cyanobacteria only reproduce asexually, with genetic recombination accomplished through transformation or conjugation. During transformation, DNA, released from donor cells, is incorporated into recipient cells. Conjugation is the process by which two cells become connected by a narrow tube, through which DNA moves from one cell into the other.

4.2.3 *Algal groups and abundance*

Algae usually make up a significant amount of the plant biomass in still waters. Larger algae, such as the filamentous algae that is often referred to as blanket weed, can grow densely in shallow water and provide distinctive habitats for other organisms. Microscopic algae are present as floating forms (phytoplankton) and sessile forms (epiphyton) attached to plant stems, rocks and even bottom substrate, providing the base for food chains that are particularly important for fish fry and zooplankton (small microscopic aquatic animals) in still-waters.

When present in high densities, phytoplankton may form an algal bloom. Blooms often result from a population explosion of phytoplankton in response to changing environmental conditions, including nutrient enrichment (eutrophication). Blooms can result in oxygen depletion and biological impacts. These blooms are often visible as a discoloration of the water; the degree of water coloration can, with experience, indicate the density of the algae in the water column. Given that most still-waters are eutrophic, it is easy to understand why it is difficult to manage still-waters that have a predisposition to algal blooms.

4.3 Macrophytes

4.3.1 Macrophyte form and function

Macrophytes are the group of plants that most fishery managers will be familiar with, as they are generally much larger than algae, have a defined structure and can be easily seen in some shape or form along the banks of most ponds, lakes, rivers and canals. The types of aquatic plant are extremely diverse, but they can be separated into four categories based on their habitat. These are riparian, emergent/marginal, floating (attached and unattached) and submersed (Fig. 4.1).

Higher plants evolved from primitive algae or algae-like ancestors, but probably on land, not in the water. Today's aquatic macrophytes are specialised forms that have recolonised aquatic environments. Because aquatic angiosperms (flowering, sexually reproducing plants) are closely related to terrestrial types, it is not surprising that the morphology and anatomy of these two groups are basically the same. However, their differences become obvious when considering the adaptations that aquatic species have evolved to survive in their watery environments.

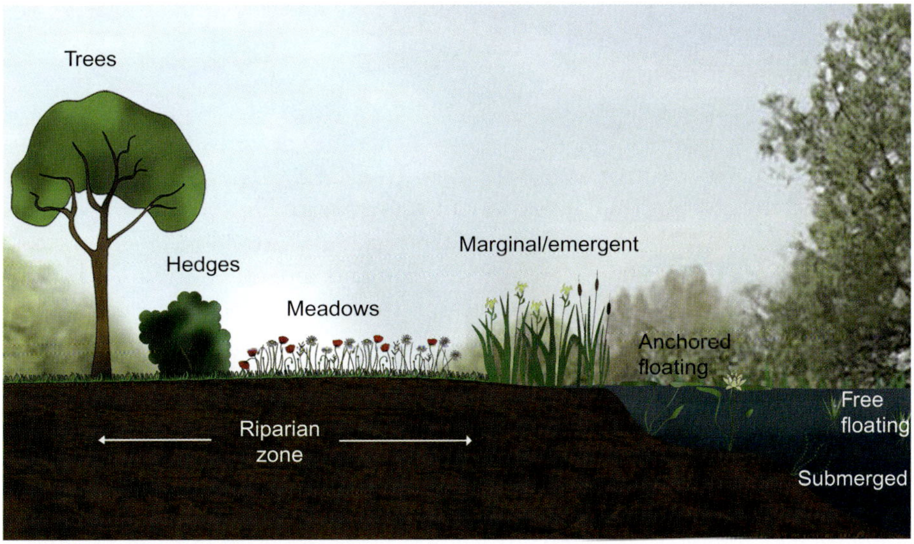

Figure 4.1 Plant succession around the edge of a still-water.

The flowers of most aquatic angiosperms must be elevated above the water for pollination to occur: entomophily is pollination by insects and anemophily is by wind; only very rarely does pollen transfer occur underwater (hydrophily). Getting and keeping flowers at the surface requires special adaptations; water lilies, for example, have waxy, bowl-shaped flowers that float by themselves, much like a small boat; some submerged plants such as water milfoil (*Myriophyllum spicatum*) develop rigid stems that protrude above the water and on which small flowers are borne.

Aquatic macrophytes tend to replace sexual reproduction by vegetative reproduction, which may be related to the difficulty in raising the flowers above the water for aerial fertilisation. Vegetative, or asexual, reproduction is a vital key to survival among the aquatic plants. Some species rarely generate viable seeds and those that are produced serve more as a 'back-up' to ensure the species' survival in the event of a disaster. Vegetative reproduction occurs primarily by stem fragmentation, which is one of the primary methods used by Canadian pondweed (*Elodea canadensis*), but some species such as duck weed (*Lemna* sp.) use the whole plant, hornwort (*Ceratophyllum demersum*) uses its shoot fragments, and specialised organs such as tubers are used by yet other species such as *Potamogeton* sp.

4.3.2 Macrophyte types

Bankside (riparian) plants

Many plants occupy the dry land of the bank, although their roots may be in the groundwater surrounding the lake or pond. Most conspicuous of the bankside plants are trees, particularly ones resistant to flooding such as willow, alder and poplar. Around small ponds trees may occupy valuable space and their leaf litter may cause water quality and siltation problems. On larger lakes they are a feature of the landscape that provide shade for fish and are aesthetically pleasing for the visitor. Various forms of grass, low shrub and wild flower are also common features on banks and may help stabilise the edges of the pond or lake (Fig. 4.2).

Emergent and marginal plants

Emergent plants are those whose roots grow underwater, but their stems and leaves are found above the water. Emergent plants are also sometimes described as 'marginals' because they grow in the shallow waters along the bank (Fig. 4.3). They provide useful habitats for many fish species and for a whole variety of invertebrate life such as dragonfly nymphs, water snails, etc., and as such are to be encouraged. A further beneficial factor is that the emergent plant stems confine water movement and aid the warming of the margins. This makes this marginal area very suitable for the establishment of dense zooplankton communities on which juvenile fish can feed. However, in a recreational fishery, they may need to be strictly controlled as they may make access to the water difficult and can snare fishing gear.

Floating plants

These are plants in which part or nearly all of the plant floats on the surface of the water. Floating unattached plants, such as duckweed, are those in which most of the plant is at or near the surface of the water. Roots, if present, hang free in the water and they are not

Figure 4.2 Grass and sedges stabilising a still-water bank.

Figure 4.3 Marginal plants in shallow water.

anchored to the bottom. Floating attached plants have leaves that float on the surface, but their stems are beneath the surface, and their roots anchor the plant in the substrate. Some floating plants can be a nuisance as they can expand rapidly over the surface of the water body, shading out the water column so plankton cannot grow and limiting access by fishermen. Floating leaved plants such as water lilies (Fig. 4.4) form valuable shade and resting places for fish.

Figure 4.4 Floating leaved plants: water lilies.

Figure 4.5 Submersed plants.

Submersed plants

Submersed plants are found when the entire plant is below the surface of the water. Their growth is affected by water clarity and they can often be the subject of overgrazing by the fish community, especially if this includes a population of carp. Submersed plants (Fig. 4.5) are particularly valuable as spawning sites for many species of fish and provide refuge for the fry once they hatch.

5 Invertebrates

Invertebrates are a large group of animals that are of vital importance to the whole aquatic environment and particularly fish. Very few fish are entirely herbivorous and only a few are entirely piscivorous, so what do the rest eat? Well, most rely on invertebrates as food for part, if not all, of their lives. Furthermore, invertebrates play an important part in the cycling of nutrients in ponds as they break down dead organic material or feed directly on plants. The ecology of freshwaters would be very different if invertebrates were not present. Invertebrates are also good indicators of the health of a still-water; the types and numbers of different species present in a water body are a useful indicator of the medium- to long-term water quality there.

Invertebrates are animals that do not possess a backbone. They comprise a very large and varied group. They are very diverse in body shape and size. This chapter looks at the main types and describes their function.

5.1 Protozoans

These are very small (microscopic) single-celled animals that carry their genetic material in a membrane-bound nucleus. They are motile and tend to be further classified depending on how they move. The most common are ciliates, flagellates and amoeboids. They are very common in freshwater where they mainly feed on bacteria, other protozoa and algae. There are also several members of this group that are common ectoparasites of fish. They are eaten by a wide variety of the smaller stages of other invertebrates. They reproduce by both sexual and asexual reproduction. Asexual reproduction takes place by simply dividing the cell into two or more numerous segments. Each segment can then grow into a new individual. These segments can also form resistant spores which can be moved by wind and water, allowing protozoans to colonize new still-waters very easily.

5.1.1 *Flagellate protozoans (Mastigophora)*

This group includes protozoans possessing one or more long flagella during their free-living stages. They are usually very small and may live as solitary individuals or occasionally in colonies.

5.1.2 *Amoeboid protozoans (Sarcodina)*

This group includes protozoans that move by amoeboid motion, including the use of pseudopodia. This group also includes protozoa that do not move at all and some that live in primitive shells made of substances secreted or collected by the animal.

5.1.3 *Ciliated protozoans (Ciliophora)*

This group includes protozoans that move using numerous hairs (cilia) often covering the whole surface of the cell. These cilia can also be used to provide a feeding current, drawing food items into the gullet. This group of animals is very large and includes numerous species.

5.2 Sponges (Porifera)

Sponges are immobile, encrusting organisms that are common and well recognized in the marine environment. There are several common freshwater examples, all belonging to the group Spongillidae. Sponges form encrusting growths, which are soft, but slightly rough to the touch. The growth is irregular in shape and perforated by numerous small holes. The sponges are filter feeders and the growth is a colony of individuals interconnected by a network of tunnels. The sponges use flagella to draw water through the tunnels to provide food particles and oxygen.

5.3 Hydras (Cnidaria)

Hydras are a very common animal and abound in most freshwaters. They are long, thin organisms with four to eight tentacles arranged in a single ring around the mouth. They attach to any firm substrate and dangle the tentacles into the water column. Any animal that brushes these tentacles is grabbed, subdued with stinging cells (cnidae) and drawn to the mouth to be eaten. They reproduce by budding, where a new individual develops as an outgrowth of the parent and eventually breaks off to form an independent individual.

This group also contains the free-swimming freshwater jellyfish. This is a very rare but distinctive creature, which is a maximum of 20 mm in size.

5.4 Platyhelminthes (flatworms)

Platyhelminthes are commonly known as flatworms, as the body is flattened and often elongated and their movements are very worm-like. Although they tend to have a definite head end it is not very developed and the body is very soft. Members of this group are often parasitic, including flukes and tapeworms, and these often have specialized organs for attaching to the host. There are also many free-living species belonging to the group Turbellaria. These are small (0.5–35 mm), flattened worms that have a characteristic gentle gliding motion over the substrate.

5.5 Roundworms (nematodes)

Commonly known as roundworms as they are slender, unsegmented worms and round in cross section, nematodes are an amazingly abundant and successful animal group, particularly in the aquatic environment. They tend to be small and a similar colour to their environment and are therefore often overlooked. They have a firm outer skin, which means their movements have a characteristic thrashing motion. They generally predate on bacteria and protozoans as well as being scavengers on dead organic material. The group also contains many species that are parasitic on numerous varieties of plants and animals.

5.6 Rotifers

Rotifers are tiny, usually microscopic, creatures with a rigid outer skin which can be contracted in length. They are often free-swimming but some species do attach to substrates. They feed using a series of ciliated tufts on the head end that look like wheels when they are in motion and led, historically, to the group being called 'wheel animalcules'. These cilia tufts create currents of water that sweep food particles into the mouth. These are then chewed in a characteristic pharynx. They mainly feed on bacteria, algae and other organic matter. When the animal is unattached the cilia tufts are also used to pull the animal along in a very characteristic motion. On the tail end they often have a small attachment, 'toes', to hold them in position when feeding.

Although these animals are very small they represent one of the most important groups in freshwaters. They can be very abundant and represent a hugely important first food for many fish fry. They are also important link between the bacteria and algae and larger invertebrates as well as young vertebrates.

5.7 Hairybacks (Gastrotricha)

Hairybacks are a common but overlooked group. They are very small with little or no colour. Characteristically they have a slug-like body shape with long sensory hairs on a rounded head. The body is covered in small cilia, hence the common name, and this gives the animal a very characteristic smooth gliding motion.

5.8 Moss animals (bryozoans)

These are colonial non-motile animals, which attach to solid objects in the water and can grow into quite large-sized colonies. The colonies consist of many branched tubes up to 1 mm in diameter; as such they are similar in some ways to the more recognizable marine corals. Individual animals are small with a crown of ciliated tentacles, which can be pulled into the colony when danger threatens. Bryozoans feed using the crown of cilia to filter the water for bacteria, protozoa, algae and organic material.

5.9 Segmented worms (annelids)

Annelids are worms that show constrictions or segmentation in their bodies. They generally have cylindrical or slightly flattened bodies. Many have hairs or bristles on the segments and this feature leads to their further classification into oligochaetes and polychaetes. This group also includes the parasitic leeches. The oligochaete worms are probably the most important subclass of this group in the aquatic environment. These are generally found in the mud and detritus of the pond bottom, and in areas like the tangles of filamentous algae floating on the surface and anchored at the edges of the pond. They are scavengers, consuming everything during their passage through the substrate; anything digestible is then absorbed in the intestine. They will also feed directly on protozoans and diatoms as well as smaller crustaceans.

Annelid worms are small but important animals playing a big role in the recycling of organic material in the pond bottom (the hydro-soil).

Leeches are also members of this group, representing a surprisingly large number of blood-feeding animals which prey on a wide variety of other animals.

5.10 Snails, limpets and mussels (molluscs)

These are soft-bodied animals, normally with a hard shell. There are two main groups represented in freshwaters: the gastropods, which have a one piece shell (snails and limpets), and the bivalves, which have a shell in two equal parts that fit closely together and are hinged on one side (mussels).

Snails are instantly recognizable with their single, usually spiralled shell. They move about in a gliding fashion on one soft, muscular foot and most have a single rasping tongue. This tongue acts as a rasp to scrape away the periphyton growing on any substrate in the still-water. They are one of the most important groups in any freshwater system.

Bivalves, such as the mussel, are enclosed between two half shells (valves) with a hinge on one side to allow the two halves to open, but clamped shut if danger approaches. When open the animal reveals two tubes: one is used to draw water in and the other to expel the water. These water currents are then used for respiration and feeding. The bivalves filter the water for food items such as suspended particles of organic material and phytoplankton. Any movement is provided by a single muscular tongue-like foot, which can be used for digging or dragging along.

5.11 Arthropods

The arthropods are a large group of freshwater invertebrates. They nearly always have paired jointed limbs and a segmented, (more or less) hard exoskeleton. The exoskeleton is dead tissue and cannot grow with the animal and must be routinely cast off in moults. The new exoskeleton is soft and can be stretched before it hardens. This also allows for significant changes in shape between each moult.

The arthropods are split into four groups: water bears, Chelicerata (spiders and mites), crustaceans and insects.

5.12 Water bears (tardigrades)

These are fairly common in small water bodies that are prone to drying. They are small, plump, tubular animals with four paired legs each with hooked claws at the end. The most remarkable feature about water bears is that they can survive complete drying, their bodies shrivelling to small cysts. When these are wet again the water bear will completely revive within a few hours.

5.13 Water spiders and mites (Chelicerata)

There are two spiders that are closely associated with water: the water spider, which is truly aquatic, and the raft spider, which is not truly aquatic but is always found in close association with water.

Water mites are quite small, usually between 0.5 and 2 mm. They have oval, plump and often brightly coloured bodies. The adults have four pairs of legs. They are a surprisingly common and diverse group, with over 300 species found in the UK.

The two most important groups of arthropods are the crustaceans and insects (Table 5.1). Although superficially very similar, there are some easy clues to tell the two groups apart.

Table 5.1 Differences between crustaceans and insects

Crustaceans	Insects
Two pairs of antennae on the head	One pair of antennae
Five or more pairs of jointed legs	Three pairs of jointed legs and often one or two pairs of wings
Abdomen often has paired appendages	Abdomen lacks appendages but often has tails

5.14 Crustaceans

There are a huge variety of crustaceans, from small members of the zooplankton such as the water flea, *Daphnia*, to the largest of freshwater invertebrates, the crayfish. They are also remarkably numerous. They all have an exoskeleton, which is a support structure (like bones) on the outside. It is hard, chitinous and jointed. This 'shell' is unable to expand, so as the crustacean grows it needs to shed its exoskeleton and grow a new one. The new shell is ready under the skin when it is shed and remains soft for a day or two as the crustacean expands, then it hardens. During this period, crustaceans are very susceptible to predation and often burrow or hide under rocks. The body consists of a variable number of segments in three regions: the head, thorax and abdomen. These regions can be fused, or sometimes enfolded in a single large plate. Most segments bear a pair of appendages. The first pair of appendages on the head is the antennae, with other appendages forming the mouthparts. Each segment on the thorax has appendages that can be developed for a variety of purposes, such as walking or swimming legs, pincers for defence, or grasping structures for obtaining prey. Paired appendages on the abdomen are often absent but if present are often formed into plates or feathery structures that beat continuously to draw water across the gills.

Crustaceans are fundamentally an aquatic group, but as with all animals they require oxygen for respiration. They have quite different methods for obtaining their oxygen and releasing carbon dioxide depending on their size. Small crustaceans can absorb oxygen directly across the body surface (e.g. *Daphnia*). Larger ones have gills that are often associated with the forelimbs (e.g. crayfish).

Their feeding varies enormously from filter feeders, which strain food from the water column, to scavenging predators like crayfish, which feed on live and dead animals.

In recent years the introduced American signal crayfish, *Pacifastacus leniusculus*, has expanded its range and occurs in considerable numbers where conditions are suitable. They can cause a significant nuisance in still-water ponds, damaging banks by their habit of burrowing into them. Furthermore, they can damage fish stocks by predating on eggs. They also pose a serious threat to the native British white-clawed crayfish.

5.15 Insects

The insects are the largest group of invertebrate animals, outnumbering all others put together. They are primarily terrestrial but some are aquatic for at least part of their life cycle, usually the immature stage (Fig. 5.1). Most insects have wings, at least in their adult forms. Like all arthropods they have a hard exoskeleton. They always have three pairs of jointed limbs on the main body and no paired appendages on the abdomen.

All insects require oxygen to respire, and they usually use a system of tracheal tubes joined to the surface through a small hole known as a spiracle. Aquatic forms have needed to modify this system in a variety of ways, as follows.

- Some breathe air through a spiracle located at the tip of the abdomen. This can be elongated into a breathing tube.

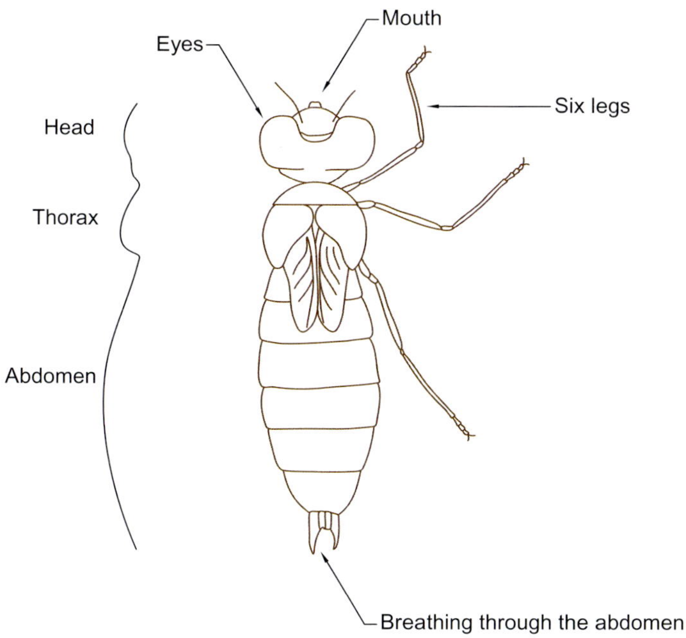

Figure 5.1 The body of an insect.

- Many nymphs and larvae have gills, which are feathery or plate-like structures, usually on the abdomen.
- Beetles and bugs breathe from a reservoir of trapped air beneath their wing cases or on hairs on their undersides. Most need to resurface to replenish this reservoir.

Insects only mature after their final moult. This means that they do not grow any further and their energy (food) requirement is drastically reduced. Some in fact do not eat at all at this adult stage. For example, many aquatic insects spend 2 or 3 years as immature underwater larvae only to emerge as winged adults, which survive for a few days to mate and then disperse to lay their eggs.

Feeding again varies enormously but insects usually have paired mouthparts that are adapted to a variety of different modes of feeding, for example cutting, chomping, piercing and sucking.

5.16 Conclusions

These major groups represent the most important members of the invertebrates and the ones that fisheries managers need to know. Examine any food chain or food web and there will be invertebrates in key roles. Without these key members of the web, vital roles will not be achieved, nutrients will not be recycled properly and the ecology of the water body will collapse. A considerable amount of fisheries management can be achieved by encouraging a good variety and abundance of invertebrates to live in the water.

It is also important to recognize the role of invertebrates in recycling dead organic material and to make sure conditions are maintained so that this work can be achieved.

The zooplankton invertebrates such as the rotifers and the various water fleas are critically important as the key link between the phytoplankton and the higher animals such as fish, which cannot feed directly on phytoplankton.

6 Fish

6.1 Introduction

Fish are a huge group of animals, represented by over 28 000 different species. They are the oldest as well as the largest group of vertebrates. They are characterised by being almost exclusively aquatic throughout their lives. They are limbless, but have two sets of paired fins as well as a variety of single fins. They breathe predominantly by using organs known as gills. They are cold-blooded (poikilothermic), which means their body temperature varies with that of their surroundings. Their skin is usually covered with scales.

Most temperate freshwater fish are called teleost ('true bone') fish, which means that they have a bony skeleton. The only fish found in temperate freshwater that do not belong to this group are the sturgeons and the lampreys. Neither of these groups is found commonly in still-waters, although a few species of sturgeons have been introduced into still-waters in recent years.

Teleost fish are extremely successful and well adapted to their environment.

6.2 Fish species

There are around 50 species of fish found for at least part of their lives in the freshwaters of the British Isles. This is quite a small number of species considering the diversity of the aquatic habitats available to them and the number of species in similar waters in continental Europe. Although many of the species are native to the British Isles, there are also many that have been successfully introduced over the centuries. Some of these have settled and become well established and are now generally considered 'native', for example the carp. Some, however, still raise concerns among conservationists about the potential impact they may have, should they become more widespread; the zander and wels catfish for example, are described as 'non-native' and are subject to regulation.

Fish species have developed and evolved to make the best use of certain habitats. Many of the species present in the British Isles are especially developed to live in rivers. These species are usually unsuited to living in still-waters as they are not adapted to this very different type of habitat. In this chapter we will concentrate on those species already adapted to the still-water, and those that are commonly stocked in still-waters with apparent success, such as the chub. The barbel has been included, as it is stocked commonly in still-waters, but it should be borne in mind that special conditions must be maintained for this species owing to its requirement for clear, oxygen-rich water.

This chapter sets out some detailed information about each species, which should allow fishery managers to understand better the species under their custody. This, along with Chapter 2 on the ecology of lakes, should allow the manager to establish the best conditions for the fish species and assess any other species for likely success in any given situation.

The fish described in this chapter are the following:

Family Cyprinidae

- Barbel
- Bream
- Silver bream
- Common carp
- Chub
- Crucian carp
- Goldfish
- Grass carp
- Gudgeon
- Minnow
- Orfe
- Roach
- Rudd
- Tench

Family Esocidae

- Pike

Family Percidae

- Perch
- Ruffe
- Zander

Family Anguillidae

- European eel

Family Gasterosteidae

- Three-spined stickleback
- Nine-spined stickleback

Family Siluridae

- Wels catfish

6.2.1 *Cyprinidae*

Barbel *(Barbus barbus) (Fig. 6.1)*

Figure 6.1 Barbel, *Barbus barbus*.

Recognition features

- Long streamlined body.
- Dorsal fin prominent with a pointed tip and slightly concave edge.
- Large cycloid scales, 55–65 along lateral line.
- The mouth is on the underside of the head and has four distinct barbels.
- Golden bronze with a dark back and creamy white underside.

The barbel is a large river fish with a long, streamlined, slightly flattened body. The juveniles are an overall brown colour but as they get larger they develop a dark brown back with golden bronze flanks and a creamy white underbelly. Adults generally grow to around 70 cm, but can get up to 90 cm and weigh in excess of 6 kg (the British rod-caught record 2008 is 9.837 kg). The barbel is native to the southern UK and mainland Europe. It is widespread and relatively common.

The barbel will generally mature after 3–5 years. Spawning takes place once a year in the late spring, usually May–July. Barbel spawn in small groups among gravel and stones in fast-flowing rivers. The adhesive eggs stay within the gravel bed until they hatch 10–15 days later. The fry tend to congregate in slow flowing areas where they feed on zooplankton. As they mature and get larger they tend to move out into the main flow of the river, feeding on invertebrates and even small fish in among clean gravel and associated vegetation.

The barbel is a very popular coarse fish, being one of the most sought-after river fish. They are a hard fighting fish that can grow very large, attracting specialist anglers.

Barbel have increasingly been stocked in still-waters over several years. They appear to grow and survive reasonably well, but there is little evidence that they can successfully spawn in most still-waters. It is vital that any stocking is with barbel obtained from a reputable fish farm. Barbel can sometimes be introduced from fish poached from the wild. Not only is this very harmful to the wild populations but evidence suggests that these fish find it very difficult to make the transition from rivers to still-waters. Generally barbel should not be stocked in shallow still-waters with a predominance of bottom feeders such as carp.

Common bream *(Abramis brama) (Fig. 6.2)*

The common bream is a large, bottom-feeding species with a deep thin body. When young it is a very silvery fish, but as it matures it becomes darker and more golden brown to green. It can grow to nearly a metre in length and weigh up to 8 kg (the British rod-caught record 2008 is 8.329 kg). The common bream is native to the UK and central Europe where it is widespread and common. It is a popular food fish in eastern Europe.

Recognition features

- Deep, laterally compressed body.
- Dorsal fin set well back.
- Long anal fin extends from middle of body to tail (24–30 branched rays).
- Deeply forked tail fin.
- Small cycloid scales, –51–60 along the lateral line.

Figure 6.2 Common bream, *Abramis brama*.

Bream generally mature after 4 years. Spawning takes place in May and June in dense vegetation in shallow water. The males develop white bumps (tubercles) over the body surface and become territorial. The females spawn at night in large groups and their sticky eggs attach to the vegetation. After 5–10 days the eggs hatch. The fry start by feeding on zooplankton such as rotifers and then larger zooplankton such as daphnia. They quickly move on to their normal adult diet of benthic invertebrates, preferentially worms, molluscs and insect larvae such as chironomid midge larvae. The adults feed by using their extendable tubular mouth to root around in the pond bottom, taking large mouthfuls and 'winnowing' out the food items. They are strongly shoaling fish from larval fry to late adulthood. Some of the very large individuals appear to be solitary or in small shoals.

The common bream is one of the most popular of the coarse fish. It responds well to anglers' baits and is readily caught. It grows quickly to a good size and survives well in many different conditions. As a bottom feeder it will disturb the bottom of the pond, causing suspended solids to cloud the water column. They are good in a mixed silver-fish population. The common bream should not be mixed with large numbers of carp as the species compete for the same food source.

Silver bream *(Abramis bjoerkna) (Fig. 6.3)*

Recognition features

- Deep laterally compressed body.
- Dorsal fin set well back.
- Long anal fin extends from middle of body to tail (19–24 branched rays).
- Deeply forked tail fin.
- Larger cycloid scales, –40–45 along the lateral line

Figure 6.3 Silver bream *Abramis bjoerkna*.

The silver bream is a small mid-water species, with a deep, compressed body. When young it is a very silvery fish but as it matures its back becomes a darker olive/grey, although it still retains a very silvery look. Its maximum size is around 20–25 cm in length (the British rod-caught record 2008 is 0.425 kg). The silver bream is native to England, central and northern Europe, where it widespread and reasonably common. In England it is generally restricted to the southeast but is becoming more widespread.

Silver bream mature at around 3–4 years old. Spawning takes place in May–July in dense vegetation in shallow water. They spawn in groups early in the morning, scattering their sticky eggs among the submerged weeds. The eggs hatch in 4–8 days and begin feeding on small planktonic invertebrates, moving to larger invertebrates and plant material as they grow. Silver bream form strong shoals with individuals of a similar size and age throughout their lives.

The silver bream is not a popular species for anglers, largely because it is commonly mistaken for a small common bream. Indeed it is very similar to a common bream of the same size, except the silver bream is slightly thicker in the body and has larger scales. They do make useful additions to a lake as they feed in a very different manner.

Common carp *(Cyprinus carpio) (Fig. 6.4)*

Figure 6.4 Common carp, *Cyprinus carpio* (mirror variety).

Recognition features

- Large, thick-set, rounded body.
- One long and one short barbel on either side of mouth.
- Large cycloid scales, –33–40 along the lateral line.
- Long slightly concave dorsal fin.

The carp is a large, bottom-feeding species with a thickset rounded body. They are usually a dark olive brown on the back, lightening to a light yellowish brown on the belly. It can grow to very large sizes, sometimes exceeding 1 m in length (the British rod-caught record 2008 is 26.9 kg). The carp is native to eastern Europe and Asia, but has now achieved worldwide distribution. It is very widespread and common in the UK, where it was first introduced in the nineteenth century.

Carp mature after 2–3 years. Spawning takes place in June to July in dense vegetation on the margins of the water body. The males develop small white lumps (tubercles) on the flanks and head. The females spawn at dawn and the sticky eggs attach to the weed. The eggs will hatch after 5–7 days and the fry start feeding on zooplankton, particularly rotifers. They quickly move to feeding on the bottom-dwelling invertebrates that make up the bulk of their diet. Their mouth is well designed for this feeding technique, being protrusible. However, carp are a very adaptable species and they can often be seen feeding on invertebrates on marginal vegetation and even off the surface. They tend to move around in small shoals of similar-sized individuals.

The carp is probably the most popular species for anglers on still-waters. It grows quickly to very large sizes and responds well to a variety of anglers' baits. It is a very adaptable species so can survive and thrive in many different conditions. As a bottom feeder, it will disturb the sediment on the bottom of the water body, causing suspended solids to cloud the water.

Carp have been bred for food over many centuries and it is one of the main global species for aquaculture in ponds. This has led to several different varieties of carp being commonly found. The most obvious of these varieties is the common carp, which is the fully scaled type. Mirror carp were bred to make it easier to remove the scales before cooking. Ornamental varieties are also common, from full-colour varieties, known as koi, to slightly gold-tinged cross breeds known as ghost carp. These are all the same species of fish, however.

Chub *(Leuciscus cephalus) (Fig. 6.5)*

Figure 6.5 Chub, *Leuciscus cephalus.*

Recognition features

- Long, rounded, thick-set body.
- Large cycloid scales, –45–48 along the lateral line.
- Dorsal and anal fin large with obvious convex edge.
- Rounded anal fin and caudal fin often with a dark edge.

The chub is a large, mid-water and surface-feeding fish with a rounded, streamlined body. When young it is a very silvery fish with an olive green back; as it gets larger its upper body darkens to a steel grey colour but retains the silvery flanks and belly. It is a large fish, attaining sizes of between 40 and 50 cm, rarely getting to up to 80 cm (the British rod-caught record 2008 is 3.9 kg). The chub is native to the UK and central Europe and is widespread and common. It is caught for food in some parts of Europe.

Chub generally mature after 3–4 years. Spawning takes place during April–June among clean gravel and stones in flowing water. They spawn in the early morning in small groups of one female with several males in attendance. The eggs hatch within 6–8 days and start to feed on benthic invertebrates. As they grow older their diet changes to include larger invertebrates, fruit, vegetation and even small fish. When young, chub move around in shoals but as they mature they become much more solitary and even appear to be somewhat territorial. In rivers they migrate significant distances to feeding and spawning areas. Little is known about their activities in still-waters.

The chub is a popular fish among anglers because it is a large, hard-fighting fish that responds well to anglers' baits. It grows quickly to a good size and appears to do very well in still-waters. Current evidence suggests that it does not successfully spawn in most still-waters, although it does spawn in large, clear-water lakes.

Crucian carp *(Carassius carassius) (Fig. 6.6)*

Figure 6.6 Crucian carp, *Carassius carassius*.

Recognition features

- Very deep bodies with marked lateral compression.
- Dorsal fin long with a marked convex edge. Caudal fin with blunt lobes.
- Large cycloid scales, 33 on the lateral line, which is usually incomplete and fades towards the tail.
- Mouth slightly upturned.
- Golden bronze colour.

The crucian carp is a medium-sized, deep-bodied fish with marked lateral compression. Young fish are a golden bronze colour, which turns darker as they get older, with mature specimens being olive brown on the back through deep bronze on the flanks to a golden yellow on the belly. Adults can attain 20–30 cm in length and 1.5 kg in weight, although many crucian carp populations remain 'stunted' with the mature adults rarely getting more than 15–20 cm and 0.5 kg (the British rod-caught record 2008 is 2.01 kg). The crucian carp is native to the UK, Europe and central Asia. There is some argument about whether it is a native to the UK and western Europe, but there seems little evidence to suggest it is introduced into these areas, where it is widespread.

Crucian carp generally mature after 2–3 years. Spawning takes place in May–June in shallow water, in thick vegetation. The fish shed sticky eggs onto the vegetation where they take between 5 and 7 days to hatch, depending on the vegetation. Once hatched, and after the yolk sac has been used, the fry tend to form loose shoals and feed on small zooplankton. As they grow they start to feed mainly on invertebrates, such as molluscs and worms on the bottom of the pond, as well as plant material. Both the juveniles and the adults tend to live in loose shoals. They tend to occupy the mid-water and bottom of still-waters. Crucian carp prefer still-waters and particularly thrive in rich, lowland still-waters with abundant vegetation. Crucian carp have a remarkable ability to survive very low dissolved oxygen situations for long periods, leading to situations where they are often the only fish species to survive. Here they form healthy populations dominated by small 'stunted' individuals. In larger waters they can grow much larger and seem to adopt a slightly different body shape.

This is a very popular angling species and is readily caught. It survives well in well-managed fisheries dominated by silver fish, but it is generally unsuitable for stocking in waters dominated by common carp.

Crucians readily hybridise with goldfish and common carp. They are then displaced by the vigorous hybrid. They are also threatened by the non-native parasitic tapeworm *Bothriocephalus acheilognathi*.

Goldfish *(Carassius auratus) (Fig. 6.7)*

Figure 6.7 Goldfish, *Carassius auratus*.

Recognition features

- Deep, slightly laterally compressed body.
- Dorsal fin long with a concave edge. The main hard fin ray at the front of the fin is strongly serrated.
- Caudal fin is deeply forked.
- Large cycloid scales, with 27–31 along the lateral line.

The goldfish is a medium–sized mid-water species with a relatively deep and laterally compressed body. Wild goldfish are a deep brown colour throughout their lives, although there are many colour variants from deep orange through to almost white. Adults can get to around 30 cm and 1 kg in size. The goldfish is native to Asia, but owing to its ornamental value it has achieved worldwide distribution, being found on most continents. It is an introduced species in the UK.

Goldfish generally mature after 2–3 years. Spawning takes place during June and July in thick weed in the shallow margins. The eggs hatch in 4–7 days, depending on the water temperature, and the young feed on zooplankton. As they grow, the diet changes to small invertebrates and plant material. Goldfish tend to form loose shoals and prefer shallow, rich ponds with abundant vegetation. They are a hardy species, however, and appear to thrive in most still-water conditions.

The goldfish has become increasingly common in still-waters in the UK and Europe. It is apparently popular with some fishermen.

Goldfish readily hybridise with common carp and crucian carp, producing vigorous offspring that can successfully outcompete the original parent stock. Stocking with goldfish or hybrid strains is advised against.

Grass carp *(Ctenopharyngodon idella) (Fig. 6.8)*

Figure 6.8 Grass carp, *Ctenopharyngodon idella*.

Recognition features

- Slender rounded body, with a large head and small eyes set low on the head.
- Large cycloid scales.
- Small dorsal fin with a convex trailing edge.
- Anal fin with a concave trailing edge.

The grass carp is a large mid-water feeding species with a solid, rounded body. It has a dark blue-black back, shading to silver on the belly. It can grow quite large, with adults reaching up to 100 cm in length (the British rod-caught record 2008 is 20.185 kg). Grass carp are originally from Asia, but have been widely distributed around the world as a method of weed control. They were introduced into England in the 1960s to control weed growth in some still-waters. Since then they have become widely distributed, but because of their specialised breeding requirements there appear to be no breeding populations.

Grass carp mature after 3–4 years. Spawning takes place in fast-flowing water in the headwaters of large rivers. Temperatures must be around 23–25°C. The eggs are unusual in that they are pelagic, being carried down river in mid-water until they hatch in the slower flows downstream. The fry start by feeding on small invertebrates, but as they get older they begin to feed on aquatic plants. Grass carp form loose shoals throughout their lives.

Grass carp are not particularly popular among anglers as they are generally difficult to catch and do not fight as hard as their size might suggest. They are not very effective in controlling weed growth as they do not feed until the waters are quite warm.

Gudgeon *(Gobio gobio) (Fig. 6.9)*

Figure 6.9 Gudgeon, *Gobio gobio*.

Recognition features

- Rounded, slightly dorsally compressed body.
- Dorsal fin large and triangular in shape.
- Large cycloid scales, –40–45 along the lateral line.
- Two well-developed barbels.

The gudgeon is a small, rounded, bottom-dwelling fish. Its back is a dark olive green to brown with lighter flanks and a silvery belly. It has dark markings along the flank and upper body. It is a very small fish, rarely reaching over 15 cm although it can grow to about 20 cm. (the British rod-caught record 2008 is 0.141 kg). Gudgeon are widely distributed throughout Europe and Asia and they are native to the UK.

Gudgeon mature at between 2 and 3 years old. They will spawn on stony or gravely bottoms, or among weed, and prefer flowing water, although they will spawn successfully in still-waters. They spawn during May or June and the eggs hatch in around 15 days. They feed on insects, molluscs, crustaceans and occasionally plant material. They are generally a solitary fish, but sometimes form into small shoals of perhaps a dozen individuals.

The gudgeon is not a popular target species for anglers, but it does do well in some still-waters where it probably makes an important contribution to the health of the lake.

Minnow *(Phoxinus phoxinus)*

> **Recognition features**
>
> - Slender rounded body.
> - Small cycloid scales, ~80–100 scales along the lateral line.
> - Dark brown or black stripe along the length of the body.

The minnow is a small mid-water species with a long rounded body. It has a brown-gold body with numerous brown and black spots down the flanks usually joining to form an obvious black line along the body. It is a small fish, rarely exceeding 10 cm in length (the British rod-caught record 2008 is 0.135 kg). The minnow is a native to the UK, Europe and Asia where it is very widespread and common.

Minnows mature after 1–2 years. Spawning takes place in June or July, usually over clean stones or gravels. The males particularly become more colourful during spawning, with a scarlet belly and black throat. The eggs take 5–10 days to hatch and immediately start feeding on small invertebrates. As they grow they continue to feed on a variety of invertebrates. Minnows generally aggregate into small, loose shoals.

The minnow is not a target fish for anglers, although it can be important as a bait fish. It does well in lakes and ponds and can be a very useful addition to the lake ecology, particularly if predatory fish are present.

Orfe *(also known as ide) (Leuciscus idus) (Fig. 6.10)*

Figure 6.10 Orfe, *Leuciscus idus.*

> **Recognition features**
>
> - Long, rounded, slightly compressed body.
> - Large cycloid scales, ~55–61 along the lateral line.
> - Dorsal and anal fin large with slightly convex edge.

The orfe is a large, mid-water and surface-feeding fish with a rounded, streamlined body. When young it is a very silvery fish, but as it gets larger its upper body darkens to a brownish grey colour. Coloured ornamental varieties are common, particularly golden and blue orfe. It is a medium-sized fish, attaining sizes of between 30 and 40 cm, although it can occasionally get up to 80 cm. (the British rod-caught record 2008 is 3.7 kg). The orfe is native to central and eastern Europe and parts of Asia, where it is widespread and common.

It is caught for food in some parts of Russia. It was introduced into England and is now widespread and common.

Orfe generally mature after 3–4 years. Spawning takes place during April and May, among weeds and clean stones in shallow water. The eggs hatch within 15–20 days and start to feed on benthic invertebrates. As they grow older their diet changes to include larger invertebrates, although they will take fruit, vegetation and even small fish. Orfe move around in shoals of similar-sized and aged individuals. They are found equally in rivers and still-waters.

The orfe is a popular fish among anglers because it is a good-sized, hard fighting fish that responds well to anglers' baits. It grows quickly, to a good size and it appears to do very well in still-waters.

Roach *(Rutilus rutilus) (Fig. 6.11)*

Figure 6.11 Roach, *Rutilus rutilus*.

Recognition features

- Slightly laterally compressed body.
- Dorsal fin directly above base of pelvic fins.
- Large silvery cycloid scales, –42–45 along the lateral line.
- Upper lip slightly prominent.
- The iris (the eye surround) is generally red.

The roach is medium-sized, mid-water fish with a relatively deep body. The young fish are silvery, but they quickly develop a dark blue–green back and light underside. Adults generally get to 20–40 cm and between 1 and 2 kg (the British rod-caught record 2008 is 1.899 kg). The roach is native to the UK and Europe, where it is very widespread and common throughout the lowland catchments.

Roach generally mature after 2 or 3 years. Spawning takes place once a year in the middle of spring, usually April–June. The roach spawn at dawn, shedding their sticky eggs on dense, submerged vegetation in the shallow areas of the lake. The eggs will hatch in 5–10 days. Once the yolk sac is used, the fry will feed on zooplankton and other small invertebrates. The juveniles prefer to live in large shoals in the shallow, marginal areas with plenty of vegetation, which provide cover and food. As they mature and get larger they move out into open water, shoaling in smaller groups of similar-sized individuals. These larger individuals feed on larger benthic invertebrates and pondweed, particularly attached algae (periphyton). They tend to occupy the mid-water to bottom of the pond.

The roach is one of the most popular coarse fish in the UK. It responds well to anglers' bait and is readily caught. It grows steadily and survives well in many different conditions. It is an excellent species in any mixed fishery as it is very adaptable. In the right conditions, it can quickly overpopulate with a very large population of small fish.

Rudd *(Scardinius erythrophthalmus) (Fig. 6.12)*

Figure 6.12 Rudd, *Scardinius erythrophthalmus.*

Recognition features

- Deep, laterally compressed body.
- Dorsal fin distinctly behind the pelvic fin base.
- Pectoral, pelvic and anal fin red.
- Large cycloid scales, –40–45 along the lateral lines.
- Distincive upturned mouth.
- Distinct keel and anal fin.

The rudd is a medium-sized, top-feeding species, with a deep, laterally compressed body. It has a dark greenish-brown back, grading through to a silvery underside. The flanks are golden brown. The iris is a deep gold with a red fleck on the upper side. The fins are normally a bright red. It can grow to 20–25 cm and 0.5–1 kg (the British rod-caught record 2008 is 2.1 kg). The rudd is native to the UK and central Europe, where it is very widespread and common.

Rudd generally mature at 2–4 years and live to 10–14 years. They spawn in late spring, usually from April to July. Rudd spawn by shedding their eggs on submerged vegetation around the margins. The eggs hatch in around 5–10 days and the fry stay within the protection of the vegetation, feeding on small zooplankton. As they grow they form compact shoals, feeding on the surface and in midwater. They prefer zooplankton, weed-dwelling invertebrates and terrestrial insects on the surface. Rudd appear always to spend their time in fairly compact shoals, no matter how large they get.

The rudd is a popular fish to catch, being attractive and readily taking bait, which is particularly important if other species are more reluctant to feed on anglers' bait. It grows steadily but rarely attains large sizes. It is a very good fish for a mixed fishery, as it does not compete directly with any other fish species.

Tench *(Tinca tinca) (Fig. 6.13)*

Figure 6.13 Tench, *Tinca tinca.*

Recognition features

- Body thickset and very slightly laterally compressed.
- Dorsal fin high and distinctly arched. All fins distinctly rounded.
- One small barbel at either edge of a small, thick-lipped mouth.
- Very small, deeply embedded scales, 95 or more along the lateral line.
- Olive green with a small red/orange eye.

The tench is a large, bottom-feeding species, with a thickset, slightly laterally compressed body. The tench is a beautiful olive to dark green throughout its life, with small orange to red eyes. Adults generally grow slowly to a maximum length of 60–70 cm and can weigh up to 7 kg. (the British rod-caught record 2008 is 6.9 kg). The tench is native to the UK and Europe, although absent from the northern parts of both. The species has been widely introduced into Asia and North America.

Tench generally mature at 3–4 years old. Owing to their slow growth, they may only be 15 cm at this age. Spawning takes place once a year in spring, usually between May and July. They spawn among aquatic plants in shallow waters, to which their sticky eggs attach. The eggs will hatch in 4–5 days, and once the yolk sac is used the fry will feed on zooplankton, before switching to grazing among the bottom of water plants on a variety of aquatic invertebrates and attached algae. Tench prefer still-waters, or very slow flowing rivers with an abundance of macrophytes. The juveniles may form loose shoals, but as they mature they become more solitary, almost appearing to have preferred territories.

The tench is one of the most popular and easily recognisable of the coarse fish in the UK and Europe. Its popularity relates to its size and its beauty. It is generally a difficult fish to catch on rod and line. It is quite adaptable to a mixed fishery, but does not thrive in waters without good communities of aquatic macrophytes.

6.2.2 Esocidae

Pike *(Esox lucius) (Fig. 6.14)*

Figure 6.14 Pike, *Esox lucius*.

Recognition features

- Long, thin streamlined body.
- Dorsal and anal fin set well back on the body.
- Very large mouth filled with many backward-pointing teeth.
- Small cycloid scales, 110–130 along the lateral line.

The pike is a large predatory fish, with a slim streamlined body and a large head with an enormous mouth full of teeth. It has a dark green, mottled back, grading through light green flanks to a creamy yellow underbelly. It is a very large fish, routinely growing to 50–100 cm and occasionally 140 cm and 30 kg (the British rod-caught record 2008 is 21.2 kg). The pike is widespread and abundant throughout the Northern Hemisphere in Europe, Asia and North America. It is a popular food fish wherever it is found.

Pike generally mature after 3–4 years, the males earlier and smaller than the females. They spawn among weed during February–May. The eggs attach to the weed and hatch after 10–15 days. When they hatch they start by feeding on invertebrates but quickly turn to small fish and then onto larger prey, including other vertebrates such as ducks and water rats. They are very partial to other pike and are, unsurprisingly, very solitary and territorial. It appears that females grow much larger than males, and it is said that any pike over 5 kg will be a female. They are an ambush predator, laying in wait for their prey and only striking at the very last minute when the prey is within one to two body lengths.

The pike is one of the most popular freshwater fish in the UK. It grows exceptionally quickly to a large size and is a ferocious fighter, particular in the first moments as it attacks the bait.

6.2.3 *Percidae*

Ruffe *(Gymnocephalus cernuus) (Fig. 6.15)*

Figure 6.15 Ruffe, *Gymnocephalus cernuus*.

Recognition features

- Single, large dorsal fin with obvious spines at the front and soft rays at the back.
- Very large eyes and mouth.
- Covered in small ctenoid scales, giving the skin a rough feel.
- Large spines at the posterior end of the operculum.

The ruffe is a small fish with an obvious spiny dorsal fin, large eyes and mouth. Its back is a dark green-brown colour with irregular dark blotches, whereas its underside is pale yellow to cream in colour. It usually grows to around 15 cm but can exceptionally grow to 25 cm (the British rod-caught record 2008 is 0.142 kg). The ruffe is native to much of Europe but was confined to eastern England, although its range has spread considerably in recent decades.

The ruffe matures after 1 to 2 years. They spawn in shallow water among stones and vegetation, mainly in April to May, although they probably spawn several times in a year. Initially the fry grow quickly on a diet of small invertebrates before moving on to larger invertebrates and other food such as fish eggs and young fry. The ruffe is a shoaling fish, moving around in groups throughout its life.

The ruffe is not a popular fish for anglers, being such a small size and attacking baits destined for other larger target species. It has numerous alternative common names such as pope, tommy and daddie.

Perch *(Perca fluviatilis) (Fig. 6.16)*

The perch is a beautiful, deep-bodied fish with four to six dark bands over olive-green flanks. The abdomen is creamy silver with prominent red pelvic and anal fins. It can grow to 20–40 cm and weigh 1–2.5 kg (the British rod-caught record 2008 is 2.523 kg). The perch is native to the UK and Europe and is now widespread and common in the UK, northern Europe and Asia. It is a popular food fish in many countries; in England it used to be canned and sold as 'perchines'.

Recognition features

- Slightly laterally compressed body.
- Two dorsal fins. The first very spiny with a dark spot to the rear.
- Heavily armoured head with obvious bony plates.
- Small ctenoid scales, –60–70 along the lateral line.
- Red pelvic and anal fins.

Figure 6.16 Perch, *Perca fluviatilis*.

Perch generally mature after 3–4 years and live to around 10 years. They spawn in early spring, usually March to May. Perch tend to spawn in the margins, producing long strings of eggs which lie wrapped around submerged vegetation. The eggs hatch after 10 days and the juveniles feed on small invertebrates as well as fish fry, including other perch fry. As they get older they start to feed on larger invertebrates and begin to feed actively on other fish. Small perch shoal in groups but as they get bigger they do become more solitary, with large individuals tending to move around in small shoals of just a few individuals.

The perch is a popular fish for coarse fisherman, being one of the most attractive of the freshwater fish species. It responds well to anglers' baits and is readily caught. As a predator it is a useful fish to control the populations of other fish species. However, they can mature at a very small size (10–15cm) and in certain circumstances they can create an abundance of small perch. It appears that perch only grow to a large size where there is a plentiful supply of small prey fish.

Zander *(Sander lucioperca) (Fig. 6.17)*

Recognition features

- Long, thick-set body.
- Two dorsal fins, the first with obvious spines.
- Large jaws with obvious teeth throughout.
- Small ctenoid scales, –80–95 along the lateral line.
- Irregular dark stripes vertically on the flanks.

Figure 6.17 Zander, *Sander lucioperca*.

The zander is a large predatory species with a long, thick body. The back is a dark brown–green with obvious dark brown vertical stripes running to the middle of the body. It grows quite large, reaching average sizes of 50–70cm, occasionally much bigger to a maximum of 120 cm (the British rod-caught record 2008 is 8.7 kg). The zander is common throughout Europe. It was deliberately introduced into England (the River Great Ouse) in 1963 and has spread widely since.

Zander generally mature after 3–5 years. Spawning takes place from April to June in among marginal vegetation. Pairs of fish spawn together at dawn and lay sticky eggs on vegetation and stones on the bottom. Unusually, both parents will guard the eggs until they hatch 5–10 days after spawning. The fry start on a diet of invertebrates but move quickly on to eat mainly of other fish. The zander is a hunter, patrolling open waters and chasing down small prey fish. Occasionally it appears to hunt as a group, although it mainly hunts as an individual.

The zander is a very popular fish with anglers, being a fast-growing, large and aggressive fish. It survives well in larger, clear still-waters and may be a useful addition to a specimen water as it prefers small fish as prey. Licences are required to stock this species.

6.2.4 Gasterosteidae

Three-spined stickleback ***(Gasterosteus aculeatus) (Fig. 6.18)***

The stickleback is one of the most common and recognisable species in freshwaters. It is a small, thin-bodied fish with an olive-brown back and silvery sides and underside. The body is covered in large bony plates and there are three obvious strong spines on the back. The stickleback is a very small fish that rarely exceeds 10 cm in length (the British rod-caught record 2008 is 7 g (0.007 kg)). Sticklebacks are widely distributed and common throughout Europe.

Sticklebacks mature after 1–2 years. Spawning takes place in March–June. They are unusual in that the male builds a nest of fibrous plant material and attracts females to spawn within it. The male also develops a vivid red throat to help attract a mate. The eggs take 10–20 days to hatch and are guarded constantly by the male. This guarding behaviour

Figure 6.18 Stickleback, *Gasterosteus aculeatus.*

continues until the fry are a few weeks old. The fry feed on small invertebrates. As they get older they continue to feed on invertebrates but may extend their diet to include fish fry. Sticklebacks tend to aggregate into small shoals, only becoming solitary and territorial during spawning times.

The stickleback is not a target species for anglers. It is common in most still-waters and will almost certainly appear without help.

Nine-spined stickleback *(Pungitius pungitius)*

This is a rare species. It is similar in most respects to the three-spined stickleback, except that it has 7–12 spines running along its back. It also goes dark with a black throat when in breeding condition.

6.2.5 Anguillidae

European eel *(Anguilla anguilla) (Fig. 6.19)*

Figure 6.19 European eel, *Anguilla anguilla.*

Recognition features

- Long, thin cylindrical body.
- Tiny cycloid scales deeply embedded in the skin.
- No pelvic fins.
- Anal, caudal and dorsal fin fused to form one long fin.
- Small apical mouth and small eyes.

The eel is a medium-sized fish with a very long, cylindrical body. The anal, caudal and dorsal fins all merge to form one long fin that goes around most of the body. It has no pelvic fins. The young stages are transparent, with pigment only becoming apparent as they become elvers in the rivers. The upper surface is generally a dark green–brown with a creamy yellow belly. As they mature and get ready to migrate, the back becomes very dark with a silvery white underside. The eel can grow to very large sizes, around 1.5 m, but more commonly they are 40–90 cm, (the British rod-caught record 2008 is 5.0 kg). The eel is widespread and common throughout western Europe and the Mediterranean. However, dramatic declines in the number of elvers returning to the European coastline mean that there are great fears for the long-term future of this species.

Little is known about the actual spawning of eels and this mystery is only slowly being revealed. It is known that the young, leaf-like larvae of the eels drift into the waters off the European coast in early spring, where they change into glass eels before running up the rivers and developing further into elvers. Once established in freshwaters they may remain for as long as 20 years, although estimates vary from 10 years to more than 40 years. Once they decide to migrate to the spawning grounds, they brook no obstacle, even going across land to reach their goal. Once they leave our shores, evidence suggests that they journey across the Atlantic Ocean to a stable area around Bermuda and the Bahamas, known as the Sargasso Sea. Here they spawn at great depth and the adults perish. The eggs hatch and drift on the Gulf Stream back to the coast of Europe, a journey that may take as long as 3 years. Once in freshwaters, they will feed on a variety of invertebrates and small fish. They are a nocturnal and solitary species, hiding during the day and roaming at night in search of food.

The eel is a very popular species among some anglers as it grows to a large size and is very difficult to tempt to take baits. However, to some anglers it is a nuisance species as it is difficult to handle and unhook.

6.2.6 *Siluridae*

Wels catfish *(Siluris glanis) (Fig. 6.20)*

Figure 6.20 Wels catfish, *Siluris glanis.*

Recognition features

- Large, flat head with a huge mouth.
- Elongated body with very smooth scaleless skin.
- Very small dorsal fin that is high on the shoulders.
- Long anal fin extending along nearly half the length of the fish.
- One pair of very long barbels, two pairs of shorter barbells.

The wels catfish is a huge, bottom-dwelling scavenger, with a large, flat head and an elongated body. Its back is normally a dark, mottled dark blue/brown black. The sides are a light reddish brown and the belly is pale. A dark mottling covers the whole body. The wels catfish can grow to a very large size, up to 3 m in length and 200 kg (the British rod-caught record 2008 is 28.1 kg). It is widespread throughout Europe, including the brackish Baltic and Black Seas. It has been introduced into several countries including England, where it is now quite widespread.

Wels catfish generally mature after 4–5 years. Spawning takes place in a shallow excavation made by the male fish. The wels catfish pair up as the male attracts females into his territory. The eggs are laid in the shallow depression and then guarded by the male until they hatch. The fry start on a diet of small invertebrates but quickly move on to eat large invertebrates, fish and other vertebrates. They are mainly scavengers, using their long barbels to hunt across the bottom for food. However, they are also a keen and active predator, taking large prey items including ducks from the surface. They are a solitary species, particularly as they get larger as they are highly cannibalistic.

The wels catfish is understandably popular with anglers, being such a large and strong fish. However, it is relatively difficult to catch and does not grow very large in most still-waters.

6.3 Fish anatomy and physiology

6.3.1 Body structure

The body form of fish reflects the sort of life they lead and the habitat they live in. The basic shape (Fig. 6.21) is adapted to moving through water with the minimum of resistance. Water is considerably denser than air, so the ability of a fish to move freely through it is dependent on it having a streamlined shape. Unsurprisingly, therefore, most freshwater fish look very similar to each other in overall body shape. The basic shape is a torpedo, with a rounded cross section, getting smaller at the front and back, with the fins placed on the outside to allow propulsion and stability. This is known as a fusiform body shape. Different shapes have adapted from this basic pattern; for example, many still-water fish have evolved towards a slightly deeper body form because they do not need to maintain position against a current in the way a river fish would. Most other alternatives to this basic body form are adaptations for other needs, such as reproduction or feeding, as we will discuss later.

There are several ways an animal may propel itself through the water, by using combinations of fins as highlighted in Figure 6.21. However, most temperate freshwater fish have evolved the successful methods of undulating their caudal fin and sometimes their trunk as the main method of propulsion.

The source of the power for this swimming method comes from the lateral body muscles lying on either side of the fish's backbone (around 40% of the body weight of most cyprinids). These muscles can be split into two groups: the red muscle and the white muscle. The white muscle makes up the larger mass and is usually closer in to the backbone.

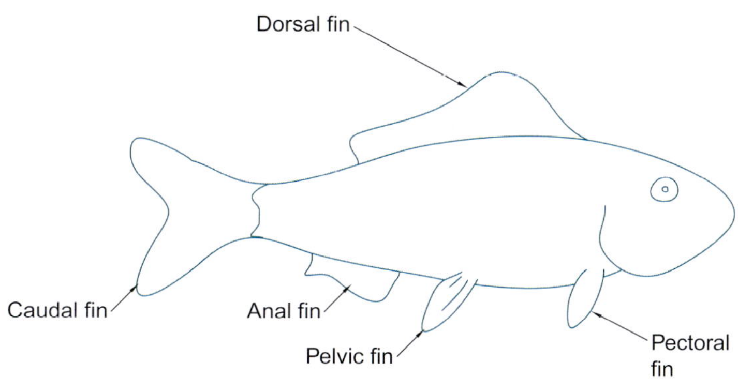

Figure 6.21 The basic shape of a fish.

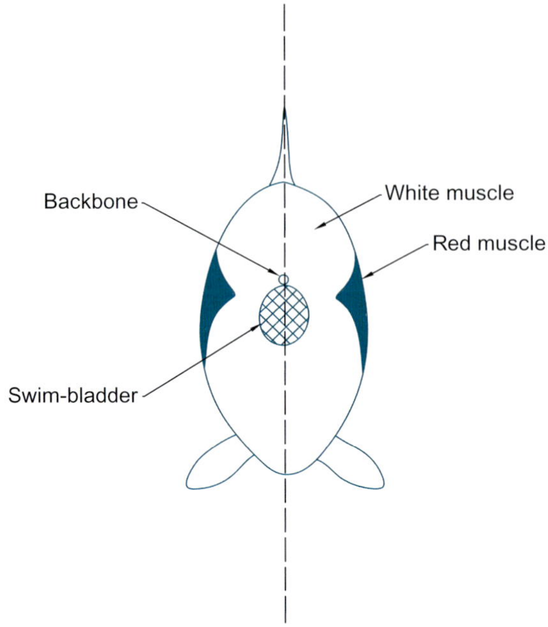

Figure 6.22 Cross section of a fish showing muscle blocks.

The red muscle is usually found in two blocks outside the white muscle, along the midline of the fish (Fig. 6.22). It is a useful generalisation to say that the white muscles are used to give fast, powerful, but brief swimming spurts, and the red muscles are used for the slow, steady, but long-lasting (sustained) swimming.

The other fins have a variety of uses, mainly to do with the stability of the swimming fish. With a fusiform body and a caudal fin powering away, there would be a tendency for the fish's head to move from side to side (yawing) and up and down (pitching) as it moves. There is also a tendency for the body of a fish to roll. These tendencies are eliminated by positioning the dorsal and anal fins on the top and bottom of the body, thus acting as keels and preventing yawing. The paired pectoral fins can be spread away from the body to prevent pitching and rolling. These paired fins are also very useful for the fish when braking and turning.

The use of the other fins decides exactly where they are placed on the body surface, depending on the lifestyle of the particular species. The various positions of the fins are highlighted in Figure 6.21.

There is another form of locomotion that is very important to fish, which is the slow and delicate movements that they may be required to perform. This is very dependent on one of the most important adaptations to life in water: the swim-bladder (Fig. 6.23). A fish's body is denser than water and therefore should sink. In fact, for some groups of fish like the sharks, this is exactly what happens. However, teleost fish have developed a swim-bladder. This is a gas-filled sac, near the centre of the body, that can be filled with air. The volume of air can be increased or decreased, depending on the depth at which the fish is swimming. This gas bladder is filled with enough air to make the fish 'neutrally buoyant' (meaning it will not sink or float). Thus a fish at rest will remain stationary in the water

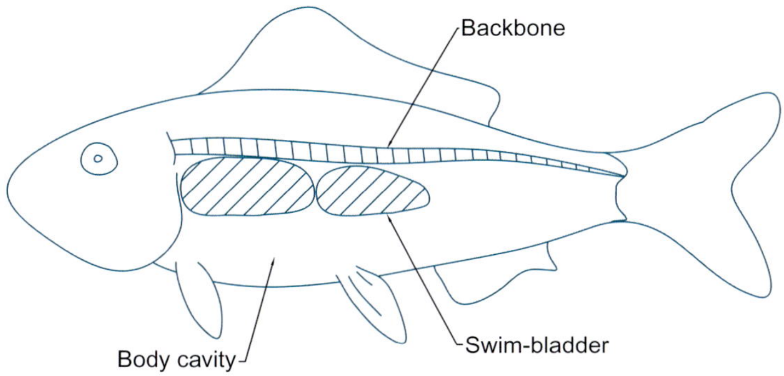

Figure 6.23 Side view of fish showing location of swim-bladder and spinal column.

column. Teleost fish have developed their various fins to be able to move them to and fro very gently. This adaptation is one of the most fundamental and important aspects of the body form of fish. It allows fish to fulfil a wide variety of roles and feeding techniques and so occupy a wide range of niches in the aquatic environment.

6.3.2 Sensory systems

How a fish understands what is happening in the world around it is very important and quite different to how we, as humans and land animals, may understand the concept. Water is quite a different environment and requires very different senses to those needed to live on land. Eyesight particularly is probably less important in the water, whereas hearing, taste and touch are much more important. We will deal with each of the sensory systems in turn before looking at how a fish understands its surroundings by using the information it can collect with its senses.

Sight

The aquatic environment has a lower light intensity than the land. Water, and the particles suspended in it, absorbs light, so light intensity decreases quickly with depth. Fish eyes therefore have adapted to use the limited light and the alternative wavelength of light available.

The overall design of the eye (Fig. 6.24) is similar to that seen in most vertebrates. It is essentially a fluid-filled ball, with a small opening at one side allowing light to enter, with a lens to focus that light onto a patch of receptor cells on the other side of the ball. Teleost fish have little or no pupil control, so they cannot alter the amount of light entering the eye. Some species have a mirror-like quality to the back of the eye, formed by special cells, called the tapetum lucidum. This allows full use of all the light that does enter the eye and is a feature of nocturnal fish such as the zander.

The lens is a spherical, clear ball held in place by muscle tissue. It can be moved backwards and forwards in the eye to focus the light available. The light passes through the middle of the eye, which is filled with a liquid known as the aqueous humour. The light is focused onto the back of the eyeball, known as the retina, which is where the cells responsible for light detection are found. In fish, the retina contains two types of receptor cell, known as

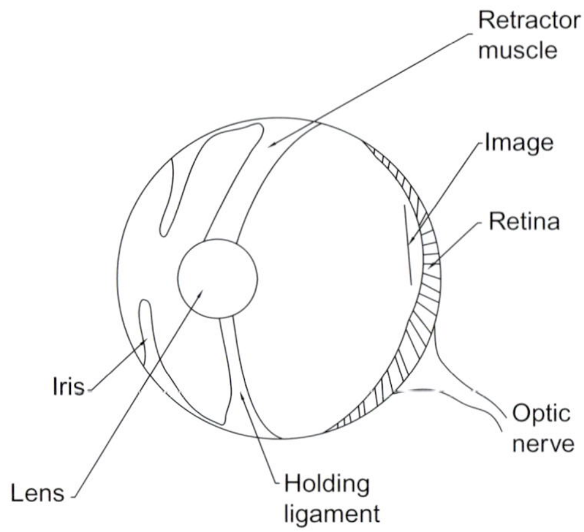

Figure 6.24 Design of the fish eye.

rods and cones, which contain the visual pigments. Most freshwater fish have cells containing all the visual pigments needed to confer full colour vision, as well as pigments that allow fish to perceive ultraviolet light.

The positioning of the eye on the fish's body is also important. Most fish have the eyes placed on the sides of the head. This allows good all-round vision, with some fish almost achieving the ability to look behind themselves at the same time as they look in front. However, this positioning limits the binocular vision (the point in front of the fish where both eyes can see the same thing). Binocular vision is very important for depth perception (telling how far something is away from the viewer), which is very useful for predators when they strike at their prey. So some predators, notably the pike, have eyes that are set to the front of the head to improve their binocular vision.

Smell

The nostrils are paired organs, usually found on either side of the top and front of the head. There are two openings into each of a fish's nostrils (olfactory sacs) so that water may enter through one opening and exit through the other. This flow is achieved by the design of the openings, allowing the water flow generated as the fish swims naturally to enter the front opening and exit through the rear opening. This current is further enhanced by the skin inside the nose being covered in fine hair-like structures (a layer called the ciliated epithelium), which beat in a uniform direction maintaining the flow of water. In all temperate freshwater teleost fish there is no connection between the nostril and mouth.

The inside of the olfactory sac is thrown into a series of folds, serving to increase the surface area of the nose significantly. These primary folds are further folded into smaller, secondary lamellae on which can be found the chemosensory cells. There are several different forms of these cells, responsible for detecting different types of chemical. There are millions of these chemically-sensitive cells in each olfactory sac.

Tests indicate there at least four groups of chemicals that stimulate the chemoreceptor cells in the nose. These probably relate to detection of food (amino acids) and pheromones, allowing fish to detect and communicate with members of their own species and to give warning of the presence of other species.

Taste

A sense very closely related to smell and one that is well developed in fish is the sense of taste. This is the primary sense for feeding so is obviously important. Among the teleost fish, the taste buds are densely packed in the mouth; in many species they come together in a group in the roof of the mouth and develop into an organ known as the palatial organ. This organ not only tastes the food but also has touch receptors and can produce protrusions that can retain food items while non-food items are spat out.

However, taste buds are not restricted to the mouth. In some species they are commonly found on small appendages that appear around the mouth, known as barbels. Furthermore, they can even be found on other areas of the body, such as the pectoral fins and even across the general body surface.

In common with other vertebrates, fish appear to be able to distinguish the four main characteristic tastes: sweet, sour, salt and bitter.

In addition to the senses of taste and smell that we would recognise as similar to those we use, some fish also have solitary chemosensory cells, similar to taste buds, scattered across the body surface. These detect a variety of chemicals and, although their purpose has not been completely established, they are likely to be used for sensing the presence of predators.

Hearing

Sound is very important underwater. Not only does sound travel five times faster in water than in air, it will also travel much further. Furthermore, sound is split into two components: particle displacement and pressure waves, both of which are important to, and can be detected by, fish using the acoustico-lateralis system. This system consists of both the inner ear in the head and the lateral line system, which is found on the trunk of the body and around the head.

The lateral line system is a long, gel-filled canal, with pores that open to the environment. The hearing system is based around a single type of mechanoreceptor cell called a hair cell (Fig. 6.25). This cell is able to detect when particle displacement sound causes movement of the gel, and thus moves the hairs. The hairs are also directionally sensitive. These cells are found in both the lateral line system and the inner ear.

Using the lateral line system along the body and around the head, fish can not only detect pressure waves, but also detect the directional source of these waves. This system will also detect movement in the water. It appears to be very useful in sensing the movement of nearby fish so is functional for shoaling fish as well as predator detection. Furthermore, when a fish is swimming, it sets up movement in the water around it similar to a bow wave on a boat. The lateral line system can detect this water movement and any changes in the bow wave, for example if the fish swims close to an object.

The inner ear can also detect pressure waves using hair cells, but it has several additional functions. The inner ear consists of a series of three semicircular canals and three sac-like organs, containing small bones. These small bones, or otoliths, are held in the sac-like organs by hair cells. Being denser than the rest of the fish, they tend to 'lag behind' if the fish moves, thus giving the fish a sense of movement. They are also affected by gravity and so give fish

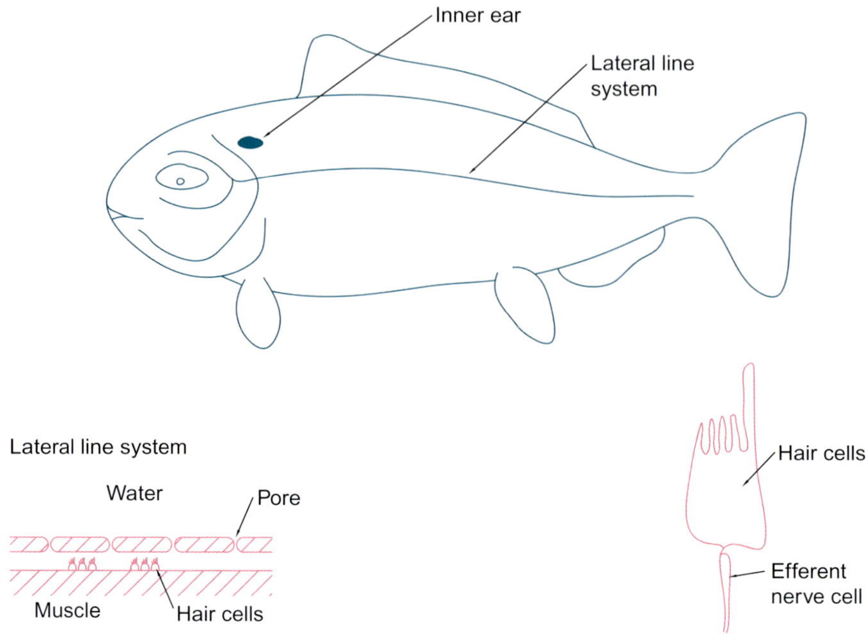

Figure 6.25 The lateral line of the fish.

a sense of up and down and of balance. Owing to the difference between the density of the otoliths and that of the fish tissue, this system can also detect the particle displacement part of a sound. However, how do fish detect the important pressure wave part of sound? Strangely, this is where the swim bladder has a function.

It is suggested, and backed up by considerable evidence, that fish use the swim-bladder to convert sound pressure into particle displacement that can be transmitted to the inner ear. The swim-bladder does this by pulsating slightly in sympathy with any sound pressure waves. This pulsating creates particle displacement in the fish tissue, which the fish can 'hear'. This works just by having the swim-bladder in reasonably close proximity to the inner ear. However, cyprinid fish, like carp, have improved on this. They have a special link of bones between their swim bladder and the inner ear called Weberian ossicles. The pulsations of the swim-bladder are transmitted directly to the inner ear along these bones. As a result, cyprinid fish have among the best hearing of any fish species.

Touch

The sense of touch is almost certainly delivered by the acoustico-lateralis system and the hair cells. These hair cells can be found on the skin, so it is likely that fish use these to sense touch.

Understanding the surroundings

In terrestrial vertebrates, vision is commonly the most important and therefore dominant sense. However, vision is more difficult in the low light intensities of the aquatic environment. Furthermore, the shifts in available light and bandwidth with increasing water depth make it even more difficult to rely on vision. For still-water fish, these problems often become even worse when the water clarity is poor, such as during an algal bloom or when there are high levels of suspended solids. Sound and water movement have very useful qualities in water that make the inner ear and lateral-line systems very important to fish. These systems not only allow them to hear sound, but also to detect very small changes in water movement, giving fish a sense of what is around them, what is moving, which direction it is in and how far away it is.

Chemical stimuli play a central role in directing many aspects of fish behaviour. This will, of course, include feeding, but also reproduction and recognition of predators, prey and members of its own species.

Fish also have a very good sense of their temporal surrounding, using various indicators to establish the time of year. For example, the pineal gland is a light-sensitive gland that is thought to play a role in timing of reproduction, growth and migration; in many species this is linked to photoperiod, which is the length of daylight during each day.

6.3.3 Digestive systems

There is much variety in the possible food items available to fish. Generally fish can be categorised as herbivorous, omnivorous or carnivorous, although it is difficult to categorise many species along these lines because they have such a varied diet. Most temperate freshwater fish are omnivorous or carnivorous. There are several truly carnivorous fish, which eat fish and water mammals as well as large invertebrates. Most other fish eat some vegetation and some invertebrates.

The function of the digestive system is to hold or detain the food item, to swallow it and then break it down physically and chemically until it is small enough and soluble enough to be absorbed into the bloodstream. This is a specialised function of the fish's anatomy, and individual species have evolved to be very efficient in obtaining and digesting particular types of food. This can be seen clearly when comparing the digestive systems of fish feeding on different things (see Fig. 6.27).

Mouth

The mouth of fish is very adaptable and is quite different in different species, giving clues to the particular diet of that particular species (Fig. 6.26). For all teleost fish, it is a two-part biting jaw. Some fish have teeth or bony plates for either seizing and holding prey items or crushing them.

Inside the mouth is the buccal cavity. As already mentioned, this has a role in tasting the food, but it also has a role in sorting and holding food, whereas non-food items are expelled.

Towards the back of the mouth are the gill rakers, whose main function is to direct the food to the back of the mouth and prevent any going into the delicate gill tissue. This is particularly important in predatory fish where most often the prey item is still alive as it is swallowed.

Fish dentition varies with species and diet. Most fish have one or more rows of teeth in their upper and lower jaws, whose form varies according to diet. An additional set of teeth, the pharyngeal teeth, are found on the pharyngeal bones in the throat. These are used for crushing and grinding the food as it is passed back to the oesophagus. These teeth look just

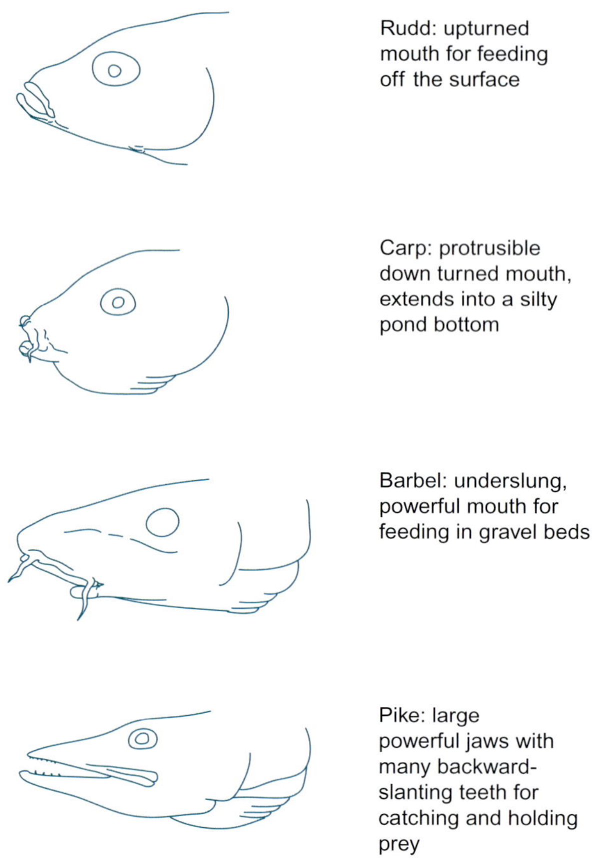

Figure 6.26 Various forms of fish mouth adapted to different types of diet.

like molar teeth and are ground against a hard plate. They appear to use the muscle of the pectoral fins for the grinding motion.

The oesophagus of fish is very short, but expandable to allow the easy passage of large food items. Mucus is produced along the oesophagus to aid the progress of the food item and to prevent damage to the lining.

Stomach (Fig. 6.27)

The stomach region shows huge variation between different species, depending on their diet. This ranges from a full musculature stomach in the pike, to a slightly expanded area of a long undifferentiated tube in the carp. It can be considered as an organ of storage, mixing and beginning breakdown of food items.

In carnivorous fish, the stomach is very distinct; it is sac-like, with very elastic but muscular walls to allow for large prey items. It is cut off from the rest of the gut by a sphincter, which retains the prey in the stomach until it is sufficiently broken down. The stomach walls secrete digestive enzymes as well as acids to aid the breakdown of the prey. The muscles of the stomach wall grind the food in peristaltic waves, thus grinding and mixing the food,

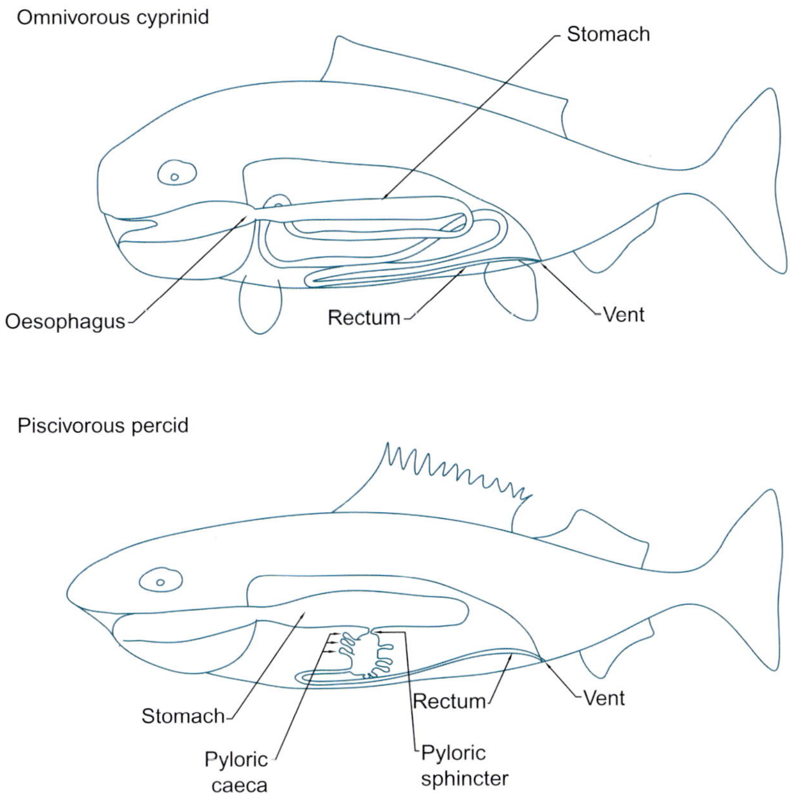

Figure 6.27 The digestive system of the fish.

aiding in its breakdown from a solid into a semi-liquid 'chyme'. This is then allowed through into the upper intestine.

In herbivorous or omnivorous fish, the stomach is much less obvious. Usually there is just a slightly expanded area of gut between the oesophagus and the first turn of the gut, which acts in a similar way to a true stomach. These species tend to concentrate on small prey items that require less work to break them down. They also tend to have well-developed pharyngeal teeth to grind up the prey before they enter the stomach.

Intestine (Fig. 6.27)

In the cyprinid group, and in herbivorous fish without a stomach, the intestine is a long, thin tube running backwards and forwards through the body cavity. Its function is to complete the breakdown of the food items and absorb all the nutrients released in the process.

In carnivorous fish with a stomach, the lower intestine is much shorter, but there are often projections from the main intestine known as pyloric caeca. These appear to serve a secretory function, releasing enzymes and other chemicals into the intestine to aid digestion.

For all types of fish, the pancreas is another organ that is involved in the secretion of enzymes and other chemicals, such as bicarbonates, which help digest the food and balance the pH.

Bile is produced in the liver and stored in the gall bladder, from where it is released in response to food in the upper gut. The bile also plays a role in neutralising the acidic chyme that comes down the intestine from the stomach.

There are significant differences in the anatomy and physiology of the alimentary tract of different fish species; this is in response to the normal diet of that species. It also seems that most fish are inefficient at digesting plant material, lacking any specialist adaptation to allow them to do so. In fact, there are very few true herbivores in temperate freshwater fish. Most species of fish are well adapted to feed on certain specific food items. Nutrition will be dealt with in Chapter 22, but it should be noted how well adapted each fish species is to its diet and this fact should not be ignored when introducing fish to an artificial diet.

6.3.4 Respiratory and circulatory systems

Most fish obtain oxygen from the surrounding water using a respiratory organ called the gills, specially evolved for the purpose. Other body surfaces are sometimes used, at particular times during the life history, or during migrations. Some freshwater species have adapted the ability to breathe air, allowing them to cope with low dissolved oxygen situations, or with water bodies drying up. In temperate water these are obviously of less concern and although some species, for example the crucian carp and the eel, show adaptations, most use the normal respiratory systems.

Compared with air, water holds very little oxygen and is very dense, so it takes a great deal of energy to pass the water over a respiratory surface. The gills therefore are very highly developed.

Gills

The gill structure is similar for all temperate freshwater fish. The gills consist of two sets of filaments attached to a hard gill arch, which is in turn supported by the skeleton of the branchial arch (Fig. 6.28). There are four sets of these branchial arches on either side of the head. The two sets of filaments are set behind the gill arch in a V-shaped row. Each of these filaments has a row of lamellae along each surface, thereby giving an enormous surface area for gaseous exchange. The surface of these lamellae is only one cell layer thick, so there is

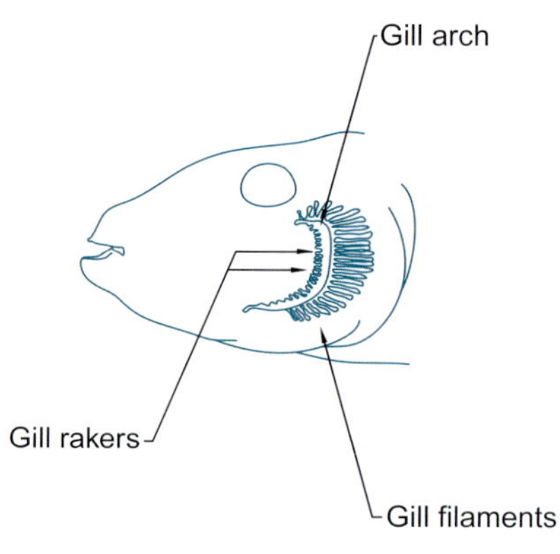

Figure 6.28 Branchial arch and gills.

only one cell between the circulatory system (bloodstream) and the external environment, thus reducing the diffusion distance to the minimum possible. The gills are supplied with blood direct from the ventral aorta through the afferent branchial artery.

In this manner the structure provides a huge surface area with a minimum diffusion distance and an excellent blood supply for carrying the respiratory gases to and from the body.

The forward part of the branchial arch has stiff, finger-like projections, which interlock with the adjacent gill arches. These are known as the gill rakers. The primary function of these gill rakers is to prevent any food items, or other particles from getting from the mouth out across the delicate gill tissue. However, in some species they do appear also to act as a method of filtering food items out of the water flowing across the gills.

Respiration

The fish needs to achieve a steady flow of water across these respiratory surfaces so that it can breathe efficiently. There are two main methods of achieving this, the branchial pump method or ram ventilation.

As already stated, it requires a good deal of effort to force the viscous and dense water across the gills. So it would seem logical to use the forward swimming motion to achieve this. Some fish do exactly this, a method known as ram ventilation. By keeping their mouths open as they swim forward, water is forced across the gills. This method is rare for still-water fish: most use the branchial pump method for all, or for the vast majority of, the time.

The branchial pump method of respiration requires water to be actively sucked in through the mouth, and then passed across the gills before being released from the operculum. To achieve a good, steady flow of clean water requires many adaptations, and the sequence of events for each 'breath' reveals these.

First, the mouth is opened and the buccal cavity is expanded. This draws water in through the mouth. To prevent water entering through the operculum, this is closed and a flap of skin along its edge acts as a valve, shutting this gap completely. The mouth is closed, although many fish also have a flap of skin, like a curtain, on the top of the mouth, which prevents water flowing back out of the mouth. The buccal cavity is contracted, and the opercular cavity is expanded. This forces the water out of the mouth, over the gills and into the opercular cavity. Now the opercular cavity is contracted and the operculum is opened, forcing water out. The operculum is then closed and the whole system can start again. This system is therefore run effectively by two pumps: the buccal cavity and the opercular cavity. By careful control of these two pumps, the water flow across the gills is almost constant.

The gills are important not only in the respiratory gas exchange, taking in oxygen and excreting carbon dioxide, but also in the excretion of other waste products. The gills of most teleost fish are the most important site for the excretion of nitrogenous waste products, usually in the form of ammonia and ammonium ions. Furthermore, they are important in maintaining the salt (ion) balance within the fish, which is achieved by the active uptake of salt ions like sodium (Na^+) and chloride (Cl^-).

6.3.5 Reproductive systems

The production of a new generation is one of the primary motivational factors for any living thing, and the various fish species have numerous adaptations to achieve this. However, in freshwater teleost fish, most species are very similar in their methods of reproduction. It is a cyclical process, usually divisible into distinct phases from maturation of one set of adults to maturation of the next generation. Among temperate freshwater species of fish, most have very distinct breeding seasons, usually confined to 2–3 weeks a year (Box 6.1). This breeding season is very closely linked to particular environmental changes that occur every year.

Most teleosts have separate male and female adults. The females lay eggs, which are fertilised outside the body (oviparous). The females are very fecund (meaning they produce lots of eggs). The eggs have a relatively short, temperature-dependent incubation period. The newly hatched larvae tend to have a very short larval stage, with a very high mortality rate, before metamorphosis into recognisable fry.

Production of eggs and sperm (milt)

The testes originate as paired organs running along the body cavity. They are usually whitish or cream coloured organs that are divided into indistinct lobes and attached to the wall of the body cavity. There is a sperm duct which leads to an opening (the urinary papilla) near the rectum.

Sperm are formed from germ cells, by a process known as spermatogenesis. This involves a period of cellular multiplication, followed by cell division, a process known as meiosis, which halves the number of chromosomes. These spermatocytes then undergo a period of growth, before metamorphosing into motile sperm cells.

The ovaries are again paired organs, although in some fish, such as the perch, one of the pair of ovaries is so reduced as to be invisible to the naked eye. Ovaries are sac-like

Box 6.1 Breeding seasons of the main British freshwater fishes

		Breeding season (month)											
English name	**Latin name**	**J**	**F**	**M**	**A**	**M**	**J**	**J**	**A**	**S**	**O**	**N**	**D**
Silver bream	*Abramis bjoerkna*					■	■	■					
Common bream	*Abramis brama*					■	■	■					
Bleak	*Alburnus alburnus*					■	■	■					
Barbel	*Barbus barbus*					■	■	■					
Goldfish	*Carassius auratus*					■	■						
Crucian carp	*Carassius carassius*					■	■						
Common carp	*Cyprinus carpio*						■	■					
Gudgeon	*Gobio gobio*					■	■						
Chub	*Leuciscus cephalus*				■	■	■						
Orfe	*Leuciscus idus*				■	■							
Dace	*Leuciscus leuciscus*			■	■	■							
Minnow	*Phoxinus phoxinus*						■	■					
Roach	*Rutilus rutilus*					■	■						
Rudd	*Scardinius erythrophthalmus*					■	■						
Tench	*Tinca tinca*					■	■	■					
Pike	*Esox lucius*		■	■	■	■	■						
Brown trout	*Salmo trutta*										■		
Bullhead	*Cottus gobio*			■	■	■							
Pumpkinseed	*Lepomis gibbosus*					■	■	■					
Ruffe	*Gymnocephalus cernuus*				■	■							
Perch	*Perca fluvatilis*				■	■	■						

structures, usually continuous with the oviduct which leads to an opening just beside the vent. The development of the eggs is slightly more complex than the sperm. It is very similar in the initial stages, with germ cells undergoing a period of multiplication. After this there is a period of growth and then cell division, which reduces the number of chromosomes (meiosis). This meiotic division is incomplete until after the egg is fertilised, when the egg finally expels the 'lost' half of the chromosomes. The oocytes undergo a period of significant growth, often swelling up to 100 times in size after the last cell division, as they accumulate yolk (vitellogenesis).

Most teleosts spawn more than once in their lives and the ovaries of these fish generally contain at least two distinct populations of eggs at different developmental stages. These ovaries contain a reservoir of germ cells at early stages of development and each successive annual egg production is grown from this reservoir.

Sex determination

Reproductive systems and patterns of behaviour are extremely diverse in fish and although temperate freshwater fish tend to have adopted a similar pattern, it is worth considering this variety, particularly in methods of sex determination.

Sex determination is largely under genetic control. For most fish this is, in common with most vertebrates, the XY system. In this system, those individuals with the same chromosomes (XX, homogametic) are female and those with different chromosomes (XY, heterogametic) are male. In other fish species, different sets of chromosomes are used to differentiate the sexes and here the male is homogametic (ZZ) and the females are heterogametic (WZ). There are also a variety of other possible genetic sex-determination systems, using multiple or even single sex chromosomes.

Although sex determination can largely be considered to be under genetic control, environmental factors can also play a role in fish. Many environmental factors, including photoperiod, salinity, social environment and rearing temperature have been shown to exert an influence on sex determination. The exact mechanism of this environmental control is unclear, but essentially fish that are genetically male or female can become functional females or males respectively. This change can be experimentally induced by treating juvenile fish with sex hormones (androgens or oestrogens). In the wild, it is rare to find a population of fish that are 50% male and 50% female, which may be caused by the influence of these environmental factors.

Reproductive pattern

There are many strategies a fish can adopt for reproduction. However, there are some general rules, which are as follows.

- Egg size and number are inversely related; that is, the fish can produce more small eggs or fewer large eggs.
- European freshwater fish usually have non-buoyant, sticky eggs.
- If any parental care is shown for the eggs or fry, then there will be fewer eggs.

Spawning

To get the best survival for the fry, it is vital that the timing of spawning is closely linked to the cycles of availability of prey consumed by the fry. Fish are usually found to have discrete spawning seasons, timed so that the eggs hatch at a time when food items are abundant. For temperate freshwater fish, spawning is therefore usually annual and timed for various intervals in the late winter, through to early summer, depending on the fish.

The timing of spawning is therefore very important, each fish needing to be ready for those few weeks when mates will be available and the conditions correct. Egg and milt production starts not long after spawning, when there should be an abundance of food, and therefore energy, that the fish can put towards the production of gonads. The timing of the production of gonads, and the efforts to ready them for the spawning season, appear to be largely reliant on environmental factors. The two most important of these environmental factors are temperature and day length; for example, the carp will start to mature its eggs as winter breaks and spring starts. The carp uses the slowly rising water temperatures, as well as the increase in the length of daylight, to recognise the start of spring, both of these environmental cues being vital for triggering a response. These responses are controlled by the release of different hormones (mostly testosterone and oestrogen), which trigger reactions within the gonads.

These signals only get the fish so far, however, with the fish producing and maturing the eggs and milt over a time period dictated by environmental changes. However, final maturation of the eggs and milt, and actual spawning, are timed by different factors. For example, the final stages of maturation of the eggs require the production of a different hormone altogether, the maturation-inducing hormone. Environmental cues help to ensure the fish are ready to spawn at a particular time of year; however, these cues are not sensitive enough for the last part of the process. An egg, once released from the female, has to be fertilised within a very short period before it is no longer viable. So the final part of the maturation is delayed until all the fish are ready and begin courtship. For this final part of the spawning cycle, conditions must be correct, not only environmental but also physical and individual. For example, goldfish will not spawn in the absence of a suitable weed substrate to spawn on (although they will use a variety of things if no weed is available, including floating carrier bags!).

The last part of spawning is controlled by a courtship ritual, specific to an individual species, and the release of sex pheromones which allow communication between the spawning individuals. These pheromones are released into the water and sensed by the other fish participating in the spawning, which lets all the individuals synchronise their own state of readiness, so egg and milt release happens at exactly the same time. The courtship behaviour that accompanies the spawning varies from species to species, but includes visual signs such as the bright red belly of the male stickleback. Contact signals, such as spawning tubercles on male cyprinids, help to excite the female during close-contact mating rituals. Some species of teleost fish also use sound as part of their mating ritual. These sounds are commonly caused by beating muscles along the swim-bladder.

Much of the spawning ritual is to do with timing the release of eggs and milt, but it will also help in choosing the best site for laying eggs. For most still-water species, the most appropriate or best sites will be fairly limited. Laying floating eggs will mean they are susceptible to many predators. Laying sinking eggs will allow the eggs to sink to the bottom of the still-water, where the mud will almost certainly smother the eggs. So for most still-water species, laying sticky eggs among marginal vegetation is the favoured option. With most of the larger cyprinids, spawning is set for mid-spring when the phytoplankton blooms are starting, with a related bloom of zooplankton, such as rotifers, which are perfect first food for the fry. Seasonal influences have led the fish to mature, and now they start to congregate in the shallows of the still-water, around submerged or marginal vegetation. A short period of warm weather will usually bring them into spawning condition and the females will start to dart through the vegetation, usually first thing in the morning. The males, which have already formed large white lumps on the head and flanks (tubercules), will dash after each female, nudging her and rubbing alongside. Usually three or four males will be following each female. This activity may go on for several hours each day, for a couple of days. Eventually, when all the fish are ready, the female will again race through the weed, shedding eggs and pheromones as she goes. Several males will race alongside her, shedding milt.

Once released, the eggs and milt come together and fertilisation occurs. At this point most of the adults of teleost fish species will have no further influence on the eggs, and there is no parental care. In fact, some cyprinids will begin eating eggs soon after spawning. An exception is the stickleback male, which builds a nest and, once the female has spawned within it, will guard the eggs until they hatch.

Development of eggs and larvae

At fertilisation the egg recombines the two sets of chromosomes found in the milt and egg. The eggs and milt were formed in a process (meiosis) that halved the number of chromosomes, so fertilisation reinstates the full number of chromosomes, half from the mother and half from the father. This fertilised, single cell starts to divide quickly, forming a noticeable bulge of cells on one side of the yolk sac. Eventually this group of cells start to differentiate and form a band around the yolk sac. As this process continues a noticeable head starts to form, and features such as eyes start to be recognisable. The time it takes for hatching of the fish is very temperature dependent, but is usually between 60 and 120 degree days (2 days at 18°C would be $2 \times 18 = 36$ degree days). When the fish first hatch, they will not be fully developed. The yolk sac will feed the embryo for a further few days as the larval fish continues to develop. Most freshwater fish spend very little time as larvae and are very vulnerable at this stage. They metamorphose quickly into little fish, commonly known as fry, and start to feed on their normal diet. The obvious exception to this is the common eel, which may spend as long as 3 years as larvae.

6.3.6 Coordination and control systems

In complex animals like fish, the control and coordination of activities is achieved either by electrical impulses in the nervous system, or by chemicals (hormones) being released into the bloodstream (the endocrine system). Although the two systems are often dealt with differently, they are essentially very similar, and closely linked.

In simple terms, the signals, whether chemical or electrical, serve to regulate the function of the organs within the fish's body. Signals are sent from the 'control centre' to the effector organ, which responds with a particular set of actions. Information about these actions is then transmitted back to the control centre which may then further modify the actions or stop sending signals.

The major difference between the nervous system and the endocrine system is one of speed: the nervous system gives rapid control over the body functions; the endocrine system gives slow control.

Nervous systems

The nervous system of fish consists of the brain and spinal cord, known as the central nervous system (CNS). It is based upon electrical activity in the basic cells of the system, the neurons. There are two major components of the nervous system, the voluntary and the automatic systems. These can be looked at as impulses that the fish has some control over (voluntary), and the automatic impulses that the fish has no control over. So, for example, hunting for prey, particularly the final snatch, is all controlled by the voluntary nervous system, whereas the beating of the heart or the movements in the stomach muscles will all be controlled by the autonomic (automatic) nervous system.

Endocrine systems

The endocrine system is made up of the various organs (glands) that secrete hormones in response to certain stimuli. These hormones will then control certain aspects of the fish's behaviour, or the development of certain organs such as the gonads. For example, the pituitary organ is at the base and rear of the brain. It is the most complex of the endocrine glands and controls the function of several other endocrine glands. Its most important functions relate to the release of gonatrophic hormones, which stimulate the growth and development of the gonads, and the release of growth hormones which stimulate growth.

6.3.7 Immune systems

All fish are subject to a wide range of possibly dangerous parasites, bacteria and viruses, collectively known as pathogens or disease-causing organisms. To protect themselves, fish have developed a very effective defence system, their immune system. This system can be split into two separate systems, the non-specific and the specific.

Non-specific immune system

As the name suggests, this part of the immune system defends against all possible invaders. It includes all the physical barriers and some chemical ones. The skin of fish is a very important part of this system. Not only does it provide an effective physical barrier, it also produces one of the best chemical defences in the animal world. The mucus of fish is not just slime, but a complex mix of mucus and chemicals that provide a chemical and physical defence against all invaders, particularly bacteria and viruses. The mucus is constantly being washed off and replaced. This serves to carry away any prospective pathogens.

Another possible entry route for pathogens is though the normal ingestion of food. To defend against this, most fish have a stomach area that has a very low pH and digestive enzymes that will kill most pathogens.

If the skin is breached by a wound, a system of events will occur that are designed to seal the wound and prevent infection. First, the wound is sealed with an inflammatory response, which also helps with osmoregulatory problems. This reaction also causes the blood vessels in the area to close up, preventing excessive bleeding. As the blood hits the outside environment, blood-clotting agents cause the blood to clot and seal the wound from the outside. Several other chemicals are released around the wound site, particularly C-reactive proteins, which will kill most pathogens, and interferon which kills viruses. White blood cells also move to the area, and actively consume and destroy any pathogens. This type of white blood cell is known as a phagocyte.

The non-specific immune system is very successful at preventing entry and dealing with wounds.

Specific immune system

This deals with specific pathogens that have established themselves on or in the fish. It is the basis for 'acquired immunity', as the fish can 'learn' to deal with specific pathogens and use this information to deal more effectively with the same pathogen later in life. This can be passed on to the next generation. It is also the basis for vaccinations.

Every organism has a unique set of proteins and chemicals that make up its body. Any invading pathogen will have slightly different proteins on its outer surface. These proteins will be unlike anything found normally within the body of the fish. These proteins are known as antigens. The fish can recognise these proteins as being alien or 'non-self', and they react by producing antibodies. These antibodies are also proteins and react to the presence of the antigens by attaching to them. Once attached, the antibody can act in one of several ways: it can attract white blood cells, particularly the phagocytes, which destroy the pathogen, it can prevent the pathogen from reproducing or it can prevent the pathogen from releasing toxins.

This system is very effective, but it may take several weeks for the fish to develop sufficient antibodies to deal with a rapidly reproducing pathogen. During this time, the fish may succumb to the pathogen. However, the real benefit of the specific immune system is only fully seen when the fish is attacked by a pathogen for the second time. If the fish survives and recovers from the first infection it develops a 'memory' of the antigens from that particular pathogen. If it is subjected to another attack from the same pathogen it can now use the memory cells to start producing antibodies straight away, thus eliminating the pathogen before it has a chance to get out of control. This is known as acquired immunity.

6.3.8 Osmoregulatory control system

Fish, and other aquatic organisms, have a particular problem relating to maintaining the amount of water in their bodies. This is due to a process known as osmosis, which is diffusion of water through a semi-permeable membrane, from a dilute to a concentrated solution. This means that fish living in freshwater are constantly taking water in across their gills and other membranes. This is because they will have body fluids that are more concentrated than the external freshwater environment. This influx of water must be dealt with, otherwise the fish will quickly die as their body fluids become too dilute.

The primary means of ridding the body of excess water is by excreting copious amounts of dilute urine, produced in the kidneys. However, this can lead to further problems. No matter how effective the kidneys are at removing the essential salts (e.g. chloride, Cl^-, and sodium, Na^+) from the urine before it is expelled, the fish can still lose too much salt. To maintain salt balance, fish have developed a salt uptake mechanism in the gills, which is highly efficient. Fish will also gain salts in their normal diet.

In short, freshwater fish will take in water across its gills and other membranes. To rid the body of excess water the kidneys have adapted to produce large quantities of dilute urine. Ammonia secretory activities are also undertaken by the gills, which use the secretion of ammonium ions to aid the active uptake of salt ions, such as sodium and chloride, which are lost during the production of the copious urine.

The regulation of the salt and water balance is an important role for the endocrine system. The salt–water balance is monitored and regulated by the endocrine system and hormones, for example prolactin, which is secreted to regulate activities in the body such as the amount of water removed by the kidneys and the level of sodium uptake by the gills.

6.4 Conclusions

Fish are remarkably well adapted to their environment. This is not surprising when one considers that fish were the first vertebrate animals to evolve and they have had such a long time to adapt to a wide variety of aquatic environments. This they have achieved by developing a wide range of body forms, physiological functions and behaviours that have enabled them to colonise all types of freshwater and marine habitats.

7 Disease-Causing Organisms

Chris Williams and Ian Wellby

7.1 Introduction

Fish disease is one area where fisheries managers must be able to understand the basic principles and problems associated with fish health, and its relevance to fisheries management. At the same time, they must be able to recognise when more knowledge than their own is required and be prepared to seek professional expertise and advice.

There are several areas where disease may impinge on the management of a fishery. These are listed below.

- The main concern of most fishery managers is obviously with the fish within the fishery. It is therefore important to understand the practices and processes necessary to prevent disease outbreaks within the fishery and what happens in the event of such outbreaks.
- Managers of a fishery often have a role in managing or overseeing the management of fish farms used to produce fish for restocking purposes. These systems require knowledge of disease in order to manage them effectively.

Box 7.1 Dynamic equilibrium

Parasite ↔ Host

The theory of dynamic equilibrium is exemplified by the huge reproductive capabilities of most parasites. Most adult parasites are capable of producing thousands of eggs or live young, often without the need for a partner. If all these new parasites survived to infect a new host, then mortalities would quickly result. Under normal circumstances, however, most of these young die because they fail to find a host, or are killed by the fish's immune system.

- Disease issues will also affect the management of a site through the rules affecting the transfer of fish from one place to another. Fishermen and fish farmers are often keen to transfer fish from one water system to another in an effort to improve their sport. Legislation has been introduced in most countries to prevent the transfer of specific diseases between catchments or geographically isolated areas and thereby prevent damage to fisheries caused by these diseases. Fishery managers must be well aware of all the relevant legislation that applies to their activities. It is important to understand the natural history of the diseases for which the controlling legislation is devised and to understand the purpose of the legislation, which has been devised to protect not only the environment but the industry as well.

The first thing a fishery manager needs to realise is that all fish in the wild will carry a variety of pathogens on and in their bodies. The aquatic environment is rich with parasites and they play an important role in the natural aquatic ecosystem. In normal circumstances the fish and its parasites exist in a balance, where the fish's immune system will regulate the pathogen numbers to a level where the damage is survivable. This is known as a dynamic equilibrium (see Box 7.1).

This chapter sets out to explain the many different types of pathogen and parasite that can affect fish, as well as to highlight some of the non-infectious diseases. The chapter begins with some principles of fish disease, followed by a discussion of some of the major groups of pathogens.

7.2 Pathogens and disease

Pathogens are a natural part of the ecology of any fish species in the wild environment. Consequently, understanding the processes that may trigger the pathogen to start causing a disease outbreak and how it may affect a fishery is a fundamental part of fishery management. Although the words 'disease' and 'parasite' often instil fear in many fishery owners or managers, it is important to realise that all fish in the wild environment will harbour pathogens. What makes an occasional occurrence of a pathogen become a disease outbreak that can decimate a fishery? The answer lies with an understanding of the balance or 'dynamic equilibrium' between the fish host, the pathogen and the environment in which they live.

A disease outbreak is seldom merely the result of a meeting between a pathogen and a host. Under most circumstances, fish and their associated parasites coexist in this state of equilibrium, where the fish's defences will regulate the pathogen numbers to a level where the damage is survivable.

Most parasites have evolved to use the fish host as a means of survival, not to kill it. This would simply be counter-productive, destroying the thing on which it depends. Similarly, fish have evolved to withstand the impact of many pathogens by means of a well-developed immune system and natural defences. Disease only usually comes about when this balance is upset, in most cases for the following reasons.

- a factor that reduces the defences of the fish host and thus a decline in its ability to kill the pathogen, for example stress or poor nutrition;
- a factor beneficial to the pathogen that makes it easier for it to locate a host, for example overcrowding of the fish.

Although the relationship between pathogen, host and environment is often a complex one, understanding some basic principles of fish health and the processes that can upset a fishery goes a long way towards minimising disease. Healthy fish require good nutrition, optimal water quality, a suitable habitat and minimal stress. Failing to meet these fundamental requirements will result in a degree of ill health and increased susceptibility to disease. Similarly, pathogens have exactly the same requirements to survive: nutrition from available hosts, optimal conditions to multiply, etc. If all these factors remain in balance then the host and parasite also remain in balance. However, if conditions become beneficial to the pathogen and not the host, then a disease outbreak may occur.

It is also worth noting that many pathogens have a very high reproductive capability, often producing thousands of offspring in the hope that just one or two individuals survive to reproduce. In most natural situations, this is exactly what happens with most juvenile parasites dying either through an inability to locate a host or as a result of the fish's defences. The result is a low level and relatively trouble-free parasite population. Upsetting this natural balance by increasing the survival of these offspring can lead to the pathogen population spiralling to epidemic proportions very quickly.

Understanding the relationship between the fish host, the pathogen and their environment is vital in assessing the health of any fishery and the potential of disease. By maintaining the dynamic equilibrium, disease will seldom result. Upset it in favour of the pathogen, and disease will often become the costly outcome.

7.3 Basic principles of parasitology

All of the organisms that are detailed in the following section are termed parasites, with viruses, bacteria and fungi being referred to by name. However, strictly speaking, all of these organisms are parasitic, in that, for at least some of their life cycle, they derive nutrition from a living organism, which is harmed in the process.

Parasites of fish and the subject of parasitology is a huge and often complex field. The number, morphology, impacts and behaviour of fish parasites are highly diverse, and despite increasing attention over recent years, we still have much to learn about this subject. Fish parasites have adapted to a parasitic existence in many weird and wonderful ways, some very simple, others very complex. To illustrate this complexity, we need only look as far as the life histories or life cycles of some common parasites. Some parasites have what is termed a 'direct' life cycle. In other words, they require only one fish host to survive and multiply. An example of this is the skin fluke *Gyrodactylus* sp., which gives birth to its live young while still attached to the fish. This very simple life cycle can occur without either the adult parasite or young leaving the fish. In contrast to this, nematodes have extremely complex 'indirect life cycles', which may involve not only fish as a host, but also birds, mammals, invertebrates and other fish, acting as 'intermediate hosts'.

Understanding the life cycle of a parasite is undoubtedly the most important part of parasitology to a fishery manager. Such information can reveal how the parasite behaves in its environment, how it may affect the fish, which size or species of fish might be affected, what the parasite requires to survive, and ultimately how it may be naturally controlled if it becomes a problem (i.e. how can the life cycle be broken). It should be noted that in the case of many parasites with complex, indirect life cycles, the absence of the intermediate hosts would result in the absence of the parasite itself.

The effect a parasite may have on its host is almost as varied as parasites themselves. Some have the ability to cause rapid debilitation and mortality if allowed to reach epidemic proportions (e.g. whitespot). Others, however, may have a more subtle impact on their host, debilitating it over a long period, resulting in a slower chronic problem in the fishery (e.g. tapeworms). Some parasites need to get to the next host to complete their life cycle and so debilitate the fish in such a way as to make it more susceptible to predation.

Equally varied is the diversity of niches that parasites have evolved in. Parasites of fish may generally be split into those that feed on the skin, fins and gills (termed ectoparasites), or those found in the intestines, body cavity or internal organs (termed endoparasites). Parasites in a fishery may also behave in specific ways, depending on their evolutionary adaptations. For example, some parasites may be 'host specific', which means they only parasitise preferred species of fish (e.g. several tapeworm species). In contrast to this, other parasites may potentially infect almost any fish that swims, and are therefore described as having very low 'host specificity' (e.g. whitespot). Understanding these processes and how parasites affect their host is a very useful part of fishery management if problems are to be avoided.

7.4 Fish pathogens

There are a wide variety of fish pathogens. These can be broken down into families of parasites to make study of them easier.

- Viruses
- Bacteria
- Fungi
- Protozoa
- Flukes (Monogenea)
- Trematodes (Digenea)
- Roundworms (Nematoda)
- Tapeworms (Cestoda)
- Spiny headed worms (Acanthocephala)
- Leeches (Hirudinea)
- Crustacea

7.4.1 *Viruses*

Viruses are extremely small organisms that, without exception, require another organism's cells to reproduce. They are therefore termed 'obligate' pathogens, because they cannot survive without another organism to live within. A virus is a very simple organism, usually made up of a protein coat, inside of which simple genetic material is contained. A few have additional, slightly more complex lipoprotein outer coats. There are many species that are extremely pathogenic to fish, and new species are being discovered every year.

Viruses actually reproduce by inserting their genetic material into a host cell, which is hijacked into reproducing more and more of this genetic material (Fig. 7.1). As a result, more and more new viruses are produced until the host cell is exhausted and dies. The new viruses are then released back into the host body to infect more cells. The virus harms the host by causing this cell death.

Diagnosis of viral organisms is very difficult and can only be achieved by indirect laboratory techniques that identify chemicals, such as antigens, in the tissues of the fish, rather than through the direct observation of the viral organism. These techniques can be difficult, expensive and sometimes inaccurate. Certainly until the particular virus comes to the attention of fish biologists through mortality events or overt disease symptoms, they may not be identifiable. It is certain that there are many fish viruses around the world that have not been discovered.

Viruses can survive for a short while away from the host and can be transferred from host to host by blood-feeding organisms, such as the fish leech. However, most are probably spread by being shed from the host directly into the water and then invading a new host. This means that viruses can be very easy to transfer from fishery to fishery, as they are extremely tolerant of environmental extremes. Therefore, although most viruses are probably transferred directly with fish movements, many viruses can be moved on damp material such as nets or mud on boots.

As there are no treatments for viral disease within a fishery, prevention is the only suitable method of avoiding outbreaks. A prophylactic approach such as the vaccination of young

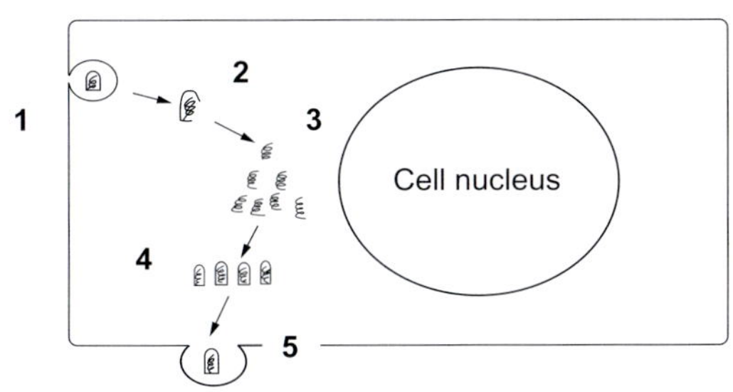

1. Penetration of the cell membrane by a virus.
2. The viral coat disintegrates.
3. The virus interferes with the normal cell function and causes the cell nucleus to create more viral RNA.
4. The new viral RNA creates more coats.
5. The virus leaves the cell.

Figure 7.1 Illustration of viral reproduction.

fish would be highly desirable. Unfortunately, vaccination methods for viruses are still very much at the experimental stage despite numerous trials. A robust biosecurity system is really the only method of protecting a fishery against viral diseases.

Important viral diseases include the following.

Spring viraemia of carp

Spring viraemia of carp (SVC) is a disease that has been known since early in the nineteenth century, although it was not until 1971 that a virus was isolated from infected carp, which confirmed the viral character of the disease. The virus is known as *Rhabdovirus carpio* and appeared to be confined to mainland Europe and Asia, until the late 1990s, when SVC was confirmed in a consignment of goldfish and koi imported into the UK from China. SVC is mainly a disease of common carp, but may also affect other cyprinids including crucian carp, goldfish, bream, tench, barbel, roach, grass-carp and silver carp with varying, but lesser, degrees of virulence. Some non-cyprinid species may also prove susceptible, such as the wels catfish.

Spring viraemia of carp is a systemic viral disease of the common carp, especially affecting young fish. The condition is almost certainly made worse in intensive culture conditions where mortality is usually very high. Although the mortalities from this disease predominantly occur in cultivation systems, it has been recorded causing significant mortalities in wild fisheries, particularly in the UK. In these cases, considerable mortalities and consequent economic loss have arisen from infections of SVC. In Europe this led to SVC being included as a List III notifiable disease by the European Union Council Directive (91/67/EEC), concerning animal health conditions governing the marketing of aquaculture animals and products. This allows the UK government to place restrictions on fish that may be infected with the virus. These restrictions are still allowed under the newer Animal Health Directive Legislation.

Fish infected with SVC show symptoms ranging from lethargy, enteritis, peritonitis, oedema, dropsy, exophthalmia, thickening of the swim bladder and pinpoint haemorrhages (petechiae) in the internal organs, skin and muscle. Other symptoms such as emaciation and erratic swimming movements are often reported. It is the authors' experience that all, some or none of the above symptoms may be seen in infected populations. SVC usually occurs in the spring as the water temperatures rise between 10°C and 15°C. It is possible to isolate the virus from fish even when the water temperature rises above 15°C, but in those cases the virus does not appear to be as pathogenic. Mortalities are usually fairly low at first, but level out to steady daily losses.

Diagnosis of SVC is by cell culture techniques, which at present are both difficult and expensive.

There are no treatments for SVC and so prevention is the only effective means of control. The mode of transmission of the disease appears to be horizontal, with the possibility of transmission by leeches and other ectoparasites, as well as piscivorous birds.

Carp pox

Carp pox is the most commonly reported of all the viral infections. It has little effect on the host fish apart from raising white, wax-like lesions that are easily seen with the naked eye. It has only very rarely been recorded causing any significant problems apart from the unappealing look of the lesions. These lesions disappear as the water temperature rises, leaving clear, apparently undamaged skin behind.

Koi herpesvirus

Koi herpesvirus (KHV) causes a severe disease and mass mortalities in populations of common carp and its ornamental varieties. This herpes variety was first identified in 1998 in carp from Israel, where mass mortalities were occurring in fish farms throughout the country. However, the methods available to identify viral infections are such that it is difficult to tell where this virus was present before that time. Certainly KHV has been identified in preserved samples from carp mortalities in the UK as early as 1996. Since the advent of more accurate diagnostic tests, the disease has been recorded around the world. The devastating losses that have occurred in intensive and extensive carp culture facilities in Europe, Asia and North America suggest that these may well be recent infections. Since 2005 the disease was made notifiable to the World Organization for Animal Health (Office International des Epizooties, OIE) and since 2007 KHV has been notifiable in the UK and some other European countries. It is recognised as one of the most serious infections impacting on carp populations.

In the UK, KHV was suspected in several mortalities, but up until 2006 only relatively few outbreaks of the disease had been confirmed in carp populations in England. During the summer of 2006, KHV disease was confirmed in carp on a total of 23 sites suffering carp mortalities. Many of these mortalities were very serious, with large losses of fish.

KHV affects common carp and all its varieties. It has also been recorded infecting other cyprinid fish, such as goldfish, but not causing disease.

Infected fish show a variety of symptoms, the most typical of which is severe gill necrosis where the gills exhibit white, grey or black eroded gills. The damaged gills are often prone to secondary fungal infections such as saprolegnia. Infected fish often have 'dry' skin, as the skin and gills exhibit decreased mucous production. There are often small patches of haemorrhages across the fins and body. The eyes may be sunken. Owing to the impact on the gills, infected fish tend to be lethargic, swim close to the water surface and gather at water inlets or waterfalls where dissolved oxygen is highest. They often exhibit a respiratory distress (increased respiratory rate, gasping at surface). When disturbed they may show sudden hyperactivity and uncoordinated movements.

Outbreaks of KHV have only been recorded in the temperature range 15–28°C. During a disease outbreak, mortality rates can be up to 95% of the carp population, even in extensive fisheries. Evidence from previous mortalities suggests that good environmental and stock management conditions help to alleviate the effect of the disease and a much lower percentage of the fish succumb to the infection.

Like most other herpes viruses, KHV can become latent within an infected host. This means that a fish that survives a disease outbreak will remain infected, but with a very small number of viral organisms within its body. This fish remains contagious and will pass on the virus to other fish, given the right conditions. This is very difficult to detect using current testing techniques.

There are no treatments for KHV so therefore prevention is the only means of control. Although there are ongoing efforts to produce a vaccine against KHV, a robust biosecurity system is really the only method of protecting a fishery against this disease.

Diagnosis of KHV can be by cell culture techniques or other laboratory-based indirect tests. These are both difficult and expensive. Wherever there are unexplained mortalities of carp, an outbreak of KHV should be considered until proved otherwise.

7.4.2 Bacteria

Bacteria are small, single-celled organisms that occur throughout the natural environment. There are thousands of different species known, but likely to be millions, and some have been associated with fish disease. Although bacteria are organisms encompassed by a cell wall, their internal structure has no boundaries. They are therefore known as prokaryotic organisms.

Bacteria reproduce by a process known as binary fission, which means the cell simply splits into two, forming two fully functioning new bacteria. This process can occur every 15–20 minutes under optimal conditions. In a 24-hour period a single bacteria can produce 15 billion offspring! However, conditions are rarely perfect.

Bacteria are very resilient and can survive freezing, drying, heating and some disinfectants. The usual treatment for bacteria in enclosed systems is the use of antibiotics. However, antibiotics can only be used under veterinary supervision and they are totally unsuitable for use in fisheries.

Important bacterial diseases include the following.

Carp ulcer disease, Aeromonas salmonicida subsp. achromogenes

One species of bacteria, and its subspecies, appears to be responsible for more fish disease problems than any other. *Aeromonas salmonicida* is a non-motile, rod-shaped bacteria, common in the aquatic environment.

The typical form of *A. salmonicida*, *A. salmonicida* subsp. *salmonicida*, which primarily affects salmonids, causes the disease termed 'furunculosis'. Furunculosis is very widespread, occurring throughout Britain, Europe and many other parts of the world. The disease affects all life stages of the infected populations, and is usually most virulent at temperatures of 10°C and over. A cold water form has also been recorded which can cause problems in fish as low as 2°C. It can infect fish in fresh and seawater. All salmonids appear to be susceptible and it has been recorded in a variety of other marine and freshwater species. *A. salmonicida* is generally a very adaptable and dangerous pathogen.

Subspecies of *A. salmonicida* are known as atypical strains and are implicated in a wide range of diseases recorded in other freshwater fish. In cyprinids, the familiar ulcerative disease of carp and goldfish is usually caused by *A. salmonicida* subsp. *achromogenes*. This disease is often referred to as 'carp erythrodermatitis' (CE), carp ulcer disease or spring/summer ulcer disease. When recorded in other species, it is known imaginatively as perch ulcer disease, roach ulcer disease, etc.

A. salmonicida subsp. *achromogenes* causes a septicaemia disease of cyprinids. Symptoms include haemorrhaging, surface ulcers of the skin, protruding scales, scale loss, exophthalmia, abdominal swelling and petechial haemorrhages on the gills. Internal examination may reveal varying amounts of fluid in the abdomen. Frequently there are haemorrhages and reddening of the gastrointestinal tract, enlargement of the spleen and mottling of the liver.

The classic feature of this disease is the deep surface ulcers. In scaled fish, the rims of the ulcers are more often jagged instead of rounded. In some instances, the ulcers can become overgrown with fungus.

Although mortalities are often low, many fish within a population may show the characteristic lesions. Injuries such as those caused by predators or angling are primary wounds that can lead to secondary bacterial infections by aeromonad species. This makes *A. salmonicida* subsp. *achromogenes* one of the most problematic disease-causing agents of coarse fisheries.

Prevention is the only legitimate method of control in a fishery situation. Antibiotics have been shown to be effective against *A. salmonicida* and its variants. However, antibiotic sensitivity tests must be performed before any antibiotics are administered. Many of the diseases discussed above have developed wide-ranging resistance to many antibiotics. Treatment with an antibiotic that the pathogen is resistant to will do considerably more harm to the fish. In a culture, protection from furunculosis and some of the variants can be achieved using commercially available furunculosis vaccines, which are generally very effective and are recommended for salmonid and cyprinid culture.

7.4.3 Fungi

Fungi are often confused with plants but are in fact a completely separate group of organisms, containing thousands of different species from complex toadstools to single-celled yeasts. Usually they feed on dead organic matter, playing an important part in the recycling of nutrients, but several have adapted to a parasitic lifestyle on fish. Most fungi that affect fish require some prior skin injury to gain a foothold. The skin injury can be a result of parasite invasion, mechanical damage or bacterial infections creating skin sores. Fungal infections typically appear like whitish cotton-wool type growths from the infected site. Such growths may also become coloured brown or green from silt and algae in the water.

Reproduction can be by sexual or asexual reproduction, which usually results in the formation of spores. These are often highly resilient and can remain dormant for years until suitable conditions arise. It is generally accepted that these spores are extremely common in water and all water should be assumed to contain spores.

Significant fungal diseases are the following.

Saprolegnia sp.

Saprolegnia is a ubiquitous organism, being found worldwide in all types of freshwater (this may be due to one truly ubiquitous organism or possibly lots of very similar but different species of fungi). It is true to say that all water should be assumed to contain spores of *Saprolegnia*, waiting for an opportunity to grow.

It will readily invade any open wound on any fish species where it grows finger-like hyphae into the damaged skin tissue. Once established, it can grow across the surface of the fish and invade undamaged skin causing extensive pathology, leading to osmoregulatory disruption. *Saprolegnia* can quickly turn a small wound into a life-threatening condition.

Treatment in enclosed waters such as ponds or tanks can be effective, but considerable care should be taken as additional skin damage can lead to a significant resurgence of the disease, even if the initial treatment was effective. In a fishery, *Saprolegnia* infections are most often associated with poor conditions, debilitated fish and large amounts of nutrients in the water. It is vital in such situations to establish the primary cause of the problem and return conditions to within optimal ranges.

Branchiomyces sp.

Branchiomyces literally means fungus infection on the gills, and this is often how the term is used. This is not strictly correct as *Branchiomyces* sp. is a specific type of fungal infection and *Saprolegnia* can also be found on the gills. *Branchiomyces* sp. has an ability to grow along the blood vessels of the gill tissue, severely disrupting respiratory function.

7.4.4 Protozoa

Protozoa are single-celled organisms that measure between 10 and 100 micrometres (one micrometre is one-thousandth of a millimetre). Although some are just visible with the naked eye, e.g. whitespot, most are microscopic. Protozoa are different from single-celled bacteria organisms because they have well organised membrane-bound structures within a cell wall (these organisms are termed eukaryotic). The protozoa are a massive and diverse group of organisms. Of the 70 000 or so species of protozoa that have been described so far, most are free-living and common in the aquatic environment. Many species, however, have evolved a parasitic lifestyle and are commonly found on fish in all types of fishery.

Most protozoa have a 'direct life-cycle' and only require the host fish to thrive and reproduce. Protozoan parasites reproduce by simply splitting in two (termed 'binary fission') although some undergo sexual or other sorts of asexual reproduction. Some protozoa may have free-living stages, others must leave the fish to reproduce and others may exist solely on one fish.

Nutrition is generally obtained directly from the fish host, although some merely hitch a ride on the fish, feeding on dissolved nutrients or bacteria in the surrounding water. As a result of this, protozoa vary considerably in their impact on fish, some causing little discomfort whereas others are highly pathogenic and capable of widespread mortality. Natural epizootics from protozoa in the wild usually signals unsuitable conditions, and many outbreaks are nature's way of regulating excessive populations, particularly in systems where there are no top predators such as pike in the water. As the protozoa are very numerous and varied in their morphology and impact, things are simplified if we split them up into the flagellates, ciliates, amoebas and Sporozoa.

Flagellates

Members of this group possess long, hair-like protrusions known as flagellae, mainly used for stability and locomotion. Although many species of flagellates may infect fish either internally or externally, few are capable of being very pathogenic. Probably the best-known flagellate protozoan to cause disease in freshwater fish is the ectoparasite *Costia* sp. This parasite is often a problem in salmonid hatcheries where fry are particularly vulnerable. With their ability to reproduce at a phenomenal rate, high numbers of *Costia* sp. can cause considerable harm and mortality. It is usually only debilitated fish that are prone to such infections, indicating unsuitable fishery conditions. Other examples of flagellates that infect freshwater fish are species of *Cryptobia* sp. and *Trypanoplasma* sp. that live in the blood. Although common, these parasites usually exist in low numbers and are seldom a problem.

Ciliates
Most ciliates have a widespread distribution and relatively low host specificity. Some species cause very little damage to the fish, whereas others can cause serious pathology and death. Ciliates as their name suggests, possess short hair-like protrusions called 'cilia' all over the body surface, which are used for propulsion and food collection. Most ciliate parasites of relevance to fish are ectoparasites, living on the skin, fins and gills of their host. Probably the best known, most widespread and pathological of all the ciliate protozoa is *Ichthyophthirius multifiliis* or 'whitespot'. This parasite burrows into the surface layers of the fish and feeds on the skin, causing small white spots to appear, hence the name. When it becomes mature it ruptures the skin to escape from the fish. The parasite then encysts on some suitable substrate and multiplies by fission before the cyst ruptures to release thousands of infective stages requiring a host. These infective stages then search for a new host. On finding one, the parasite invades the fish's skin once more and the life cycle is complete. With a temperature-dependent life cycle, large numbers can soon accumulate if conditions allow.

Amoebas
These are very simple organisms that move with a flowing, jelly-like (amoeboid) movement. Few of them are particularly pathogenic and most live on the skin and gills of fish. They are of little relevance to the fishery manager.

Sporozoa
Although the taxonomy of these organisms has recently been revised, this group still contains the microspora and myxozoa, both spore-forming endoparasites. These parasites are very tiny although the multicellular spores that they form are often quite large and visible to the naked eye. Little is known about many areas of the Sporozoa, including the life cycles and impact to the fish host. However, it is generally considered that the myxozoa require an invertebrate intermediate host (often an oligochaete worm) and cause inflammation to the target organs in which they reside. All myxozoa possess two or more polar capsules in their spore while the microspora possess only one. Probably the best-known diseases caused by a myxosporean are 'proliferative kidney disease' (PKD) and whirling disease, both very serious and life-threatening conditions of salmonids. In the case of whirling disease, the parasite attacks the cartilaginous structure of the skull tissue as the fry grows; the resulting damage to the growing brain causes the characteristic whirling motion in infected fish. As a result it represents little danger to adult fish.

Different Sporozoa are common on coarse fish but have rarely been implicated in any disease outbreak.

7.4.5 Monogenea

These organisms are part of a larger group known as the trematodes although they are more commonly called flukes. Monogenean flukes are small (from 0.01 to 1 mm), worm-like creatures that attach to the external surfaces of the fish by a variety of suckers or hooks on their posterior end. They generally feed on the epithelium of the host fish and only require one host to complete their life cycle.

The aggressive attachment and feeding behaviour of some monogeneans can cause irritation and significant pathology to heavily infected fish. All monogeneans have a direct life cycle and can reproduce rapidly on individual fish. Of the monogeneans that parasitise freshwater fish, there are three different groups that are of particular interest: *Dactylogyrus* sp., *Gyrodactylus* sp. and *Diplozoon* sp. Representatives of these groups are widespread in freshwater, the two most common being the gill fluke *Dactylogyrus* sp. and the skin fluke *Gyrodactylus* sp.

Gyrodactylus sp. are live-bearers and have a very high reproductive capacity. They are also known as the 'generation worm', because when viewed through a microscope three or more generations of the parasite can often be seen within one individual. The salmonid parasite *Gyrodactylus salaris* is a spectacular example of the potential pathogenic effect of these parasites.

In contrast to species of *Gyrodactylus, Dactylogyrus* sp. produces eggs that are shed into the water to hatch and develop into a ciliated, motile stage that then locates and reattaches to a fish. They are distinguished from *Gyrodactylus* by the presence of four black eye spots. In small numbers, both parasites are of little concern and are commonly found on all fish species. However, if allowed to reach high numbers they have the ability to cause irritation, respiratory distress, excessive mucus, haemorrhaging and even mortality.

The third monogenean of interest is the 'twin worm', *Diplozoon* sp. When young, this fascinating parasite attaches and feeds on the gills of its host by means of clamps. Shortly after attaching, the young parasite seeks out another juvenile diplozoon (termed the diporpa). They fuse together in the middle in a 'copulatory embrace' which lasts the rest of their lives and forms the adult parasite, hence the name 'twin worm'. These parasites are very large, being clearly visible to the naked eye, and have a very impressive array of claspers to attach to the gill. Although these attachment structures look impressive there is seldom much pathology associated with their presence and heavy infections are rare.

7.4.6 Digenea

The other half of the trematode flukes belong to the group called the digenea. Again, they are small, worm-like organisms, but they are very different to other trematodes in their appearance and effect on the host. Most digenea possess two suckers for attachment, one positioned at the posterior end and the other in the middle of the body. Although there are exceptions to this (e.g. the blood fluke *Sanguinicola inermis*), such characteristics can be seen in most digenea, which also provide a useful identification feature. All digenea are endoparasites and vary considerably in their form, although most are elongated and measure between 0.5 and 10 mm.

Unlike the monogenea, digenea have complex life cycles requiring several intermediate hosts. The fish can act as the host for the adult fluke (definitive host) or as the intermediate stage (intermediate host), where it can be found in most organs of the fish's body. Those that live in the gut are seldom highly pathogenic unless in high numbers.

Their appearance is often deceiving. Those that use fish as an intermediate host often have a piscivorous bird as a final host. To complete their life cycle therefore, they affect the fish in a way that debilitates it, increasing the chances of it being eaten by the definitive bird. Such changes are often subtle but very effective.

Some examples of this behaviour can be seen with the eye fluke, *Diplostomum* sp. that, by impacting the fish's eyesight, reduces its ability to escape predation. Another nice example is that of the parasite *Posthodiplostomum cuticola*, which causes the condition 'blackspot'.

By covering the fish in tiny black spots, each containing a fluke, it makes them more visible to the eye of piscivorous birds and thus more likely to be eaten. Although this is good news for the parasite, badly affected fish in a fishery are undesirable. Because of this 'indirect' effect on the fish host, digenea seldom cause mortality unless found in extremely large numbers.

7.4.7 Nematoda

As their common name 'roundworm' or 'threadworm' suggests, the nematodes are generally long, slender worms with a round cross section. Nematodes have complex life cycles involving two or more intermediate hosts. Fish may act as either definitive or intermediate hosts to nematodes and they may occur in all organs of the body. Although many nematodes are relatively harmless to their hosts, much is still to be learnt about many nematode species and their impact on the fish that they infect.

Perhaps the best example of a nematode impacting a fishery is the eel nematode *Anguillicola crassus*. Since its introduction into the UK in the mid-1980s, this parasite has rapidly spread throughout many catchments in the country and caused significant problems to our eel stocks. Living in the swim-bladder, it causes reduced swimming performance and, as a result, it is likely to reduce the success of the migration to the spawning grounds of the Sargasso Sea. Adult parasites can attain a size in excess of 15 mm and have a characteristic black colour, the result of congealed blood on which it feeds. Its indirect life cycle requires intermediate hosts for its completion.

Other nematode species parasitic in freshwater fish infect the abdominal cavity, internal organs and intestine, ranging in size from just a few millimetres up to 5 cm. Perhaps the most common species include the white nematode worm *Cystidicola farionis*, which is found on the swim bladder wall of salmonids such as rainbow and brown trout. This parasite is often commented on, as it is clearly visible when the fish is gutted. The red nematode *Philometra* sp. is often found in the abdominal cavity of cyprinids, *Camallanus lacustrus* is found commonly in the gut of perch, and *Skrjabillanus tincae* is found in the viscera of tench. These parasites are seldom very pathogenic.

7.4.8 Cestoda

Commonly known as the 'tapeworms' or 'flatworms', cestodes are a widespread group of endoparasites. Most tapeworms that parasitise fish are long, segmented and ribbon-like parasites or slightly shorter and unsegmented in form. Leading an exclusively endoparasitic lifestyle, most tapeworms require several intermediate hosts to complete their complex life cycle. Invertebrates, such as oligochaete worms and copepods, often act as their intermediate host; whereas other fish, birds and mammals usually act as the final or definitive hosts. Interestingly, fish may act as either definitive or intermediate hosts to tapeworms. Fish infected with adult worms usually harbour the parasites in the intestinal tract, whereas parasites in the intermediate fish host are usually located in the body cavity or internal organs.

Tapeworms found in the gut of fish attach by means of a characteristic head or 'scolex', often armed with hooks, suckers or grooves. It is these structures that are useful for identification. Once attached, they feed by absorbing nutrients across their body wall from the surrounding gut. It is the attachment behaviour and removal of essential nutrients from their host that causes most problems to infected fish. Heavy infections can cause nutritional deficiencies, gut pathology and even blockage of the gut, leading to reduced growth rates, condition loss and even mortality. Fish with very heavy infections may even possess tapeworms protruding from the anus into the water. It is usually impossible to identify fish with tapeworm infections and a post mortem by an experienced fish health expert is often required to diagnose problems accurately. Although most intestinal tapeworm infections can be subtle, tapeworms that use fish as intermediate hosts are usually more obvious. The following examples are both tapeworms of relevance to fisheries and illustrate the two different types discussed.

Bothriocephalus acheilognathi

Bothriocephalus acheilognathi, also known as the Asian tapeworm, uses fish as the definitive host, and has the potential to cause problems to any infected fishery. This parasite was introduced into the UK in the early 1970s and can cause considerable problems to carp, which is its preferred host. Small fish are particularly susceptible to infections. This is because the copepods that act as the intermediate hosts to the parasite make up a large part of the juvenile fish's diet. Attaching by a heart-shaped scolex with grooves, or 'bothria', it can attain a length of 20 cm or more. Small fish have been known to harbour in excess of 100 worms, resulting in severe gut blockage, pathology and nutritional deficiency. Eggs are shed from the fish gut into the water, where they are eaten by a copepod intermediate host and then enter fish when these are ingested, completing the life cycle. A gut full of writhing tapeworms is perhaps the most repulsive, yet fascinating, sight to greet the fish pathologist during a post mortem. *Bothriocephalus acheilognathi* is a Category 2 parasite under Section 30 of the Salmon and Freshwater Fisheries Act (1975) (England and Wales).

Ligula intestinalis

In contrast to the Asian tapeworm, *Ligula intestinalis* infects fish at an intermediate stage of its life cycle, usually roach and bream. The effect of the parasite on the fish host, therefore, is to increase the chances of the infected fish being eaten by the definitive host, in this case a piscivorous bird. Fish suffering 'ligulosis' often appear very fat and bloated as a result of the large size of the parasite and its favoured location in the body cavity. The sight of a heavily ligulosed fish can often be very dramatic, with large numbers of parasites cramming themselves into the relatively small body space available. Results of heavy infections include organ displacement, castration of the fish host, and a loss of buoyancy control. Fish suffering from this condition tend to swim close to the water surface and are less capable of escaping the efforts of piscivorous birds. Although the occasional occurrence of *Ligula* is common and of little concern, heavy infections in a fishery can cause a problem. Such infections often indicate an abundance of piscivorous bird activity.

7.4.9 Acanthocephala

The Acanthocephala, or 'spiny headed worms', represent some of the most spectacular parasites of freshwater fish, possessing a proboscis heavily armoured with numerous spines. All Acanthocephala are found in the intestine of their hosts and measure between 5 and 15 mm. As with tapeworms, the Acanthocephala have a complex life cycle requiring intermediate hosts for its completion. Fish act as the final or definitive host to most acanthocephalans, with the freshwater shrimps *Gammarus* sp. and *Asellus* sp. commonly providing the intermediate role. Differentiated into separate sexes, acanthocephalans reproduce sexually, with eggs shed into the water. These are eaten by the intermediate host and attach to the gut of the fish when the shrimp is ingested. Most acanthocephalans affect their host shrimp in some way to make it more obvious to passing fish, for example colour and behavioural changes.

Acanthocephalans are relatively widespread parasites and occur in most types of fishery where freshwater shrimps may occur. With an obvious and impressive spined proboscis, it is easy to see the way these parasites attach in the gut wall of their host. This aggressive attachment appendage is a useful identification structure but can lead to substantial pathology in heavy infections. Perforation of the gut by heavy acanthocephalan infections has been recorded in some species.

There are about ten species of acanthocephalan in the UK, some more common than others. Of all these the species *Pomphorhynchus laevis* is considered particularly harmful and as such is on the list of Category 2 parasites (see Chapter 27). Unlike the other species of acanthocephalan, *P. laevis*, also known as the 'yellow peril', has a large bulb behind its spiny head. Its severity is often associated with its ability to perforate the gut wall of the fish host, leading to intestinal dysfunction and peritonitis.

Generally, small numbers of acanthocephalans have little detrimental effect on their host. However, in large numbers, nutritional deficiency, gut pathology and even blockage can occur. Owing to the complexity of their life cycle, it is the occurrence and abundance of the intermediate hosts that determine infection levels. It may therefore be stated that fisheries containing heavily infected fish also have abundant shrimp populations. Controlling these intermediate stages is often the only way to relieve acanthocephalan problems successfully.

7.4.10 Hirudinea

There are few people who do not have an unpleasant mental image when the word leech is mentioned. Anyone who has had even a passing association with the freshwater environment is likely to have come across this particular parasite, either on fish or on stones and rocks. Leeches are a type of annelid worm, which have developed the ability to suck blood from their host. In a similar way to species that affect humans, fish leeches attach by means of suckers and feed off blood, swelling in size as they feed. Once full, they detach, leaving a resultant mark or haemorrhage at the site of attachment. There are only two common species of leech found in freshwaters that have the ability to parasitise fish, namely *Piscicola geometra* and *Hemiclepsis marginata*. Both cause mechanical damage associated with attachment and feeding.

When attached to the fish, commonly along the underside and fin bases, leeches are easy to see, measuring up to 2 cm in size. They are brown or black and are often banded. Fish that have either been sulking on the bottom of a fishery, or have remained relatively still for long periods (e.g. during the winter months), will harbour the occasional leech. However, such infections seldom reach problem levels without imbalance to the fishery.

As is the case with many fish parasites, low-level infections are of little concern and healthy fish can easily tolerate the occasional leech. However, heavy infections can be very debilitating, resulting in lesions, haemorrhaging and secondary infections. As with the Crustacea, leeches can also aid the transmission of other disease organisms including bacteria, protozoa and viruses including SVC.

7.4.11 Crustacea

There are few groups of fish parasites as numerous and diverse as the Crustacea. With thousands of species, crustaceans have adapted to a parasitic mode of existence in many ways and as a result have varying impacts on their hosts. Most parasitic Crustacea are relatively large and ectoparasitic, often being visible with the naked eye. They are also much specialised in their attachment and feeding behaviour, and many species are commonly associated with disease problems in fisheries. They are characterised by a hard exoskeleton, often numerous swimming appendages and well-developed feeding apparatus. Crustacea in general are a very widespread group of organisms, common in the aquatic environment. The best-known and widespread example of a crustacean parasite is the fish louse, *Argulus* sp. Crustacea have relatively simple, direct life cycles, requiring just one host for their completion. Sexual reproduction leads to either the release of eggs or free-living stages that grow and become parasitic and re-attach to the fish to mature as adults.

Of all their characteristics, it is probably the attachment behaviour and the associated appendages that are most interesting and diverse. Different species use varying numbers of hooks, claws, anchors and spines that often burrow deep into the host tissue. Such attachment is often quite harmful and can cause considerable discomfort. In addition to this, the grazing behaviour of many species, feeding on mucus, blood and host tissue, adds to such pathology. This invasion of the host skin often leads to secondary bacterial and fungal infections.

Many crustacean parasites can cause serious problems in high numbers, although the occasional parasite on a large fish is of little concern and will cause little discomfort. The onset of heavy infections in a fishery, however, usually indicates imbalance and suboptimal conditions. Most Crustacea have very high reproductive capabilities, but in a balanced environment natural mortality of juvenile stages is high, resulting in relatively stable, low-level populations. In overcrowded and stressful conditions, however, the numbers of surviving parasites can increase dramatically and lead to a disease outbreak.

Typical symptoms of crustacean parasites include irritation, inflammation at the site of attachment, haemorrhaging and excess mucus, leading to lesions, lethargy and death with heavier infections.

The most important crustacean parasites are the following.

Argulus sp.

Probably the best-known and most widespread crustacean parasite of freshwater fish is the 'fish-louse' or 'carp-louse', *Argulus* sp. With its characteristic eyespots and paired suckers, argulus are easily recognisable on the fish skin and fins. When observed in situ, the fish louse almost walks over the fish surface, feeding as it goes. More common in still waters than running waters, *Argulus* numbers can build up rapidly if conditions allow and cause severe condition-loss and mortality to infected fish. Female parasites leave the fish to deposit egg strips on a suitable substrate, commonly weed stems and rocks. On hatching, juvenile *Argulus* are directly parasitic, and require a host within a few days to survive. Under natural conditions this results in only the odd parasite surviving. However, high stocking densities, warm conditions and plentiful weed growth for egg attachment are conducive to rising infections. Unlike many other crustaceans, adult *Argulus* are good swimmers and commonly transfer from fish to fish. *Argulus* is so common that there are probably few types of fishery in the country that have not at some time harboured this parasite. It is therefore not the presence of this parasite that brings the onset of disease, but an imbalance in the fishery and its management. An important factor of crustacean parasites is their ability to act as vectors to other disease-causing organisms, including bacteria, protozoa and viruses such as SVC.

Ergasilus sieboldi

Other important crustacean parasites of freshwater fish are members of the Ergasildae, commonly known as the 'gill maggots'. Most of these species have been introduced into the UK from abroad, and as a result hold potential to cause problems to our native fish that have limited defences against them. *E. sieboldi* is probably the most destructive of all the ergasilids, living and feeding on the gills of its host. As with all the Crustacea, *E. sieboldi* has impressive attachment appendages, namely a large pair of hooked arms. These firmly grasp the gill filaments of infected fish, and allow the parasite to move up and down each gill, feeding as it goes. Being such a delicate organ, any significant damage to the gills can result in severe debilitation, respiratory distress and mortality. The life cycle of *E. sieboldi* is direct, with only adult females being parasitic. Relatively large white eggs sacks protrude from the back of the parasite when gravid, appearing almost maggot-like in appearance. Juveniles hatch from these eggs into the water, attaching to a fish only after several free-living stages. As a result of this, it is not just the fish within a fishery that are infected, but the fishery as a whole. *E. sieboldi*, as with many other species of the ergasilids, are Category 2 parasites under Section 30 of the Salmon and Freshwater Fisheries Act (1975) (England and Wales).

7.4.12 Non-infectious diseases

Non-infectious diseases are those diseases that do not involve a pathogen and are therefore not transmittable from one fish to another. A wide-range of these have been documented including tumours (neoplasias) and developmental diseases.

These sorts of problems are usually noticed as deformities. Common problems include the following:

- the spinal deformities scoliosis and lordosis, which result in the backbone being curved horizontally or vertically;
- deformed opercula;
- swollen abdomen;
- obvious external tumours.

If these problems arise in a substantial percentage (greater than 40%) of any group of fish on a fish farm, then the fish should be destroyed and the diet and broodstock changed for the next batch of fish.

These sorts of problems are unlikely to affect fisheries greatly, except the occasional interesting deformity that may be reported by an angler. There is no cure for any of these diseases and if it is causing distress then the fish should be destroyed.

In addition to tumours and physical deformities, there are several conditions that may affect fish as a result of environmental problems. Although these are not going to be covered in detail, probably one of the most well known is termed 'gas bubble disease'. This condition is the result of supersaturated gases (usually nitrogen) leaving solution and expanding within blood vessels, gill lamellae, skin and fins. This condition is usually more of a problem in fish culture, particularly when bore hole water is used as a source. In a fishery this problem is rare.

7.4.13 Novel pathogens

One of the most serious problems in fisheries today is the introduction of novel pathogens into the country, and then their widespread dissemination. In the natural environment a pathogen and its host have usually been in contact for many thousands of generations. In that time, by a process of natural selection, the host fishes most able to withstand the pathogen have been able to breed and produce offspring. As the pathogens have evolved new ways of attacking the host fish, so the hosts have evolved new ways of protecting themselves. This form of adaptive evolution is a very elegant way of making sure that hosts and parasites both survive and contribute to the ecosystem's rich diversity. However, it means that immune systems are somewhat specific to the types of pathogen that the population has been exposed to. If a novel pathogen is introduced into that system and none of the immune systems work on it, then the pathogen will be able to reproduce and spread without any control. This situation quickly leads to a disease outbreak, which the fish are unable to respond to.

8 Mammals and Birds

No section on the living components of a still-water would be complete without the inclusion of mammals and birds. Although these animals are often not considered when looking at the general wildlife of a still-water, many play a very important part in the ecology of the system. Both birds and mammals represent some of the more important nuisance species and so will be dealt with in more detail later, in Chapter 20.

8.1 Birds

Birds are defined as bipedal, warm-blooded, vertebrate animals that lay eggs. They are often characterised as animals with feathers, a beak without teeth and a lightweight but strong skeleton. All birds have forelimbs adapted as wings and most can fly.

In still-waters, birds fall mainly into two groups: the fish eaters, such as herons and kingfishers, and the grazers, such as most ducks and geese. Birds are a natural component of the aquatic environment and, although some are occasional visitors, many rely heavily upon still-waters to breed, eat and rest.

8.1.1 Ducks

There are many resident species of duck in the UK, with many more that are frequent and infrequent visitors. Most of these spend some, if not all, of their time on still-waters. The presence of ducks is of course entirely natural, and under natural conditions they frequent specific still-waters at levels that can be supported by its biological carrying capacity. They generally feed on weed and the invertebrates that live on the weed. This behaviour again should be considered as helpful to the aims and objectives of the fishery manager.

Problems only occur when numbers increase because the ducks are exploiting an opportunity created by certain man-made situations. These will be dealt with in Chapter 20.

Sawbill ducks

Sawbill ducks are so named because of their long, serrated bill (Fig. 8.1). They are highly specialised in catching fish. The species in this group are as follows:

- the goosander (*Mergus merganser merganser* L.);
- the red-breasted merganser (*Mergus serrator serrator* L.).

The impact of sawbill ducks on riverine habitats and fish species has been researched and is understood to exist. However, sawbills are generally only considered to frequent larger still-waters such as reservoirs and lakes on anything like a regular basis. Where they occur, fish are their only dietary item, so predation will occur.

Grebes

Grebes are numerous in species, with worldwide distribution. As such, they have their own family (Podicipedidae). There are five species that are resident in the UK, all of which frequent still-waters to a lesser or greater extent. The five species are listed below:

- black-necked grebe (*Podiceps nigicollis*);
- great crested grebe (*Podiceps cistatus*);
- little grebe (*Tachybabtus ruficollis*);
- red-necked grebe (*Podiceps grisegena*);
- Slavonian grebe (*Podiceps auritus*).

Figure 8.1 Sawbill duck beak.

8.1.2 Geese

There are many geese species that are resident in and migratory to the UK. However, although most of them will frequent still-waters at various stages, there are two that spend a significant amount of their time on still-waters. These are the following:

- the Canada goose;
- the greylag goose.

Canada goose

The Canada goose (*Branta canadensis*) is a large goose (Fig. 8.2) that is a non-native species. It was first introduced into the UK by King Charles II, to enhance his waterfowl collection at St James Park, in 1665. Since then the species has done very well, fitting into a niche not formerly occupied by one of the UK's native geese.

Canada geese are highly gregarious and as such form large flocks. They can be seen in the sky in large 'V' formations (skeins) as they move around the country between their preferred habitats. They also graze in these large numbers, known as gaggles, and as grazers spend a considerable time on grassland. Unfortunately, as this grassland can be parkland around a lake, or riparian vegetation around a small pond, these geese can cause serious damage to vegetation when their numbers overwhelm the ability of the habitat to support their presence.

The Canada goose is a quarry species and as such can be shot during the appropriate 'open' season; however, since 2005 it has been possible under a general licence to shoot Canada geese during the close season. The license expressly details the purpose of the licence, which it states is 'for the purpose of preventing the spread of disease and for preventing serious damage to livestock, foodstuffs for livestock, crops, vegetables, fruit, growing timber, fisheries, or inland waters' and 'for the purpose of preserving public health and safety'.

Figure 8.2 Canada goose.

Figure 8.3 Greylag goose.

Greylag goose

The greylag goose (*Anser anser*) is the largest of the UK's native geese (Fig. 8.3). However, many of the flocks that are found across England are perhaps best described as semi-domesticated; indeed, our farmyard geese are direct descendants of the greylag. The truly wild flocks of this goose are to be found north of the border in Scotland. Although the greylag is a grazing animal, in common with the Canada goose, there are a number of differences in their behaviour.

Greylag geese are often present as semi-domesticated flocks and for this reason do not migrate to breed. Given this fact, large and increasing numbers can become problematic from a fishery perspective.

The greylag goose, like the Canada goose, is a quarry species and can be shot during the appropriate 'open' season; however, fisheries managers should always view destructive measures as a last resort and undertake other control methods. Protection of the habitat is an integral part of this and is described in more detail in section 18.2.

8.1.3 Heron

The heron (*Ardea cinerea*) (Fig. 8.4) is a common visitor to still-waters in the UK. It is a large bird, perfectly designed for stealth and stalking its prey. It should be considered by all fisheries managers as being one of nature's predators that has a rightful place, asserting a downward pressure as a top predator on the fish stock. Herons in still-waters generally take the smaller fish that are less well-able to escape predation. They certainly will not generally take fish of a size that would interest most anglers. The selection pressure they exert therefore should be considered helpful to the aims of a fishery manager.

Figure 8.4 Heron.

8.1.4 Other birds

Although still-waters can provide suitable habitats for many specialised species, such as kingfishers, reed warblers, water rails, coots and moorhens, the aquatic habitat and the surrounding terrestrial land make for an attractive environment for many of the UK's commonest species of other groups of birds, such as songbirds, finches, warblers, tits, etc. The number and diversity of species can often be increased by enhancing and creating specialised habitats (Chapter 23). Anglers often welcome the diversity and interest most birds bring to a still-water.

8.2 Mammals

Mammals are defined as warm-blooded, vertebrate animals. They usually have a hairy skin and they feed their offspring from special glands that produce milk.

In still-waters there are two that are associated with fisheries: the fast-recovering and recolonising native species the otter, and the non-native mink. Both these species are covered in detail in Chapter 20. There are a few other species, however, that are worthy of our attention, for example the water vole.

8.2.1 Water vole

The water vole (*Avicola terrestis*) is a member of the vole family (Microtidae). Although the water vole is found throughout the UK, its distribution is extremely limited today compared with historic populations. Since 1990 the species has suffered a 90% decrease in sites where they had previously been recorded.

Water vole conservation has perhaps been caught out, as these animals were once common. The decline is thought to be a result of the following factors:

- loss and fragmentation of habitat;
- disturbance of riparian habitats;
- predation by mink;
- pollution of watercourses and poisoning by rodenticides.

The water vole is a small aquatic mammal that is native to the UK. It hunts and forages during both day and night, and indeed spends most of its life doing just that; it consumes up to 80% of its body weight daily.

It is often mistakenly identified for a brown rat (a distant cousin). The reverse is equally true, with a rat being mistakenly identified as a water vole. Rats tend to choose habitats that are nearer to human habitation and sources of food derived from human waste and pollution, although this might, of course, be left over from an angler's provisions, including leftover bait.

Although significant progress has been made in improving water quality, this has not had the same result on the water vole population as it has on that of the otter.

The water vole has, since 1998, been partly protected under Schedule 5 of the Wildlife and Countryside Act (1981). However, in 2008 it was afforded full protection under Schedule 9 of the Wildlife and Countryside Act (1981).

So the water vole is now fully protected by regulation. As previously stated, although regulation sets the scene, it is actions to improve habitat and decrease predation by mink that have the potential to increase the distribution and number of water voles. Individuals and organisations not directly involved with water vole conservation are best placed to do their bit by improving both terrestrial and aquatic habitats and implementing strategies to reduce mink numbers.

There will be a varying but considerable number of other mammals that will frequent a still-water and its surrounding terrestrial habitat. These may include the following:

- Fox
- Badger
- Deer
- Rabbit
- Stoat
- Weasel
- Brown rat
- Mice
- Shrew

Although these mammals are of no direct interest to the fisheries manager, their presence is an indicator of the biodiversity of the fishery. This can serve as an indicator of the quality of the habitat and may suggest the need for enhancement. Again, most anglers like to see these species of animals when they are fishing.

Part II
Management of Still-Water Fisheries

Part II describes the various procedures involved in acquiring a fishery, either by building a new one or by taking over an ongoing concern. It also sets out the day-to-day management of the fishery by considering the technical, social and constructional aspects of management.

9 Developing and Preparing a Fishery

The dream of many anglers, angling clubs, individuals and fishery managers is to manage their own fishery. Surprisingly this is an achievable dream for many as there are so many routes to being the manager of a fishery. You could start from the very beginning and develop a green-field site where you get the opportunity to build exactly what you want. Another way is to take on the management and control of an existing fishery or still-water. This chapter deals in the main with the development of a green-field site, but all the techniques and lessons are equally applicable to taking control of an existing fishery.

9.1 Developing a green-field site

9.1.1 Introduction

Often developing a fishery from scratch will require the development of a green-field site. However, before any development takes place a suitable site must be found. In many respects this can be the most difficult aspect of the entire procedure of developing and creating a fishery.

A fishery needs land and adequate good-quality water. The UK is a very densely populated island. In the past, its freshwater resources have been adequate for supporting the population with drinking water, and water for agriculture, industry and recreation. However, even in the past, stresses on the resources were such as to result in severe pollution and local shortages in supply. Now, with changing climate and growing population pressures, the quality and quantity of water is constantly under pressure in parts of the UK as rainfall appears to be declining and becoming less predictable. Whatever the cause and the degree of impact that this will have on the environment, it looks certain that the current trend towards heightened water awareness by the general public is set to continue for the foreseeable future. It is this awareness that will drive those organisations responsible for water regulation to be increasingly vigilant when considering applications for abstraction and discharge consent. The need for continuity and protection for the potable supply carries immense political weight.

Obtaining a plot of land suitable for development is not all doom and gloom, far from it. As this is still a wet island, standing waters occur frequently in lowland areas. It is, however, probably fair to say that any future developments will need to have an increasing regard for their potential impact upon the environment.

The bureaucracy involved in turning a green-field site into a fishery is considerable; however, before considering this it may be pertinent to consider some aspects of locating a suitable parcel of land for development (Fig. 9.1).

9.1.2 Finding and identifying a green-field site suitable for a fishery

Before identifying a suitable site and obtaining some form of tenure (licence, lease or purchase) on it, one should consider exactly what is needed, both in terms of what is essential and what is desirable in the site.

Ultimately the type of fishery that is to be developed will be the main criterion for deciding the suitability of the site, although there may be some flexibility to modify the type of fishery to suit the site. For example, certain water-quality criteria are vital in determining whether a fishery would be suitable for trout. Certainly water quality is not a factor that should be compromised: generally if it is not suitable to begin with, in practice it will be difficult to improve at a later stage.

Finding the right site in a convenient location is very important. Only investigate sites within a predefined radius of your base. Consideration should be given about what is a reasonable distance to travel on a daily basis. Of course, this decision is a matter of personal choice, but it stands to reason that, as most people live in the urban environment, then most will have to travel to rural areas to find and locate a green-field site. There is little point in coming to an agreement to develop a site many miles from home only to find that the strain of travelling is too great to sustain.

The actual finding of a potential site requires a great deal of personal effort. It is a very time-consuming procedure. Although some fishery owners will surely report that they just 'came across' their site, most sites are found after a great deal of time and effort. Time and effort are required because there is nowhere that one can go to find out where a defined green-field site awaiting development is situated. There are many ways of obtaining information that will make the task of finding such a site easier. These may include speaking to local people; success may often depend on word of mouth and who you talk too. However, local knowledge should not be overlooked.

There can be no doubt that the best method in condensing the search area is the use of Ordnance Survey maps and the ability to read and understand contour lines. If this is coupled with a map showing geological and hydrological features, much of the initial detective work

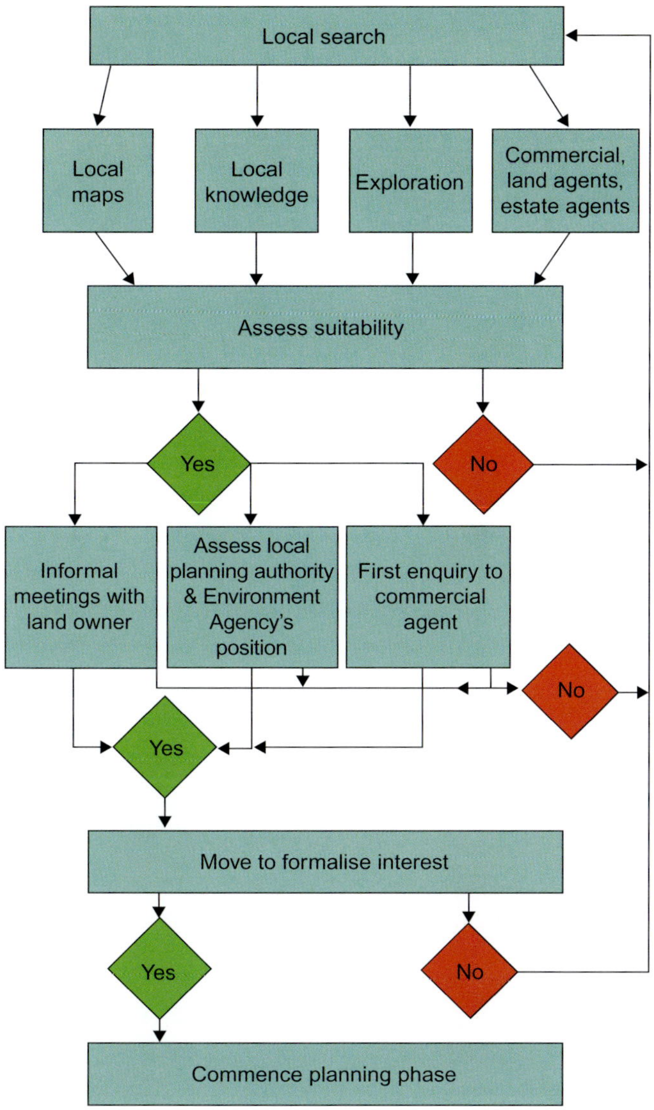

Figure 9.1 Procedures for locating suitable land for development.

can be done from the comfort of your own home. The Ordnance Survey's 'pathfinder' 1:25 000 series of maps is well suited. They are available from bookshops within the local area to be investigated. The area should be near to a water supply from either running or spring/groundwater sources; it should have an impermeable substrate and be accessible. If the area broadly fits these criteria then it will merit further investigation, and it is time for a site visit. On-line searches using geographical engines such as Google Earth or Google Map can also help refine the search and clarify the setting around an identified area.

Many of the classic signs of a suitable site will only ever be found by close observation in areas of potential site development (Fig. 9.2). These may include the following:

- standing surface water;
- boggy appearance;
- *Juncus* (soft rush) and *Carex* (pond sedge)-type grass tufts or other aquatic plants;
- nearby water source.

9.2 Site assessment of a suitable green-field site for a fishery

Once a potential site has been located, a decision needs to be taken about the site's suitability for the proposed use. To decide this some further information needs to be drawn together for consideration. It is suggested that the following factors be considered when assessing a green-field site for its suitability for a particular type of fishery (Box 9.1):

9.2.1 *Access*

The site will not be viable if access makes either the construction or the future management of the facility difficult and expensive to undertake. It is advisable to take a very detailed look

Figure 9.2 A potential site for development.

Box 9.1 Factors to be considered when assessing a green-field site for its suitability for a particular type of fishery

- Access
- Site security (including bio-security)
- Site availability
- Nature conservation interests
- Soil types
- Water supply
- Water quality
- Topography

at this aspect early in the site visit. The laying and maintenance of a hard surface suitable for vehicular access can often be a very limiting factor both logistically and financially.

9.2.2 *Site security*

The potential security of the site is easily overlooked. There can be little point in successfully creating a fishery only to see the fruits of one's labours disappearing by theft (poaching) and predation (herons, cormorants, mink, etc.). An isolated location will make the security of the site very difficult. The theft of fish is widespread, even from angling waters where fishermen are in attendance. A near-ideal site might of course have accommodation on site or very nearby.

Every site is unique. It is important to consider site and stock security as one of the contributing and deciding factors in a decision about suitability for the purpose. In addition to security from theft, the fishery must fulfil the requirements of the fishery manager's biosecurity risk assessment; this is dealt with in far more depth in Chapter 20.

9.2.3 *Site availability*

There is little point in formulating plans to develop and construct a fishery if the potential land is not commercially available. This needs to be confirmed at an early stage. Local knowledge can be an advantage: a question at a local pub, or to a passer-by, as to the ownership of the potential plot of land may be helpful. It is advisable to keep the amount of information departed to a minimum: enquire the name of the owner without giving your reason for wanting the information, other than to contact them. It is not unusual to find a piece of land that the local inhabitants cannot identify an owner for, as often with suitable sites they are poor agricultural land that is left fallow for year on year.

The local farmer may well be your next best option. Do your research first as they might turn out to be the owner. An additional source of ownership information is through the Land Registry.

9.2.4 *Nature conservation*

Increasingly, many parts of the countryside are being protected by various conservation-status designations. It is important to establish as early as possible if any of these designations affect the potential site, or indeed if any are up- or downstream. The designation can affect whether development would be allowed and, if it were, whether the management regime would be required to coexist with the conservation status.

The development of an area that has been designated with a special status will as a minimum undoubtedly involve outside bodies such as national and local conservation bodies. If the area is designated as a Site of Special Scientific Interest (SSSI), then Natural England (www.naturalengland.org.uk) will need to be consulted.

9.2.5 *Soil types*

Fortunately, the tell-tale signs that may lead to a site investigation in the first instance can also be indicators of the suitability of the soil for holding water, namely visibly wet and swampy ground. Fisheries rely on a natural bottom and sides to the pond or lake, therefore the material that is left after construction must be impermeable to keep seepage to a minimum.

The suitability of soil for a fishery is more involved than just its ability to retain water. The soil richness will directly reflect in the fertility of the pond and thus its ability to support a balanced aquatic ecology, including a fish community. The best way to establish suitability is to perform a soil survey.

Soil surveys should be done as early as possible. Soil sampling can be carried out in two ways, which are dependent on the analysis required from the sample. They are:

- disturbed samples;
- undisturbed samples.

Disturbed samples do not represent exactly how the soil lies in its natural state (before sampling), whereas an undisturbed sample represents exactly how the soil lies in its natural state.

The disturbed sample is used for the simple tests: those that can be performed at the site while making the initial site visit. The disturbed-sample method also includes a simple test, which can be done at the site to establish soil texture and its ability to retain water (Fig. 9.3).

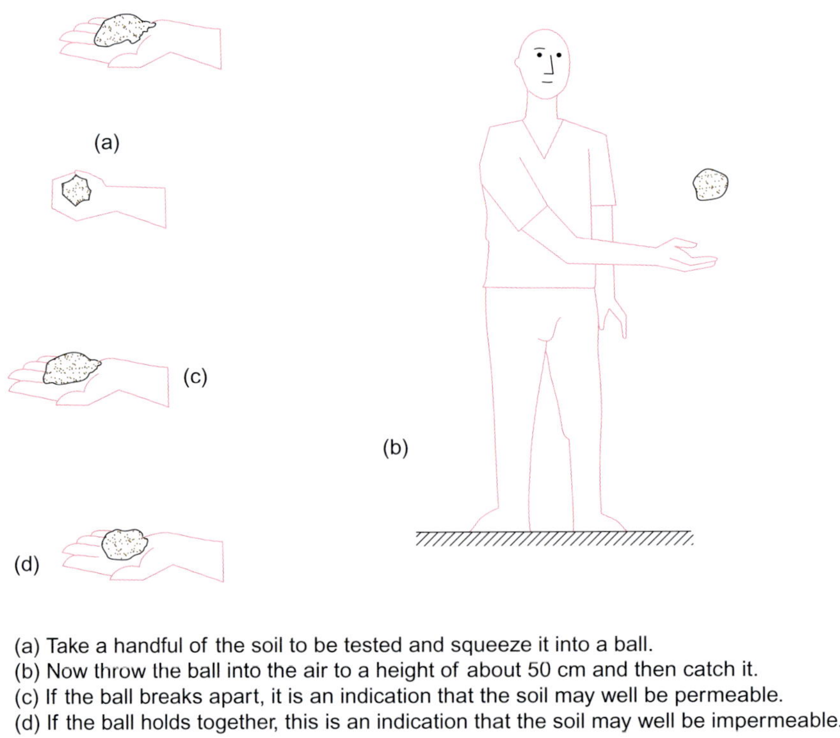

Figure 9.3 Four stages of a simple soil evaluation.

It follows that the undisturbed sample is used for the more sophisticated tests, which are more suited to laboratory investigation, such as chemical analysis. The undisturbed sample requires a more involved procedure to obtain it. Before developing a green-field site for a fishery, it is suggested that both types of sample will be required and the appropriate analyses performed.

The initial survey may be termed the 'reconnaissance survey'. It should be quick and provide a general idea of the soil varieties present and where they are found. A survey like this should be conducted by digging several open pits (Fig. 9.4) and then examining the exposed soil profile. From this it should be possible to determine what parts of the site are suitable for pond construction (those having good impermeable soil), and those parts that are not suitable (gravel outcrops, thick layers of organic material, etc.).

Close inspection of the open pit will reveal the soil profile. This comprises the soil horizon (layers of different types). It will be worthwhile making a sketch or taking a photograph, and taking a sample of each layer. This is not quite the same as an undisturbed sample, as the soil is only lying in its natural state in the horizontal plane.

It is worth noting here that should soil be required to construct a dam or dyke, it needs to have the cohesive properties or plasticity index of 15. Soils that contain too much sand or gravel should not be used because they will not hold water. Clays and silty clays are best. Soils that contain at least 20% clay by volume are suitable for making impounding banks.

Figure 9.4 A test pit.

It is clear that a soil's ability to retain water and to act as a nutrient base is important to the creation of ponds and lakes for recreational fisheries.

9.2.6 Water supply and quantity

Quantity

Although we call the ponds and lakes we use for our fisheries 'still-waters', the water in them changes constantly. The amount of time a mass of water remains in the still-water is called the 'retention time'. This can vary according to the nature of the water body. For example, a lake created by damming a river will have a short retention time compared with an isolated pond on a clay soil. The amount of water in any still-water is the result of a balance between water coming into the water body (gains) and water leaving the water body (losses). It can also vary seasonally, with much shorter retention times in the rainy autumn and winter seasons than in the summer.

On first impression, water quantity may appear to be a straightforward criterion. Indeed, if the supply is consistent and obviously of suitable volume, this may be true. Where the supply is more limited it may be necessary to estimate how much water will be required. The amount of water that will be required to fill a fishery should be worked out in cubic metres (m^3).

Gains

- Inflow from rivers or streams: usually the main source of water. If water is to be deviated or abstracted from a nearby river, an abstraction licence will be required unless there is already a permit in place.
- Seepage from ground water: especially in low-lying swampy areas, or ponds located in valley bottoms. The rates of seepage in or out of the pond depend on the nature of the soil and bedrock in which it is located.
- Pumping from ground waters: a locally important source of water although abstraction licences will have to be applied for.
- Local rainfall and surface flow in the lake catchment: usually a minor condition but can be seasonally important in areas liable to flash flooding.

For small ponds with a generous water supply (an annual flow more than six times the pond's volume) that flows all year, volume alone (in cubic metres) will be sufficient to determine the adequacy of water supply. For larger ponds with a less generous water supply (an annual flow less than three times the pond's volume), it may be prudent to calculate the seepage (Table 9.1) and evaporation factors (Table 9.2) as well.

Losses

- Outflow into rivers or streams: this can be controlled by a series of flow control structures.
- Seepage into groundwater.
- Evaporation: losses by evaporation are variable and depend on air temperature and wind.
- Abstractions: usually not important in the types of still-water that are of interest.

Table 9.1 Calculation of seepage rates through different soil types

Natural soil type	Seepage losses (mm/day)
Sand	25.00–250
Sandy loam	13.00–75
Loam	8.00–20
Clayey loam	2.50–15
Loamy clay	0.25–5
Clay	1.25–10

Table 9.2 Calculation of evaporation rates

Month and evaporation rate (mm)		Methodology
April	The evaporation rate for your area each month is available from the Meteorological Office	Add the rates (in millimetres) for each month. Divide the amount (in millimetres) by 1000 to express the evaporation in metres. Multiply this value (in metres) by the surface area (in square metres) to find the total water losses by evaporation (in cubic metres).
May		
June		
July		
August		
September		

Evaporation in a dry summer month can account for a 200-mm loss in a 0.4-hectare pond (equivalent to about 7 mm/day). In a fishery, it is the annual flow that is in many ways the limiting factor. In a still-water, a guide might be that a minimum of twice the water volume required should be available as the annual flow.

The estimation of the quantity of water available is generally dependent on flow. It is important that a fairly accurate estimation of water flow can be gained under field conditions. There are several ways to measure the amount of water in a stream or channel. The method that you choose may depend on several factors:

- the accuracy of the result needed;
- the quantity of water present in the stream or channel to be measured;
- the site access;
- the equipment available.

Table 9.3 lists the various methods of assessing water flow.

Under normal circumstances an initial site investigation should be done before approaching the landowner. It follows that, before making such an approach, one would wish to have a very good idea about the continuity and volume of the flow. The float and cross-section method will give the required degree of accuracy and is fairly simple and quick to perform with the minimum of equipment.

Type of supply

An initial investigation should quickly reveal the source of the water supplying the potential still-water. The three likely sources are:

Table 9.3 Assessing water flow

Method	Water flow	Accuracy	Remarks	Equipment
Quick and rough	Small	Approximation	For a quick estimate	None
Bucket	Very small	Very high	Most accurate of all methods	Dam, pipe, buckets, 1-l bottle, watch
Float	Small to large	Low to medium	Best for streams with calm water	Float, stakes, line, measuring stick, sheet, watch
Float and cross-section	Small to large	Medium	Best for streams with calm water	Float, stakes, line, measuring stick, record sheet, watch
Dye stain and cross-section	Small to large	Medium	Best for streams with calm water	Dye, stakes, line, measuring stick, record sheet, watch
Weir triangular	Does not vary greatly (114 l/s) or smaller or does vary greatly from small to large	High	For recording flow over a period of time	Wood, sheet metal or corrugated roof sheeting; tools for working with wood or metal; shovel, pick, line level, measuring stick
Weir rectangular	Does not vary greatly and is greater than 114 l/s	High		

- surface water;
- springs;
- groundwater.

All provide very different types of water quality. Perhaps the most important point to note at this stage is that a groundwater supply will require either drilling a borehole or excavating the pond or lake down into the water table. Excavating into the water table can only ever be a viable option where the level of the water table remains stable during the summer months.

To develop a borehole, initial test drilling is needed before a decision is made for one to be sunk permanently. This can be a very expensive operation, requiring specialised equipment and operatives. It will, if settled upon as a method of supply, also require pumping, and the associated costs will have to be met as part of any running cost. So water quality potentially may well have a trade-off against additional cost.

Before any drilling is undertaken, the Environment Agency must be approached as there are many areas where the sinking of a borehole will not be allowed. The way forward is to have an informal discussion with the water resource function of the Environment Agency. Establishing a borehole may be considered as having an impact on the aquifer. This is particularly true in the southeast England, where a large percentage of the potable supply in some areas comes from an aquifer, which may struggle to meet the demands placed upon it.

9.2.7 Water quality

Having confirmed that the soil can retain the water, and that there is enough water, we must ascertain that the quality of the water is suitable to support fish in a still-water fishery. The quality of water can be simply tested and involves taking a sample in a clean, dry plastic bottle. The sample can be tested with a commercially available test kit. Some parameters such as dissolved oxygen, pH and temperature can very easily be taken with small electronic probes. Once you have analysed the water sample, you then need to be able to understand what the analysis means in terms of creating and sustaining your fishery. This is dealt with in depth in Chapter 14.

A single sample test is only a snapshot in time, indicating water quality at the time of sampling. For this reason further evidence about the viability of the supply, particularly upstream pollution, will prove extremely useful. In many instances there may be some previous history of poor water quality, so the fishery manager, having observed the site over many months, will be able to determine that there is an adequate water supply.

In the absence of previous knowledge it is possible to gauge frequency and severity of pollution events by sampling the animals living there (biotic taxa) as an environmental indicator. Further information on the use of invertebrates to assess water and habitat quality is included in Chapter 14.

Larger carriers (rivers and streams) may have had similar sampling done by the Environment Agency. It is always worth a call to your local Environment Agency fisheries officer or area biological appraisal teams to ask if any information is available. Also, remember that any upstream fishery can impose a potential disease problems. The Environment Agency may also have record of any known risks in the catchment.

The topography of the potential site is the factor that dictates the ease with which a pond or lake can be constructed. Ideally it should suit low-cost construction and should not be subject to risk of flooding. An ideal site will also facilitate the emptying of the fishery without pumping (that is, by gravity). The objective must be to exploit the particular advantages of a given site, while ameliorating the disadvantages, in order to maximise the most cost-effective construction method.

9.3 Obtaining the site

Having decided that the site has potential, it is advisable to define the precise objectives of the scheme. This will enable a plan to be drawn together, which should be used as a guide once the site has been obtained. The plan can also form the basis of a management plan that could be presented to a financial institution to raise any funds that may be required. It will be almost impossible to progress past this stage on a commercial basis without a set of comprehensive plans and accompanying cash flow and budgets.

A set of ground plans showing the site layout with access, etc. will be required by the planning authorities. The Environment Agency will, if appropriate, require fairly concise details about abstraction and flow through (discharge) as well as details of any impoundment and associated structures such as dams. However, before spending money on these aspects, the purchase or lease of the site must be secured first.

The owner will have to be approached. To a large extent, how the approach is made will depend upon the person making the approach and those being approached. A personal approach can often be beneficial but many may be ill at ease in doing this, so an initial well-written letter may be better. Obviously, a lot depends on how it is envisaged that the recipient will react.

The buyer should be armed with a strategy, and know all of their options including what they are prepared to pay for the site; then, and only then, should the approach be made. It may also be advisable to have put the finances in place or at least know where they are coming from.

9.3.1 *Options for securing the site*

You must also know your preferred option for securing the site. These include the following:

- a shake of the hand;
- a lease;
- a licence;
- purchase.

A shake of the hand

A shake of the hand or gentlemen's agreement is a quick and easy way to reach an agreement to establish a right to construct your fishery on the proposed site. There is, however, one obvious major disadvantage. There is no security of tenure. Your being on the site is dependent solely on an agreement made orally. Furthermore, the agreement is generally made with a specific person and this may be negated should the ownership change.

Any time and effort involved in a 'shake of the hand' agreement should be minimal compared with the return. This does not sound like a situation very likely to apply to the development of a fishery.

Leasehold

A leasehold will give a predefined period of security. A renewal clause could (perhaps) be written in that allows the tenant the right of renewal at the end of that period. The period, wording, renewal, etc. of a lease is a matter of negotiation.

It is advisable to consult a solicitor. Remember that, although a lease may give the tenant security, it will be in the mind of the landlord's solicitor to ensure security, protection and indemnity for his client, the landowner.

It should also be remembered that the person or persons signing the lease are responsible for all the terms contained therein for the period of the lease, and sometimes on surrender. This will be regardless of any circumstances that may affect the operation the fishery.

It is given these harsh realities that many potential fishery managers will decide for various reasons that the risk outweighs either their confidence/ability, or the potential gain, or a combination of these and other factors.

It should be noted that before giving up it is always worth trying to negotiate a lease with a 'get-out' clause in the areas that are of concern. An example of this would be if the inflow of water was only about the minimum required and no history of seasonal and annual flows was available; then it may be possible to sign a long lease, with an option or 'get-out' clause that covered the volume of water. Of course, it may well be with such a marginal site that it would be questionable if the development cost would not be outweighed by the risk.

Licence

Licences have grown in popularity in recent years. The legal definition of the difference between a lease and a licence is often to do with security and length of tenure. A licence is generally for a shorter term. It confers a permission to perform the activities detailed in the document for the defined period of time. It does not carry the legal weight of a lease; it is not as easy to rely on the document to resolve disputes in law between the two parties.

It is believed that the popularity of a licence has occurred because it can be put into place more easily without involving solicitors. It may be that there is a connection here with the recent trend for landowners to employ land agents to deal with such matters as negotiating land deals. The land agent will produce a suitable licence without involving a solicitor. When solicitors are involved the favoured document appears to be a lease. The lease or tenancy is a lengthier document and is thus more expensive to produce and be advised upon by the legal profession. This is not necessarily a bad thing.

Purchase

The purchase of a potential site is the dream of many who would wish to own a fishery. In most cases, however, even if there is available capital there will not be sufficient money to complete the purchase, the development and the set-up costs.

It is assumed that most of those wishing to create a fishery in the UK will have to do so within conventional methods and that any financier would wish to see a realistic 'return on investment'. If a presentation can be made to a bank or other source of finance that shows there to be sufficient margin to cover the purchase, set up, running costs (start-up to opening), then if the return on investment is sufficient compared with the risk and investors may be interested.

Commercial mortgages are available for land purchase; they would generally be made in stage payments in-line with site development. This shortfall and lag in funding would need to be bridged (financed from another source).

When to sign on the dotted line

Although the agreement of the owner of the potential site for development has been obtained, as yet other permission to develop the fishery will not have been obtained. Signing of any binding agreement at this stage would be very risky. It may be useful, however, to have a solicitor draw up a letter of intent. A letter of intent will lay down an agreement of both parties that if certain criteria are met, then there is a contract. This should be a simple document that states that, if the necessary permissions are forthcoming, the interested party will have first option on a lease or purchase, etc.

Avoid signing any binding agreement until permission has been granted to undertake the proposed activity by all the relevant authorities. This is stressed, as there can be little point in having local planning consent if you are refused abstraction and discharge consents.

It is surprising how interest is suddenly awakened once the word's 'pond' and 'fish' are mentioned in some quarters. At this stage it is advisable to keep all your information very close to your chest.

Submitting and obtaining permissions

All the effort up to this stage will have been for nothing unless the relevant authorities and permissions can be gained. It should be pointed out that part of this 'seeking permission' procedure could be done as soon as the potential site has been found and identified as suitable. However, although it is possible to put in formal planning permission for a site that is not owned, such action is unlikely to win the appreciation of the site owner. The procedures required (shown diagrammatically in Fig. 9.5) are somewhat lengthy and involved; they are, however, within the scope of the layperson who has the patience to work through the forms with determination and attention to detail.

Plans must be formalised before approaching the regulatory bodies. At this stage a decision must be taken whether to involve a land surveyor; the surveyor would draw up the site

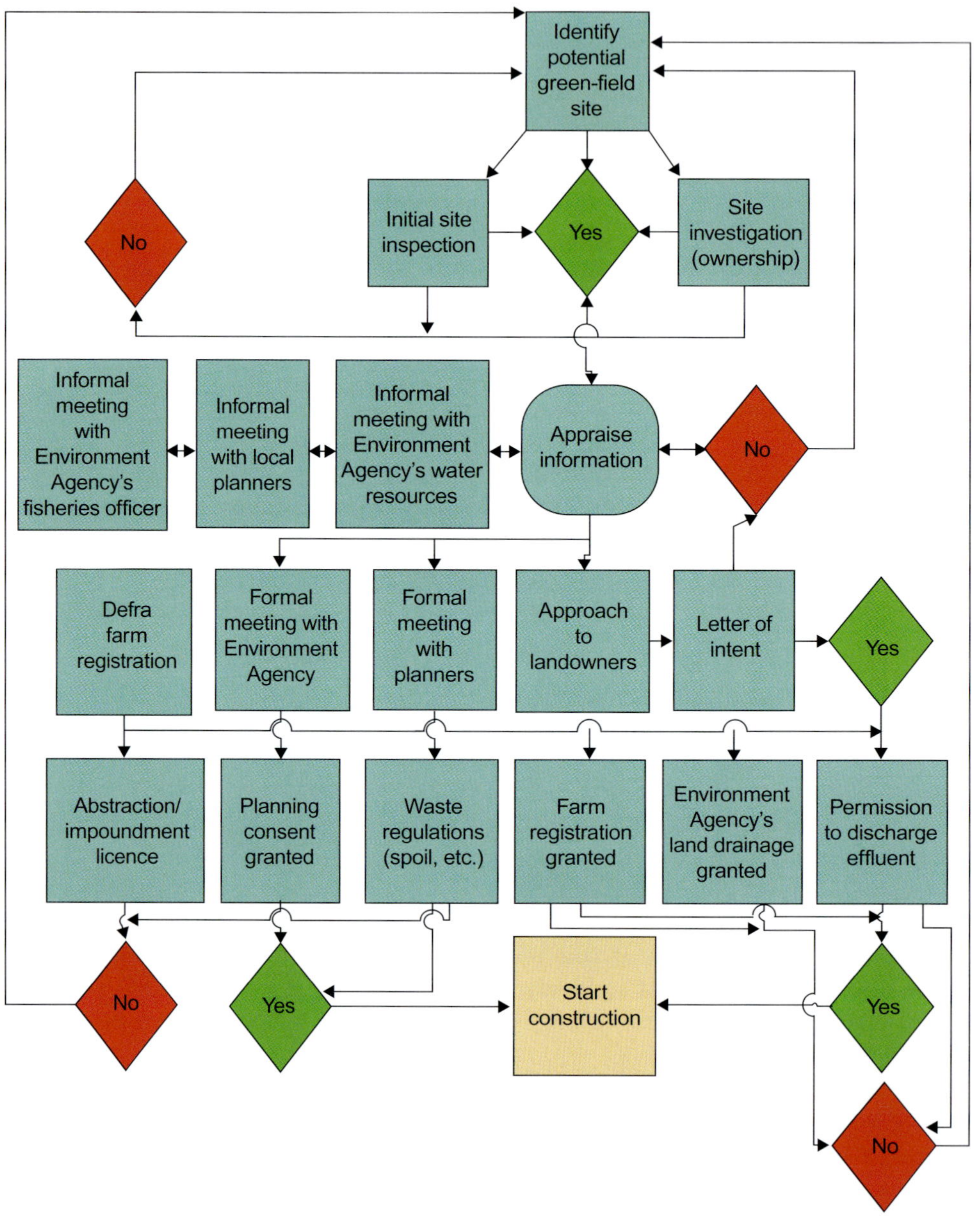

Figure 9.5 Procedures for obtaining formal permissions for site development.

plan. Depending on the size of the site, or perhaps more correctly the size of the proposed development, the cost can vary from a few hundred pounds to perhaps a thousand or more. Cost is also dependent on the firm or individual contracted. There is perhaps a better chance in negotiating a lower fee with an individual than a firm of land surveyors. Those prepared for a bout of hard-nosed negotiation may be pleasantly surprised at the discounted 'fixed rate' deals that may be achieved.

Regardless of who draws up a plan of the site, it must represent the vision of those who are to run and manage the fishery. Obviously this vision should consider expert opinion about water supply, siting and construction of the fishery and, importantly, the ongoing access to and around the site.

Increasingly, plans for fisheries must consider the environmental impact that the development may have. If it is possible to create some useful habitats while developing the site, this should be shown on the plan. Local planning authorities now have conservation officers who will be made aware of planning consents; their opinions can carry significant weight. With the involvement of a local planning authority's conservation officer there may come, as the expression goes, 'out of the woodwork', a host of other interested bodies, which often include single-interest voluntary groups.

Before considering the various stages involved in obtaining the relevant permissions and consents, it is useful to compile a list of the information required for submission to the controlling authorities. This stage of the process can be very costly in terms of both time and money, particularly if information required by the specific authority is omitted from the original submission and a subsequent resubmission has to be made. Many of these authorities and regulators only meet at predefined times, and not to suit the applicant.

It is very useful to have informal meetings and conversations with the relevant authorities to gain a firm appreciation of their requirements. A brief breakdown of the likely requirements is given in Table 9.4.

It cannot be stressed too strongly that informal meetings and discussions should be used to obtain a feel for exactly what information is required and in what way its presentation may affect any outcome. Human nature will dictate that a planning committee may have far more trouble in rejecting an application that has been submitted in a manner in which the local planning authority itself has requested, and that has considered aspects that were highlighted by them during informal discussions. That is not to say that even an application fulfilling all requirements may not be rejected: there is often no consistency. It is suggested that in most cases where a total rejection is given, this will have been made known at an informal level.

It should be acknowledged that there will be areas and circumstances where no level of negotiation, flexibility or finance will make it possible to submit a plan that will get approval. From a financial viewpoint, the sooner this is realised and a swift withdrawal made the better.

9.4 Determining the cost of development

Regardless of the information required by the regulative bodies, at this stage it is suggested that a clear idea of the cost of development should be in place. The following factors should be included:

Table 9.4 Formal planning requirements for a green-field site development

Type of permission, authority, licence or consent	Regulatory body	Detail
Planning consent	Local government planning authority	Authority to build the pond farm may be required on a defined parcel of land. Contact through the local council.
Licence to impound water	Environment Agency	Any structure such as a weir or dam across a water channel or course will require an impoundment licence. Larger impoundments (>5.5 million gallons) require a certificate from a Panel 1 civil engineer under the Reservoirs Act.
Licence to abstract water	Environment Agency	All abstraction requires a licence issued under the Water Act 1963. Any existing abstraction licence will require variation, if there is to be a 'material' change in the type of use or practice. The Environment Agency will levy a charge under the Water Resources Act 1991. This may vary according to circumstances.
Land drainage consent	Environment Agency	Any work that affects 'main river', a watercourse under the jurisdiction of the Environment Agency, or work on the banks within 8 m of the river. The work will require consent under the Land Drainage Act 1976. Additionally, under Section 28 of the Act for works such as a dam, weir or culvert or other unclassified watercourses, consent is required.

- development time and scale;
- number of ponds and size (excavation);
- total area of water (ratio of water to land);
- total water requirement (abstraction);
- hard surface access (materials);
- stocking;
- usage: number of anglers, run as a club or on a 'commercial' basis;
- return on investment (finance).

Unless the budget and projected profit and loss account support the proposed project, there can be no point in continuing with a commercial enterprise. The project figures will need to be reworked. If, after further investigation and appropriate adjustments to the plans and costs, the project is still not viable, it should be abandoned. There can be many reasons for not being able to bring a budget within the required limits. These could include, for example, the landowner requiring an unrealistic amount of money for the land or the cost of excavation. Some sites will require more soil to be removed than others; as the ratio of soil (metres depth) to the volume of water (cubic metres) increases, so will the cost. An example of this might be the difference in creating a pond on a slope where a 'cut and fill' method could be used, compared with excavating a pond on a flat area of land. The pond created by cut and fill would have a lower ratio of soil moved/water area and volume created, and would thus be less expensive to construct. It must also be considered that if the soil has to be removed from site, then it will undoubtedly be financially limiting. Whatever the cause, it is far better to 'jump ship' and get one's feet back on dry land, so to speak, reappraise the overall ambitions and objectives, amend the plan and then start the search for a green-field site again. A site does not have to be developed in one attempt: the time-scale can be increased, which will spread the cost. Unfortunately, delay in building also delays completion, which can then affect cash flow, so the spreadsheet will have to be revisited.

10 The Construction of a Still-Water Fishery

The development and planning stage of a project to construct a still-water fishery are extremely important because once a fishery has been constructed (ponds or lakes have been excavated), the options to change one's mind are very much reduced, especially if construction has gone ahead with a limited budget.

10.1 Planning phase of fishery construction

Development and planning of the fishery must be the first phase to be completed before the manager moves on to the actual construction. Once construction has started, putting right an error or correcting a calculation can be very costly as the movement of material to form a still-water is one of the most costly parts of the budget. Putting a mistake right often requires expenditure of 50% or more on the initial cost to rework work that has already been done. This section highlights some of the areas needing extra attention. In writing this chapter, it has been assumed that the site for development has been or will be acquired (see Chapter 9).

Hopefully, the type of fishery to be created will have been central to the process of obtaining the green-field site, and the site is suitable for its intended use as a fishery. The obvious first step in the development process is to put together a working plan, which should bring together the objectives and aims of the owner (the owner can of course be an individual, partnership, company or organisation). The aims and objectives will be the driver for all that follows, to ensure that the plan delivers the fishery that is desired.

10.1.1 Aims and objectives

Examples of the aims and objectives of a still-water fishery are given in Box 10.1.

The points listed in Box 10.1 are far from exhaustive but should give the manager a good indication of the things that need to be considered. Aims and objectives should always reflect the desires of the owner.

Once the aims and objectives have been decided, they should be committed to the plan that sets out to deliver them. The next stage must be to provide a methodology for delivering the aims and objectives, which will largely be achieved by how the fishery is constructed. The plan should initially consider this from a very basic level: attention to detail comes a little later once the construction phase is planned. Some of the areas listed under the aims and objectives will have been considered very early on in the process. Certainly many will have been considered as part of obtaining the development site and in obtaining planning consent. For example most local planning authorities will require information about usage, numbers of anglers and vehicle movements to the fishery on a daily, weekly and annual basis.

The number of actual ponds and their sizes will already have formed part of the planning application and the subsequent consent. It would be unwise for the construction to vary in any significant detail from the plans that were part of that approval process.

It hardly needs pointing out that the water bodies that will make up the fishery are central to all of the process. As such, early consideration of how the fishery is built around the water element is crucial, as turning holes in the ground into a still-water fishery is one thing, turning it into a fishery fit for your purpose is another. If the project is to remain and be delivered on budget, then everything needs to be included in the budget at this early development stage. So it is important to decide how each individual water body will be laid out.

We know how important plants are in forming a balanced aquatic ecology. We also know that fish respond positively to living in as near-optimal conditions as possible. So, as part of the development planning, the fisheries manager should ensure that as many plants are established initially as is practically and economically possible. There is, however, little point planting in an area where anglers will fish, as the impact from angling pressure will have a detrimental effect on the establishment of the plants.

10.1.2 Diagrammatic map of a still-water fishery

It is suggested that a diagrammatic map of the still-water be drawn, showing the pond footprint and essential features. This can now be annotated to include the planting and the positioning of the fishing stations.

To this should be added, as required, details of both vehicular and pedestrian access in terms of car parking and footpaths. An example of a drawn working plan is shown in

Box 10.1 The aims and objectives of a still-water fishery

I. What is the intended use of the fishery?
  (i) Coarse
    - Match
    - General/pleasure
    - Specimen
  (ii) Trout
    - Put and take
    - Catch and release

II. Who and how will the fishery be used?
  (i) Commercial
    - Syndicated
    - Day ticket/season ticket
    - Commercial membership
  (ii) Private
    - Membership
    - Syndicated ownership

III. What is the size of the project and what provisions need to be made?
  (i) Access
    - Car parking
      - Where will car be parked versus drive anywhere
    - Pedestrian access
      - Footpaths
      - Formal versus informal.
  (ii) Usage
    - How many anglers at any one time?
      - Size of matches?
      - Weekday/weekend usage?
    - How many fishing stations will be required to facilitate the angler numbers?
      - Fish anywhere
      - Defined swim stations
      - Formal/informal fishing platform, etc.
    - Number and size of ponds/lakes
      - Fewer, larger water bodies versus more, smaller ones
      - How will this affect the landscape?

IV. What are the characteristics of the still-water and facilities?
  (i) Water quality
    - Source
      - Quantity: is abstraction required?
      - Quality: chemical and physical parameters
  (ii) Outlet
    - Discharge: is consent required?
    - Online versus offline
    - Groundwater
  (iii) Facilities
    - Clubhouse
    - Toilets
    - Tackle shop
    - Canteen facilities

Figure 10.1. Details of how to establish aquatic plants, build fishing stations, paths and car parks are discussed later in this chapter.

The map will also have to include the constraining features surrounding the proposed site, such as existing underground pipework or mature trees that may be protected. Other flora and fauna may also have to be considered, depending on the site, particularly if the area is inhabited by protected species; in which case an ecological survey of the area may also be required. A further important point is locating the specific point of ingress to and egress from the site to satisfy the highways department that the entrance will be safe and not cause difficulties to other traffic using the highway.

Other considerations relating to planning (where consent will be required from the Environment Agency) include the proximity of the site to existing fluvial systems. If either the excavation or the storage of the spoil is within an existing floodplain then a flood risk

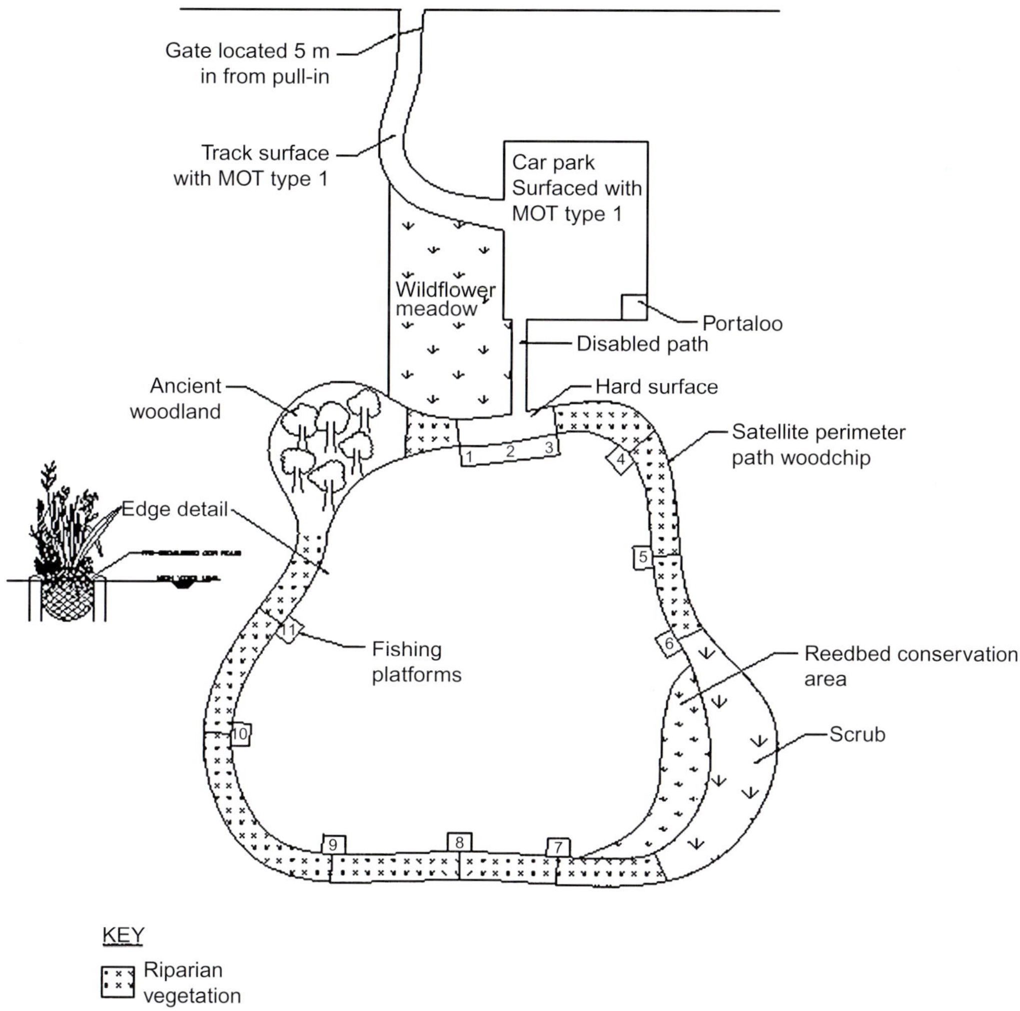

Figure 10.1 First draft of your fishery design.

assessment will have to be performed, for which the Environment Agency Flood Defence Office should be contacted for advice. Consent will also be required if the site is within 25 metres of a road or where material is being exported from the site.

Biosecurity and how to protect the fishery from associated risks must also be considered when creating a still-water fishery. As described in more detail in Chapter 20, these risks come in many forms, from the spread of disease within a fishery to invasive species colonising and outcompeting native flora and fauna. It is worth noting that many aspects of biosecurity can be dealt with most efficiently during the construction phase.

10.1.3 Building a budget

The budget is very much a 'grey' area as a component of the development plan, as it is important to have an accurate idea of cost but accurate and representative costs cannot be arrived at until the exact nature of what is required is understood and planned. For example, a cost for fishing platforms can only be factored into the budget once it is known how many are required, what they are made of, whether they will be part of the main build and installed by the contractor or will they be put in afterwards. It is a good idea to build a spreadsheet using 'ball park' figures, replacing each category with an accurate figure once the costing is obtained from an outside source.

It is important very early in the life of a project to understand how much finance is required to deliver the aims and objectives. This is because one must 'cut one's cloth' to suite a finite availability of funds or indeed allow enough warning of the need to raise additional funding. The cost of actually excavating and creating the fishery will often not be known until the project is put out to tender on a commercial basis.

10.2 Construction phase

This section does not attempt to describe the details of how to excavate a still-water fishery. Rather, it describes the process required to ensure that the construction phase of the project goes smoothly. Although an assumption is made that those with an interest in this section are not intending to undertake the practical construction themselves, an outline of the types of method are given, as the greater the understanding of the process the greater will be the chance of improved communication between all concerned. So some consideration should be given to how the work will be actually done and some advice should be sought from an experienced person or professional to get a feel for what might be the best way to approach the construction of your fishery.

10.2.1 Water retention

Another factor that must be considered both before and during construction is that of the water retention qualities of the substrate in which the still-water is excavated. Finding out after construction that water seeps through the substrate is too late! Where the still-water has been excavated in a clay substrate, any permeable material or fissures in the clay must be covered with a layer of clay compressed into place. Where lakes are excavated in permeable areas, membranes must be used to stop water seepage. There are several liners and

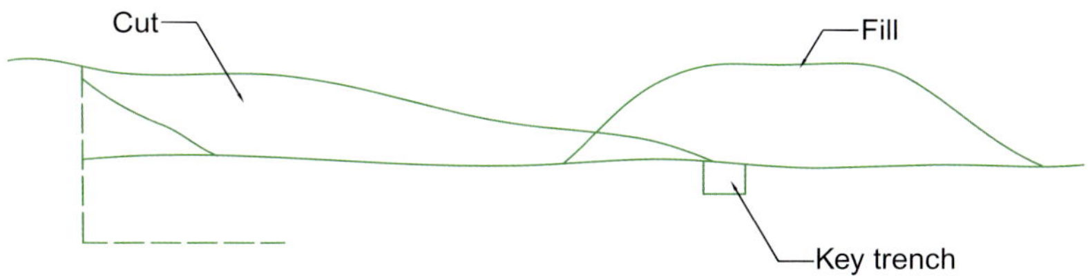

Figure 10.2 Principle of cut-and-fill approach to construction.

membranes commercially available, but these will significantly increase the cost of the project.

On sloping sites a 'cut and fill' method (Fig. 10.2) may be applicable; this involves using the excavated material to build up the height of the banks on the downhill side. As this section of material will be retaining water, it is imperative that it is impermeable and securely 'keyed' in to the lower layer. Using this method can be beneficial in some instances as it will significantly reduce costs.

10.2.2 Machinery for construction

In most cases the machinery used in the construction of lakes takes the form of a 360° hydraulic excavator, which can work from either inside or outside the perimeter of the area to be excavated, depending on site conditions. As the excavation of the spoil proceeds, it will have to be moved, and earth-moving equipment will be necessary to transport it to the allocated location. In some circumstances, where the topography is suitable, the use of a bulldozer may well be the most appropriate equipment to form the basin of your still-water, relying on a 360°^{0} excavator to perform the bank profiling.

10.2.3 Inlet and outlet structures

The control of in- and out-flows of water into a still-water fishery can be a critical component of their management. Correct control of water flow can enhance the ecology of the fishery, whereas poor control has the potential to cause damage. This section does not attempt to discuss the reasons for implementing these controls, but examines the types of structure that can be used to allow the fisheries manager the degree of flexibility required, in applying the flow control that is needed.

For many still-waters, the control structures can be either relatively simple or quite complex from an engineering perspective. For example, a downpipe (Fig. 10.3) is a simple and inexpensive method of controlling the water level, whereas a sluice or penstock (Fig. 10.4) can be far more complex and expensive to install.

Several control structures suitable for still-water fishery inlets and outlets are discussed. These are as follows:

- downpipes;
- weirs;
- penstocks;

- slot and board sluices;
- spillways;
- sluice gates.

It should be noted that many of these terms are very colloquial and are often interchangeable or incorrectly used. There can often be a requirement to fix a grille to the structure, to protect against the inflow of flotsam and of course living elements of the system such as fish.

Downpipe

The downpipe can be the most simple and inexpensive structure to install to control water flows both in and out of a still-water. It has several advantages, but also some serous disadvantages in some circumstances. Benefits of a downpipe control structure might include the following:

- their ease of installation;
- the low cost of installation;
- their low maintenance;
- their ease of operation;
- the ease of adding a grille;
- they do not leak;
- they facilitate drain down;
- they provide flexible control of water level.

However, there are several disadvantages that should be considered. These might include the following:

- they can be prone to erosion and cavitation;
- they can block easily in heavy flows;
- security concerns in open access and public areas.

To ensure that the downpipe is robust in operation, the design should include an upright, which allows the pipe to be chained in the vertical or near-vertical position, to ensure that it cannot be tampered with. A guide to installation is shown in Figure 10.5. It should also be noted that a correctly functioning downpipe has no inherent predisposition to leak.

A derivation of the downpipe is a simple outflow pipe that is put through the bank at a predefined height to set the water level. It is very cost efficient but has very few benefits. Caution should be given to cavitation around a pipe placed straight into soil with low clay content.

Weir

A weir is often incorrectly described in a manner that does not necessarily convey its true technical use, which is to control flows in streams. In the riverine environment these structures are more correctly termed 'fixed-crest weirs' when they are uncontrolled and are used to retain water at a defined level, thus controlling the depth. Another form, the gauging weir, is used on rivers to measure the flow past a fixed point. Weirs are used in still-waters

Figure 10.3 A downpipe.

Figure 10.4 A penstock.

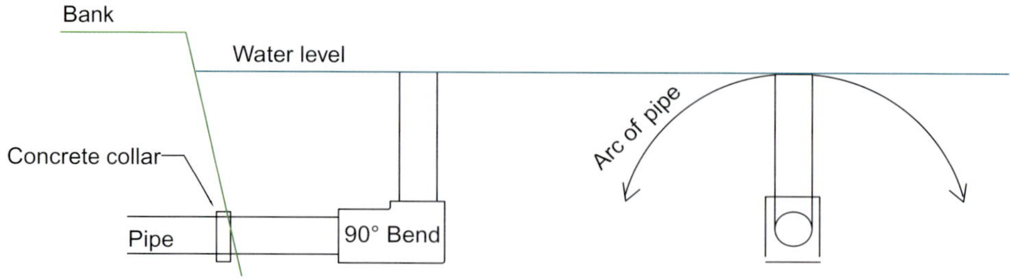

Figure 10.5 Installation of a downpipe.

Figure 10.6 A fixed weir.

in their simpler form (Fig. 10.6), to set the outflow level and to act as an inlet from a stream or as an outflow into a stream. Benefits of a fixed weir might include the following:

- they are of a straightforward design;
- they can be built from timber;
- they are not prone to blocking;
- when used as an inlet control, they can act as a silt trap;
- they are very secure;
- their maintenance is low when built with 'hard' materials.

Potential disadvantages of controlling water flow with a fixed weir might include the following:

- they can be prone to erosion where the structure meets the bank;
- they do not facilitate drain down;
- they have no facility to control water level.

Penstock

The term 'penstock' dates from the days of water mills and intricate streams through water meadows. A penstock is a type of sluice gate that is generally assumed to have a mechanical gear chain to facilitate ease of opening, and here we will assume this to be the case. So although they were often installed on mill races, they were also installed on mill ponds, thus the association with still-waters. A penstock is a gate valve. By design, they require careful installation, as they are very precisely engineered and require that the construction is level, parallel and that the foundation is solid (Fig. 10.7). Benefits of a penstock might include the following:

- they are very robust;
- they are easy to operate;
- they can facilitate drain down.
- they give good control of water level

Potential disadvantages of controlling water flow with a penstock might include the following:

- they require regular maintenance;
- they are not secure, but can be made tamper proof;
- they are a relatively costly form of control.

Slot and board sluice or weir

This term describes a weir-like structure that allows the control of water level by adding or removing boards as required. The boards are retained in a slot running down either side (Fig. 10.8). These structures can be relatively simple, or very complex in design and construction. The relative complexity would generally be considered a factor of the amount of water that flows over it and the volume of the impoundment. Benefits of a slot and board sluice might include the following:

- their ease of operation;
- they can facilitate partial drain down;
- they give good control of water level.

Potential disadvantages of controlling water flow with a slot and board sluice might include the following:

- they require regular maintenance;
- they are not secure and in some locations can be prone to unauthorised adjustment.

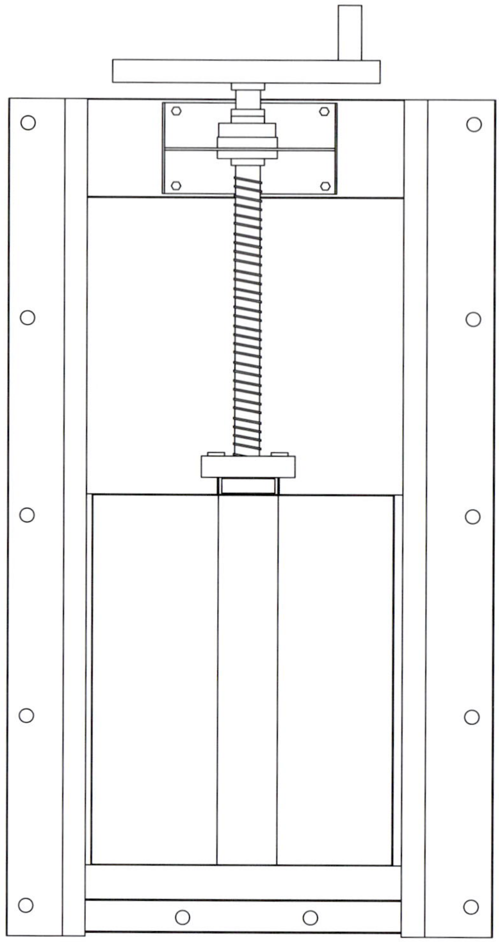

Figure 10.7 The design of a penstock.

Sluice gate

A sluice gate is a term that covers many different designs of control structure, although most of these are specifically designed for use on rivers and canals (lock gates). It is not surprising to note that traditionally a sluice gate sits at the end of a sluice channel, which forms part of the complex water engineering system found at the sites of historic watermills.

A sluice describes any of a very specific range of structures that control the flow of water: from a pipe with a valve, to a solid sheet of material impeding or impounding a flow which can, by some means, be physically removed to allow a flow.

Spillway

A spillway is placed on an area behind a reservoir dam to deflect water away from the control structure, to avoid scour at the back of the bank. Areas where there is a drop in water level as it leaves a still-water over an outflow structure are also often termed a spillway.

Figure 10.8 Simple weir with wooden boards.

Any discussion of control structures for impounding volumes of water that bring the still-water under the Reservoirs Act (see section 26.2) and any associated channels including spillways are beyond the scope of this text.

A spillway is a suitable term to apply to any still-water where the outflowing water takes a comparatively large drop in height over a relatively short length of channel or area (Fig. 10.9). Where this occurs, scour to the bank of the recipient water may occur and may require that the bank is armoured with rock rolls or pre-established coir-fibre rolls.

Consent

Any structure that is part of a system to abstract or discharge from, or into, an open system considered to be 'main' channel, may require consent from the regulatory body, the Environment Agency, under the Land Drainage Act (see section 26.1.8). Consent may also be required to abstract and discharge water from and back into a watercourse; the Environment Agency is also responsible for this legislation.

10.2.4 Bank profiles

When constructing the still-water, careful consideration must be given to attaining the correct bank profiles. This incorporates the positioning of the angling stations, marginal vegetation and access. Ideally, the profile of the bank should be gently sloping to reduce the risks of erosion associated with steep-sided banks. Marginal plants will also play a part in stabilising the bank and, during the construction phase, the banks should be profiled to accommodate this vegetation. A shallow shelf is also an advantage in promoting marginal aquatic plant growth. Correct profiling will promote good environmental conditions both above and below the water level and as such is an essential element of the ecology, which will in turn determine the overall success of the fishery.

Figure 10.9 A spillway.

10.2.5 Facilities

Paths

Paths are the first defence against habitat damage. Anglers can be considered a little like sheep in respect to paths: if you give them one to follow, they will follow it. As part of path creation you may need to make access to the adjacent areas difficult (that is, roped off, etc.), as these will be potential short cuts. The angler will soon pick up on the fact that a level and dry path is far better than wet, muddy feet from a walk along a rutted, worn walkway, even though it may be a little longer

Paths set back from the water's edge that link the fishing areas (swims) will reduce bankside disturbance and allow the important riparian and marginal vegetation to flourish.

The construction of a path is critical to how it will look and, as importantly, how costly it will be and how long it will last. When considering the installation of a path their use for disabled access should also be considered. Indeed the Disability Discrimination Act (see section 25.5) makes it a requirement for this to be considered. Criteria for paths suitable for disabled access are shown in Box 10.2.

Construction of paths can be carried out using several materials, including the following:

- trodden earth/grass path;
- woodchip;

Box 10.2 Criteria for paths suitable for disabled access

Guidelines for paths suitable to facilitate disabled access:
Suitable surface consisting of a material that can be compacted and remain so;
Width of paths:
- 1.8 m: suitable for two wheelchairs or two pedestrians on foot;
- 1.5 m: suitable for one wheelchair plus one pedestrian alongside;
- 1.2 m: acceptable where there is limited space.

Level or shallow incline:
- Minimum: 1:12;
- Best practice: 1:15;
- Ideal gradient: 1:20.

Avoid acute cambers:
- Recommended maximum: 1:50;
- Preferred: 1:100.

- tarmac;
- aggregate (gravel);
- granular fill;
- road planings.

Trodden earth/grass path
A simple trodden path (Fig. 10.10) is perhaps the simplest form of path and may be all that is required where pedestrian traffic will be light and/or not in wet periods. The benefits are the following:

- they are natural looking;
- they provide uncommon habitat to increase biodiversity;
- they are low maintenance if used lightly.

However, the disadvantages can be significant and might include the following:

- they require high maintenance if heavily used;
- they can be slippery in the wet;
- they are dirty underfoot.

Woodchip
Woodchip has become very popular as a material for paths because it is thought to have 'green' credentials. This is not disputed; however, woodchip is often laid in a manner that makes it not very sustainable. It is also not suitable in areas of a fishery that are prone to flooding as it floats away. Woodchip will provide a very suitable surface providing it is laid on a compacted sub-base (Fig. 10.11), or will only be subjected to very light pedestrian use. It has several benefits, which might include the following:

- it provides a soft 'green' finish;
- it is made up of recyclable material, often available as a by-product from work on the fishery;

- it is easy to maintain;
- it is relatively non-slip.

The use of woodchip, even when laid on a sub-base, does have some potential disadvantages, which might include the following:

- there is a tendency for the woodchip to migrate from the path;
- it is prone to rotting in damp conditions;
- it may not be wheelchair friendly.

Figure 10.10 Simple access by grass path.

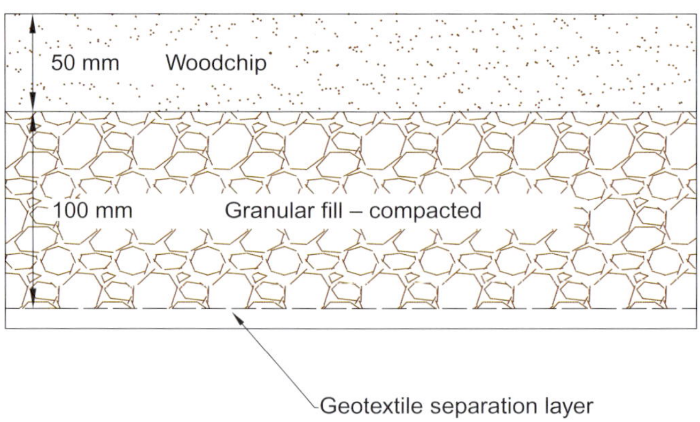

Figure 10.11 Woodchip on a compacted base.

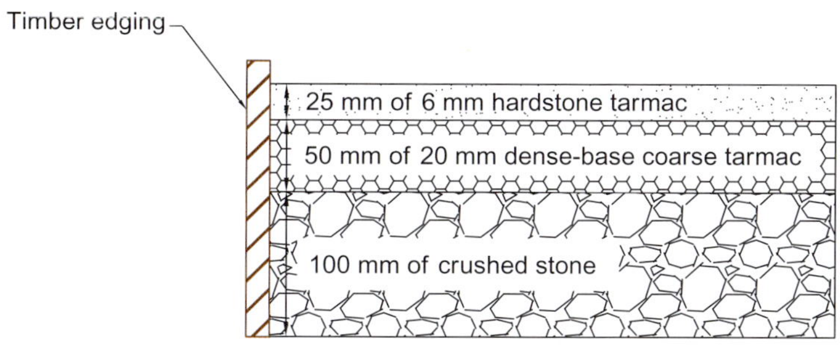

Figure 10.12 A tarmac path.

Tarmac/asphalt

Tarmac will provide a high-quality path (see Fig. 10.12) but one that comes at a comparatively high price due to the cost of the materials and the requirement for the path to be professionally constructed. In most fisheries, a tarmac path will be hard wearing. Being a hard surface it may, of course, be prone to icing in cold weather.

The benefits that tarmac paths might have are the following:

- they are easily adaptable for disabled access and are wheelchair friendly;
- they are hard wearing;
- they require little maintenance.

A professionally constructed tarmac path does have several disadvantages that may need to be considered. These include the following:

- they can become slippery with mud and ice;
- they need to be maintained;
- they are prone to damage from abuse;
- they are costly to construct and repair;
- they are environmentally questionable as the oil base may leach hydrocarbons.

Aggregate

Aggregate paths are constructed from various-sized stone or gravel. Aggregate paths suffer the same problem as woodchip in that the material is prone to migration from the path to the surrounding area. It has some benefits, which are the following:

- it is free draining;
- it is easy to construct;
- it is non-slip.

Aggregate paths do, however, have several significant disadvantages, not least the fact that the small, dense individual gravels are prone to compaction into the natural base, so some form of separating layer, such as a geo-textile, is required (Fig. 10.13). The disadvantages are the following:

- the material is prone to migration;
- they are not suitable to facilitate disabled access;
- the material is prone to undesirable compaction.

Granular fill

Granular fill material such as MOT type 1 (Box 10.3) requires a competent standard of installation, but is within the scope of a group of enthusiastic amateurs, providing attention is paid to details of installation and the project is well planned. All paths constructed of granular fill are prone to damage from standing water, so it is helpful if the specific design can elevate the path slightly above the surrounding ground (Fig. 10.14). This allows the water to drain off, rather than through, the path, avoiding any negative impact upon the compaction, which is a key element in ensuring that the path gives a long life with minimum maintenance. The potential benefits of a granular fill might include the following:

- it is suitable to facilitate disabled access;
- its ease of maintenance;
- its value for money;

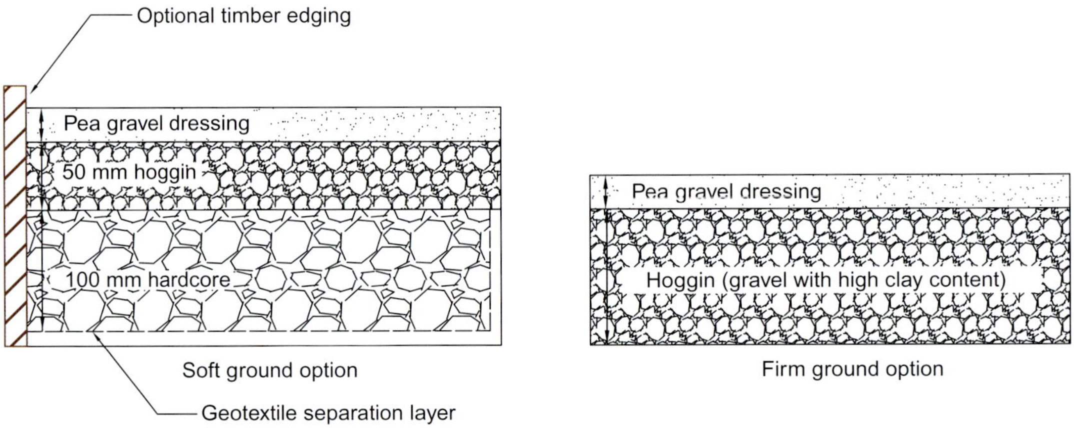

Figure 10.13 An aggregate path.

Box 10.3 Specifications for MOT Type 1

MOT Type 1

A granular, graded material with a maximum permitted particle size of 75 mm down to dust. Type 1 material must be a crushed material to provide a stable sub-base for road surfaces, paths, car parks, etc. It is known as such because the standard specification meets a Department of Transport standard set down for sub-base material to be used for highway works.

Type 1 is also known as 'scalpings' and is often made from crushed limestone, although that can change regionally to reflect the local geology.

- its good life expectancy;
- it is non-slip.

Disadvantages of granular fill might include the following:

- it is prone to damage from laying water;
- it can be prone to damage in wet conditions;
- it requires regular maintenance to keep it in good condition.

Road planings

Road planings are simply recycled tarmac that has been removed from a road before resurfacing. If a road resurfacing operation is undertaken near the fishery and you note that it is happening, it may be an opportunity to obtain this material at a very beneficial price, although this often will be on a 'cash in hand' basis. Although road planings are tarmac, it is not relaid in the way that the tarmac was laid originally (Fig. 10.15). Effectively, the material is compacted at the right concentration so that the path appears seamless. Once a seamless appearance has been achieved, it will indicate that the compaction is sufficient and that the life expectancy of the path will be acceptable. The benefits of this material are the following:

- it is cost effective;
- it is suitable to facilitate disabled access;
- it is easy to maintain.

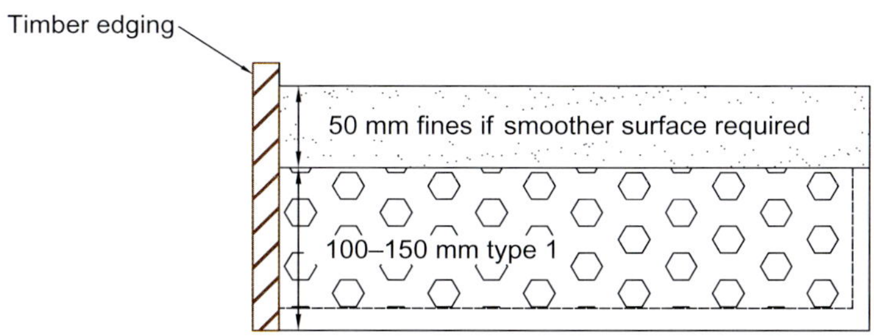

Figure 10.14 Computer-aided design of a granular fill path.

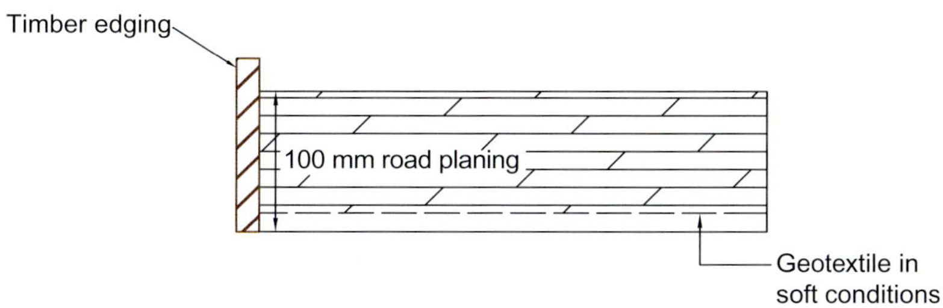

Figure 10.15 Computer-aided design of road planings.

The disadvantages might include the following:

- it is prone to powdering due to cold weather;
- it may be prone to lifting in hot weather;
- it is a 'hard' solution;
- it is environmentally questionable as the oil base may leach hydrocarbons.

Car parks

Good car parking should be provided with well-compacted and firm surfaces. If the areas are likely to become muddy and unstable in wet-weather conditions, then some form of protection must be put in place. This can either take the form of a reinforcement layer, such as a geo-textile or geo-web, or alternatively a layer of hardcore or stone chippings. This will give a firm surface, which over time will compact. The car park should meet the needs of the planned number of anglers, but some extra space should always be allowed for competition days or expansion.

Toilet facilities

Some consideration should be given to the provision of toilet facilities on the site. Where there is a clubhouse, the plumbing and pipe work may well be in place. Where there is no building it can be a problem, although there are commercially available biological and chemical solutions that do not need drains and pipe work. Chemical toilets require regular attendance and emptying, but the biological ones turn the waste into compost naturally. Toilet facilities should generally be gender specific with disabled access.

Shelter

The provision of some sort of shelter at the site is also desirable in the form of a simple shed or clubhouse. This is an area where anglers can meet, have something to eat or drink and discuss the fishing. For trout anglers this is also somewhere they can weigh their catch and leave a 'catch return' for the fishery manager.

Provision of angling swims

The generic term 'swim' is considered an area of bank from which angling takes place. From a management perspective, having anglers fishing from any area of bank as may be their whim may pose several difficulties, which might include the following:

- damage to marginal habitat;
- disturbance of sensitive habitat;
- retardation of habitat development;
- arguments between anglers.

It follows that there may be significant benefit in managing where angling takes place, by a range of options from simply marking out an area of bank on a very informal basis, to constructing an engineered structure such as a fishing platform. The most common options are the following:

- reinforced grass area;
- angling platform;

- concrete base;
- granular fill base.

Defined constructed swims have many functions. Increasingly, one of these is to provide safe and suitable areas for disabled anglers to fish. The basic requirements of a fishing area to meet this requirement are shown in Box 10.4.

Reinforced grass area

A reinforced grass area provides support to the root structure of the grass, as well as facilitating angling. Simple wear and tear from pedestrian access can be erosive. Whatever method is adopted to reinforce the area, it is important to ensure that it is free draining, as any standing water on grass that comes under use will suffer a 'puddling effect' from anglers feet.

The first and simplest method of providing a reinforced grass swim is to build up a base above the surrounding ground level and to sow a suitable topsoil (which may need to be imported) with a suitable grass seed (Box 10.5). Having sown the seed, this is protected with a turf reinforcing mat.

Box 10.4 Fishing swims suitable for disabled anglers

British Disabled Angling Association recommended dimensions:
Single platform: 1.8 m × 1.8 m;
double platform: 4.0 m × 1.8 m;
coaching training platform: 8.0 m × 1.8 m;
handrails: 950–1000 mm height, safety rail at 0.5 m required;
handrail diameter–45–50 mm;
knee bar: 450–500 mm;
wheelchair bump stops: 150 mm height;
approach to platform: suitable inclined path;
height above water: 300–600 mm.

Box 10.5 Suitable grass mixes for reinforced grass fishing swims

Grass seed mix for seeding fishing swim surfaces.
AGA: Mix 10;
35% chewing fescue;
25% slender creeping red fescue;
15% perennial ryegrass;
10% smooth stalked meadow grass;
5% brown top bent.

AGA 10 mix produces a dense sward with excellent binding qualities and recovery.
sowing rate: 35 g/m^2;
overseeding rate: 20 g/m^2.

The method of construction, including a free-draining base, is shown in Figure 10.16, whereas the finished, freshly constructed article is shown in Figure 10.17.

The benefits of this type of fishing swim might be the following:

- the cost benefit of construction;
- the ease of maintenance;
- they constitute a 'soft' solution;
- they are sustainable, providing use is not heavy.

The disadvantages might include the following:

- poor resistance to wear and tear if heavily used or continuous over prolonged periods;
- they can be damaged by rod rests;
- they can be slippery;
- the grass requires sunlight regularly across the entire surface.

The second method to protect a grass fishing area is to construct a base in a similar way to the method above. However, in this method the soil is retained on the surface of the swim

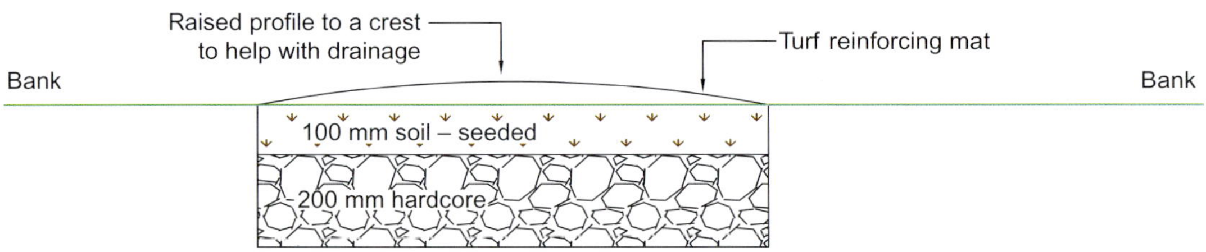

Figure 10.16 Computer-aided design of a reinforced grass swim.

Figure 10.17 A reinforced grass swim surfaced with a turf reinforcing mat.

Figure 10.18 Cellular soil protection.

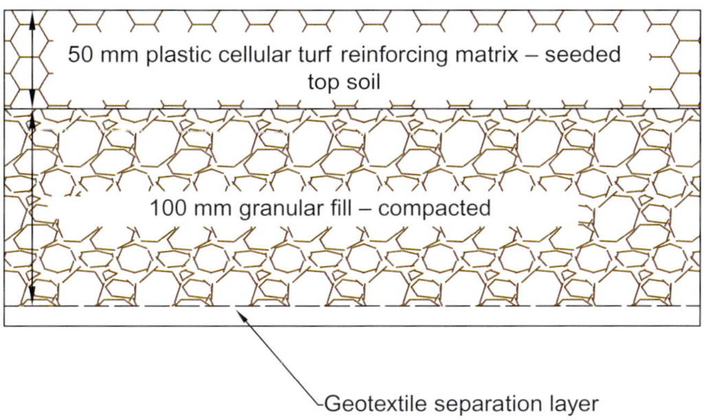

Figure 10.19 Computer-aided design of CELL material on swim.

base with a plastic cellular structure (Fig. 10.18). The installation is relatively simple and is shown in Figure 10.19, whereas the finished article is shown in Figure 10.20. The benefits of this solution might include the following:

- it provides a solid soil retaining matrix;
- it provides good protection against erosion from human use;
- it is easy to maintain.

The disadvantages of this method might be the following:

- it requires regular maintenance;
- it can be damaged by angling tackle including rod rests;

Figure 10.20 A cell-finished swim.

- it can be slippery;
- the grass requires sunlight regularly across the entire surface.

Angling platform

An angling platform provides a formal, highly defined area from which angling takes place. For the purpose of this text, a platform will be assumed to be built above existing ground level and to have a support structure based on a post (leg) design.

The design of a platform can be very simple. Indeed the best platform designs are, although some may be more involved and/or of heavier construction.

A reason contributing to the popularity of platforms is that they allow the angler to get right over the water–land interface. For this reason, angling platforms are often cantilevered over the water (Fig. 10.21). When choosing a cantilevered method it is important that the runners (the supports running from the land over the water) are of sufficient dimensions so that they do not bend or warp, either during construction or later.

Angling platforms are most often made from either a treated softwood timber or a hardwood such as seasoned oak. However, it is also becoming more common to use recycled plastic material. Softwood, if tanalised and well constructed, should give 20 years' service, whereas oak will give a longer life, with the fixings becoming the weak link. Recycled plastic should give an almost indefinite life; again, the fixings are the weak link. Softwoods are the least costly, whereas recycled plastic is the most expensive, by a ratio of 4 : 1 compared with tanalised softwood with oak sitting at about 2.5 : 1 relative to a softwood solution.

The benefits of angling platforms might include the following:

- they increase access to the water to facilitate ease of netting and general fish handling;
- they isolate the fishing area from the surrounding area;
- they can be made very safe;
- they can be made comparatively non-slip in the wet;

Figure 10.21 A cantilevered platform.

- they are excellent at facilitating disabled access;
- they are easy to maintain.

The disadvantages might include the following:

- they can be slippery in icy conditions;
- they can be prone to vandalism unless heavily constructed: try to avoid nails;
- they require some maintenance;
- there are health and safety implications as they are built structures, so they must be maintained in good condition.

A design for a standard 2 m × 2 m angling platform with cantilevered front elevation is shown in Figure 10.22. A photograph showing the finished article is shown in Figure 10.23. Integrated angling platforms are particularly suited to providing fishing access to the disabled angler. The platform may become particularly useful when connected to a walkway such as the one shown in Figure 10.24.

The benefits of a combined platform walkway might include the following:

- greater control of angler movement;
- they provide good access for disabled anglers;
- they can be made non-slip, other than in icy weather;
- they are safe, providing kick bars and low rails are included;
- they can be levelled into an existing path to provide continuity of levels.

Any disadvantages might be broadly the same as an angling platform without a walkway detailed above.

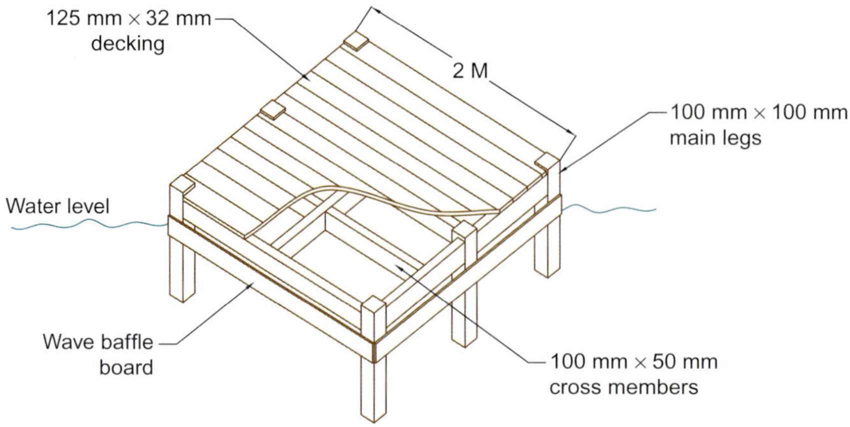

Figure 10.22 Computer-aided design of a 2 m × 2 m fishing platform.

Figure 10.23 A finished cantilevered fishing platform.

Concrete base

A concrete-based fishing swim may be required in an area that is prone to flooding (Fig. 10.25) or vandalism. It is obviously not as environmentally sensitive or 'green' as other solutions, but in certain circumstances it may be the only option to ensure a long service life.

Getting the materials to the site is often a logistical problem, but once this is covered the construction is simple. The golden rules are to ensure that the work is adequately shuttered and that the concrete depth is suitable for the intended use and surrounding substrate. In wet areas, float the concrete over a rubble base (Fig. 10.26).

Figure 10.24 A platform with walkway.

Figure 10.25 A concrete swim.

The benefits of concrete bases might include the following:

- they are robust against physical damage;
- they have a long service life;
- they are low maintenance;
- they are resistant to laying water and other harsh environmental conditions.

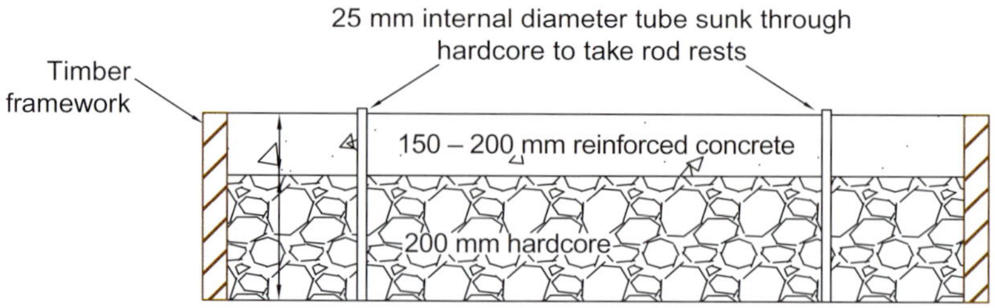

Figure 10.26 Computer-aided design of a concrete swim.

The disadvantages might include the following:

- the logistical difficulties with materials and equipment;
- they are not a 'green' or 'soft' solution;
- they are not very aesthetically pleasing;
- it is difficult to use angling tackle such as rod rests on them.

Granular fill base

A granular fill base is an aggregate-based material such as MOT type 1. Woodchip may be used as the fill or surfacing material. Both of these materials need to be laid in either a self-draining area or on a base material that will encourage free draining. Neither material tolerates standing water, although for different reasons.

Aggregate-based granular fill

It is recommended that a granular aggregate-based fill sits slightly proud of the surrounding land to ensure that standing water does not impact upon it (Fig. 10.27). Transporting materials to site for construction can be difficult. However, as this is very much site dependent, it is a matter for the overall assessment that the fishery manager must make. A newly finished compacted swim (Fig. 10.28) looks very enticing; however, it can require regular maintenance if subject to heavy usage, although the benefits may well outweigh the disadvantages.

The benefits of aggregate based granular fill swims might include the following:

- they facilitate disabled access;
- they have a non-slip surface;
- they are dry underfoot;'

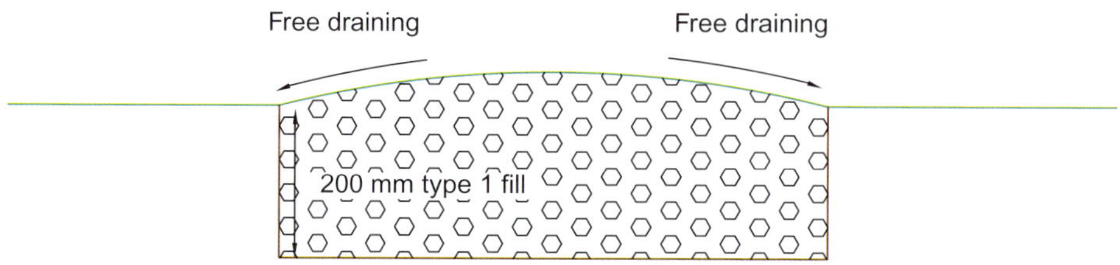

Figure 10.27 Computer-aided design of a granular fill swim.

Figure 10.28 A finished granular fill swim.

- they can be incorporated into a path;
- they naturalise very quickly.

The disadvantages of this type of fishing swim might include the following:

- they may require regular maintenance;
- transport of materials to site may be difficult
- they will accept angling tackle such as rod rests but these will damage the surface.

Woodchip-based granular fill

It will be seen from Figures 10.29 and 10.30 that the design of a swim using either woodchip-based granular fill or aggregate-based granular fill is very similar. However, the reason for creating a separate section is that they are very different in service, as highlighted under the benefits and disadvantages. Both methods, when weathered, look very natural and generally blend in well. Woodchip is very susceptible to migration, particularly if the area can be prone to flooding. Potential benefits and possible disadvantages that might exist are shown below.

The benefits include the following:

- they provide a natural 'soft' and 'green' solution;
- they are easy to maintain;
- they have a non-slip surface;

The disadvantages include the following:

- the woodchip has a tendency to migrate;
- anglers tackle such as rod rests will disrupt the base below the woodchip;
- they do not facilitate disabled access well if wood chip is too deep (over 75 mm);
- they are not as easy as other methods to dovetail into a path.

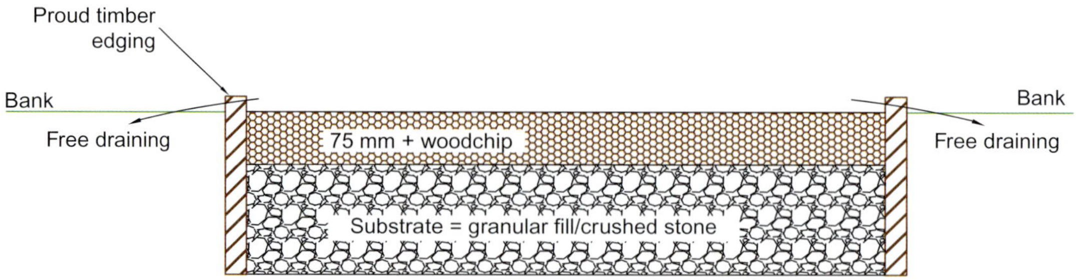

Figure 10.29 Computer-aided design of a woodchip-surfaced swim.

Figure 10.30 A woodchip swim.

10.3 Management

The first step is to decide whether you will manage the project yourself or appoint a project manager. If your decision is to appoint a project manager, then do so before you appoint your contractor who will perform the construction.

The decision whether to appoint a project manager should be based simply on your experience: have you or anyone in your organisation done this before? If so, did you learn from the process and do you know as much about still-water construction as those that will do the work for you? If the answer to any of these questions is no, then find and appoint a competent person to perform these tasks. This is not as difficult as it may seem. A competent person may be a fisheries consultant with experience of constructing ponds and lakes. Equally, it could be professional quantity surveyor or someone who, although they may have limited specific knowledge about still-water construction, would have the knowledge of the processes required to manage the contractor and the commercial aspects of the contract. There are many firms of independent surveyors who are available for project management.

Once you have appointed a project manager and agreed their role, you can appoint a contractor. Try to avoid the club member, acquaintance or friend who can do this for 'a cut-price rate'. Equally, avoid the highly recommended firm; you can fall back to this source after you know what the commercial 'going' rate for the work is. Put the work out for tender to at least three companies. Your development plan will help to provide a set of specifications to accompany the drawn plans. For this process to work, it is important that each contractor tenders on a 'like for like' basis. This will allow the received tenders to be assessed fairly and, importantly, to highlight any potential difficulties or misunderstandings. The tender process should require that the tender is submitted in logical, individually costed sections as shown in Box 10.6.

It should be noted that work should be done on a fixed-price basis and that accepting hourly or daily rates for doing the work should be avoided.

Box 10.6 Examples of fixed costs for construction

Bill of Quantity

Mobilisation

Welfare
- canteen;
- toilet;
- trying room;
- site office.

Subsistence
- travelling;
- accommodation;
- meals.

Security
- fencing;
- signage.

Excavation
- equipment;
- labour.

Planting
- creating vegetated margins;
- lily beds;
- submerged planting.

Waste

Demobilisation

Other
- uplift %;
- day rates.

Box 10.7 Items that should be included in the invitation to tender

Training and qualifications of the staff;
Method statements and risk assessments;
References and details of similar projects undertaken;
Financial standing: turnover, etc. (a credit reference should be independently obtained);
Key personnel: CVs;
Insurance details:

- public liability insurance: £5 000 000;
- employers liability insurance: £10 000 000;
- professional indemnity (if any design included): £1 000 000.

Project timings: lead-time from order to mobilisation, how long from mobilisation to completion.

The cheapest tender is not always going to be the one to accept, but a comparison of several costs helps to eliminate tenders that are much too cheap or much too expensive.

Items that should be included in the invitation to tender are given in Box 10.7.

A compulsory part of the tendering process is attending a pre-tender site meeting to ensure that the perspective contractors fully understand what is required of them. If the project will be awarded subject to a formal contract, then this needs to be specified.

Suitable 'short-form' contracts are available as a download from the Internet. Although these are not free, they are far less expensive than having a solicitor draw up a specific contract, although it may well be sensible to issue this contract through your professional project manager. If you do not have a project manager or other competent person, then consider getting professional advice for at least this aspect of your project.

The tender process, although securing a reliable cost for the project, should also draw out the experience and suitability of the contractor to complete the project. Before awarding the contract, ensure that you have met the contractor's project manager and the person who will supervise the day-to-day work, as it is important that there is a good working relationship between the client's project manager and the contractor.

10.4 After construction

In an ideal situation, a contract will have been let, and the client (you) will end up with your dream still-water fishery meeting all of your original aims and objectives, together with those that were added during construction. However, this will not often be the case as budgets, timing and a host of other reasons will cause projects to be delivered in varying stages of completion, some with fishing stations and fully planted, some with nothing more than an excavated hole in the ground.

After the construction phase is completed, the banks will have to be quickly consolidated to avoid initial erosion from wind and wave action. The simplest way to protect the banks is to plant them with marginal aquatic species. These will not only act as a buffer against erosion, but will begin to create the marginal habitat that is so critical in turning a hole into a still-water fishery.

Alternatively, bioengineering techniques such as the use of pre-planted coir rolls have proved extremely effective at providing an instant solution to erosion problems. Other techniques, such as willow spiling, also offer environmentally sensitive solutions to erosion control, promoting biodiversity and encouraging bank compaction (see Chapter 23).

Any remaining spoil should be used to profile the surrounding area sympathetically. This should then be covered, where appropriate, with a layer of topsoil, which can be seeded with a commercially available, easily manageable grass or wild-flower mix. Areas of the site should be designated for habitat creation, with a variety of locally indigenous species planted including trees and shrubs. Over time, these habitats will harbour insects and other invertebrates, which in turn encourage biodiversity at the site and provide a valuable source of nutrition for the fish.

10.5 Security

Site security is an important consideration and some form of perimeter fence is recommended. This can take a variety of forms and could, for example, be a particularly dense bramble hedge. Site security is important for several reasons and, aside from unwanted trespassers and poachers, there are biosecurity risks to consider (See Chapter 20). Sadly, some form of perimeter fence is now essential to a well-managed fishery, to stop intruders coming onto the property and stealing valuable equipment and stock.

If there is just one entrance/exit to the fishery, with a lockable gate, this allows you to monitor comings and goings and secure the fishery at night. It is also advisable to keep all equipment under lock and key, especially overnight.

10.6 Taking control of an existing fishery

Another method of obtaining a fishery is to take control of an existing still-water. This may happen in any of the ways already described in Chapter 9. Essentially, the procedures for obtaining an existing still-water or fishery do not differ from the development of a green field site, except in the permissions needed and the fact that much of the development of the still-water will already have been done.

A final way of obtaining control over a still-water fishery is to be employed by the owner as the fishery manager. This also relates to the many volunteer fisheries managers out there who do splendid work for the angling clubs who own the fishing rights on so many fisheries.

When considering taking on an existing fishery, it is important to be able to assess what the fishery represents. This assessment will obviously include the overall size, shape, location, etc. that have already been discussed in the previous chapter. However, any evaluation should include an assessment of the fish stock, to see what species are present, in what numbers and in what condition. Further evaluation should also include some assessment of the water quality and an assessment of the invertebrate life that can tell us so much about the condition of the still-water. This assessment should also run into a long-term monitoring programme if the lake is taken on. On the business side, it is also important to obtain some record of the numbers using the fishery and their requirements in terms of type of fishing and facilities.

11 Fish Stock Assessment

11.1 Introduction

A fishery manager needs to know what species are present in his fishery, how many of each species, the overall weight (biomass) of fish, how quickly they are growing and the general health status of the fish. The problem is how to obtain this information. It is difficult to see through the water into a normal fishery so the fish cannot be counted. It is almost always impractical to drain all the water out of the pond and count and weigh each individual fish. So we must, of necessity, sample the population to gain enough information to allow us to estimate the total number of fish, etc. This is a perfectly valid approach and can often give reasonably accurate answers, which can then be used to plan and evaluate the management of the fishery.

Given that the estimation of fish stock levels in the still-water environment can be difficult, it is sometimes a good idea to use indirect methods of fish stock assessment. Indirect methods use records of things like the clarity of the water, the presence of aquatic plants and the status of invertebrate populations to give information on the fish stocks. For example good water quality, luxuriant growths of soft, submerged plant species and a diverse invertebrate population will indicate 'well-balanced' fish stocks that are in equilibrium with their environment. This information gives a good impression of general health of the fish stock and the stock density relative to the carrying capacity of the pond. However, this type of information is not sufficient if you need to determine feeding rates or assess the need for supplementary stocking.

11.2 Planning

Before beginning any fish stock survey, very clear objectives must be set out and considerable thought must be given to how to achieve these. It is very easy to waste money and time through ill-conceived assessments that will not provide the information required.

The assessment process generally consists of two steps: (1) sampling the fish and (2) interpreting the results of the sampling to convert them into an estimate.

11.3 Sampling the fish

11.3.1 Capture techniques

Most fish stock assessments require that a sample of fish be caught, so obviously one of the most important decisions is how to go about catching the fish. This is not as straightforward as might be imagined. Each method has its benefits and problems and these need to be clearly understood before you start. Furthermore, each method will have specific problems associated with particular fisheries and again these need to be clearly understood before you embark on the survey.

Seine netting

One of the most traditional survey methods is seine netting, employing long-established equipment and techniques that have changed little in the general principles applied. The seine net is known to date back to ancient Greek and Egyptian times. Nowadays, natural fibres have been replaced with synthetic materials in the manufacture of nets, which has not only extended their useful life but had an impact on the efficiency of capture.

The seine net consists of a single sheet of meshed netting that is held vertically in the water by floats attached to the 'head rope' or 'cork line', which is fastened to its upper edge, and weights attached to the 'foot rope' or 'lead line' (Fig. 11.1). When deployed, the net should form a wall between the surface of the water and the bed, encircling any fish which are inside the wall. Once the net is laid and both ends are on the bank then both ends of the net are slowly pulled ashore, making certain the lead line maintains contact with the bed and the cork line does not sink below the surface so that fish cannot escape.

The main problem associated with seine netting from the shore is the need for a good bed profile over which to set and haul the net. Snags can prevent the net from being landed and silt and excessive weed growth can also prevent the seine net from working efficiently.

Figure 11.1 Use of a seine net.

The main advantages of seine nets are that they are relatively cheap and easy to use. A trained and experienced operator can deploy a net in most circumstances and catch fish. The seine net will also catch a wide variety of the fish present, as it is the least selective of all the capture techniques.

Furthermore, seine netting, especially with micromesh nets, is often the only way to capture the smaller representatives of populations in shallow freshwaters.

Trawl nets

Trawl nets are similar to seine nets but are 'trawled' behind a moving boat. Although more commonly associated with saltwater fishing, trawling can be successfully employed in surveying larger freshwater lakes. The major constraints are the cost, the difficulty of deriving quantitative assessments of the fish population under study and the damage trawl nets can do to the bottom.

Gill and trammel nets

Gill nets and trammels both work on the same principle of entangling the fish that swim into them. The gill net is a single piece of monofilament net suspended between a cork line and lead line (Fig. 11.2). Fish swim into the net and the gauge is set so that it catches them just behind the head and the fish cannot escape. A trammel net is a modification of a gill net consisting of three layers of net. A slack, small mesh, inner panel of netting is sandwiched between two outer layers of netting, which are taut and have a larger mesh size. Again the fish are entangled within the netting.

The main problems with these types of net are that they are very selective for the size of fish, with mesh shape and material type also influencing the efficiency of capture. Gill nets do not allow the fish to be sampled and released afterwards, as they are almost invariably fatal to the fish caught. Even fish that survive initial capture are often so badly damaged

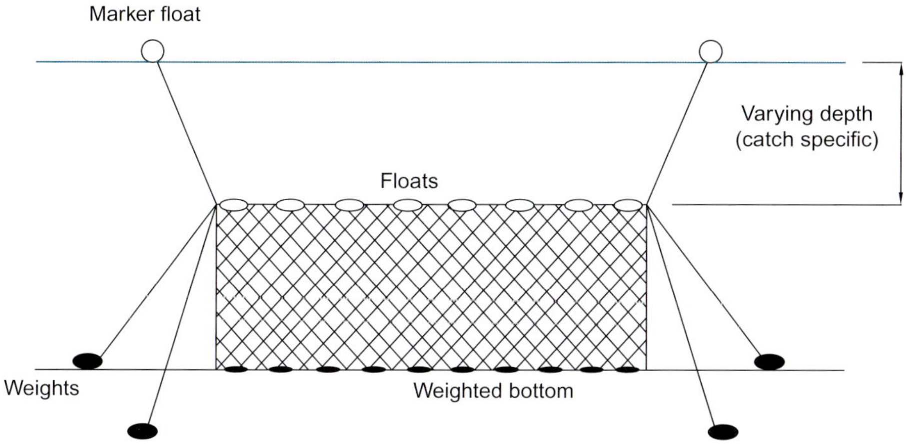

Figure 11.2 Features of a gill net.

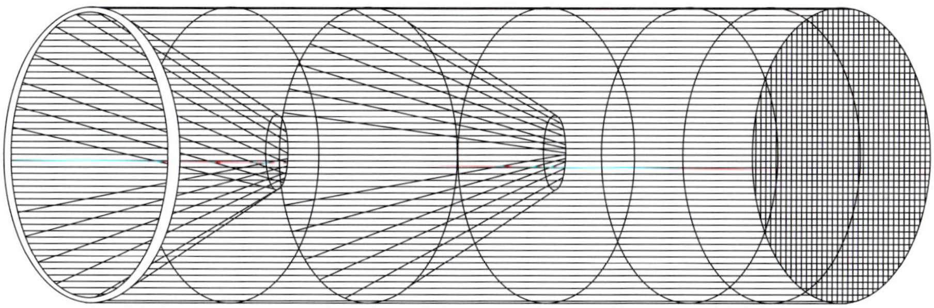

Figure 11.3 Drum traps.

that secondary infections of the wounds inflicted eventually kill the fish. Furthermore, there is a good chance of catching non-target species such as diving birds and mammals. Results for gill-nets are also difficult to interpret, especially for estimates of absolute abundance, and have to be used mainly as a relative method.

The main advantages for these methods are their ease of use.

Trapping

Traps fall into two broad categories: baited and unbaited. Bait is used to induce certain species to enter a trap and, as such, they are quite selective. Baited simple drum traps (Fig. 11.3) are mainly used as a means to capture species for food, or to remove nuisance organisms such as signal crayfish. Unbaited traps are used either to intercept migratory species, or species undergoing local movements. Such traps are often combined with leader nets or fences, the commonest example of this in the UK being the fyke net. Fyke nets can be used in still-waters where they consist of cylindrical or cone-shaped netting bags mounted on rings. These rings get smaller as you move down the net. Each has a tapered entry net, which

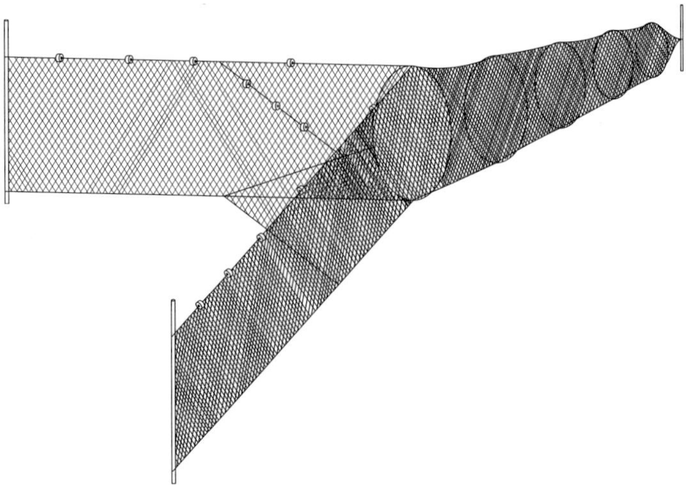

Figure 11.4 Features of a fyke net.

allows entry into each section but no exit. They usually have wings or leaders, which guide the fish towards the entrance of the bags (Fig. 11.4). Fyke nets are fixed on the bottom by anchors, ballast or stakes. The application of trapping methods for population assessments is limited by several factors, not least being that they are absolutely dependent on the behaviour of the fish in relation to the type of trap used. For example, fyke nets select only fish that are actively moving around and therefore susceptible to the trap. They are also mesh selective in that fish smaller than the size captured by the mesh of the leaders escape. Furthermore, they cannot be left unattended as they may catch non-target species such as otters or diving birds.

Fyke traps are relatively easy to use and can work well. For example when there is an individual target species and age structure, growth or condition are the assessment criteria. For example, the Lake Windermere perch (*Perca fluviatilis*) trap has been a valuable aid in the long-term study of the lake. Disregarding the difficulties of quantitative appraisal, traps remain popular for studying juvenile fish, which may be below the size that can be taken by other methods.

Electrofishing

The electrofishing survey technique uses electricity to stun the fish, allowing them to be retained to be weighed, measured, aged and then returned to the water.

The electrofishing machine (Fig. 11.5) creates an electric field in the water that interferes with the central nervous system of the fish. This stuns the fish and enables them to be scooped up in a hand net. Electrofishing can work effectively in a variety of habitats but it is considered best suited to sampling from flowing waters. However it can be used effectively on still-waters. The type and method of electrofishing depends on the size of the water and the water's ability to conduct electric current. Either alternating or direct current

Figure 11.5 Electrofishing.

(AC or DC) is used, depending on circumstances and operated either from a boat or using a backpack unit. The DC current works best in waters with a high potential to conduct electricity (hard waters). AC current is most effective in waters with a lower potential to conduct electricity (soft waters). Where appropriate, DC current is the preferred type of current because it has the least potential to harm individual fish. Sampling with electrofishing gear is efficient, quick and relatively safe if properly conducted by trained personnel. For electrofishing, the effective fishing range from the probe (anode) depends on biological effects such as species and length of fish as well as physical effects like conductivity of water or substrate type.

The main problem with electrofishing is that it is selective. This is because different fish species are more susceptible to the electric current than others and the size of the fish is very important. The longer the fish then the greater the potential difference and the more effect the electrofishing has. Furthermore, in still-waters the depth of the water and the width is very important. If the water is too deep then the fish may just swim down away from the anodes and evade capture or they may move away from the direction of the boat and so again evade capture. Even so, it remains one of the more effective methods for sampling areas with a lot of weed growth and marginal areas. It is worth noting that electrofishing can do significant harm to fish if used incorrectly. If the current is set too high the fish muscles can spasm in such an extreme manner that the backbone is broken or the artery below the backbone is ruptured. It is difficult, if not impossible, to spot this damage to the fish when it occurs, so it is not apparent to the operator at the time. It is therefore very important to set the current as low as possible to catch the fish and to reduce the amount of time individual fish are in the electrical field.

Fishermen's catch surveys

The use of information from fishermen can give access to a vast amount of data, even if the primary reason the fish are caught is for leisure. Surveys based on fishermen's catch returns have the advantage of being able to run over much longer timescales than might be possible with scientific sampling programmes. There have been numerous studies using this form of data collection, which have proved extremely beneficial in examining long-term trends. For example, trends in species composition over an 11-year period in the lower reaches of the River Trent showed that the fishery changed from a roach/dace to a bream/chub community over a period of a few years. This type of data would be hard to substantiate without the involvement of the fishing community.

This method is also a useful tool when sampling small, still-water commercial fisheries, for example sampling during a competition match. When the fish are in the keepnets it allows for relatively easy representative population samples to be recorded after weighing. Scales can also be easily removed for the purposes of ageing. This method can be effective in gaining a relatively broad range of data in a short time-frame to give a 'snapshot' of fish populations at a given time.

There are disadvantages with using catch data, however, the most notable being the difficulty in estimating abundance. There will also be species bias and the quantification of effort, not to mention the fact that healthy fish will intrinsically feed more than unhealthy fish so large sections of the actual population may not be represented.

11.3.2 Non-capture methods

It is possible to gather some good information without actually catching or handling the fish. These are called non-capture techniques and involve some high-technology but very useful equipment.

Counters

This technique is used primarily for electronically counting migratory salmonids in rivers. As the fish pass over a structure with electrodes attached they are electronically counted. The technique can be very efficient, with over 90% of the fish passing over the structure being successfully counted. The method is only really applicable to migratory river species, but can be adapted to some still-water situations.

Hydroacoustics

Hydroacoustics is a general term for the study and application of sound in water, usually by means of sonar. The practice is used for the assessment and monitoring of underwater physical and biological characteristics. Sonar provides accurate underwater topographic information, which is valuable in population surveys, for example assessing where in the water column individual fish populations prefer.

When assessing fish numbers, however, the method relies on the sound signal being bounced back from the swimbladder of the fish. These signals are analysed using complex data analysis techniques, and confidence in the results can be limited. The advantage of the method is that it provides access to fish populations that cannot be easily sampled in other ways such as in large deep lakes. Sonar equipment can successfully count fish although it is recognised to work most effectively in deep, open water with few species present.

Species and size separation is very problematic but the development of multiple beam systems is making progress in this area. The technique uses either classical vertical sonar beaming, as in the sea, in deep lakes or reservoirs, or a new technique of horizontal beaming for rivers and shallow reservoirs. The development of horizontal beaming techniques has demonstrated the power of scientific echo-sounding as an ecological tool for estimating the abundance and size structure of fish communities in large lowland rivers, for which no alternative survey technique of such power exists.

Figure 11.6 shows the output from one survey using an Eagle echo-sounder. This display shows the fish marked on the screen and the depth at which they were located.

This technologically advanced method is obviously prohibitively expensive in most private fisheries, although it is a valuable environmental tool. Although the research using this method is essential in advancing ecological survey techniques, fishery managers must rely on more practical methodologies.

Direct observation

The observer can either watch and count fish directly or use remote control camera systems. Fish shoals in shallow waters have been successfully counted from boats, and size information can also be easily obtained. This method is somewhat crude, however, with notable constraints such as water clarity, cost and training. There is scope to use the technique in certain situations, for example where more traditional sampling techniques are inefficient or result in unacceptable disturbance of the fish population. In still-water fishery situations, however, observation alone cannot be relied upon to give representative assessments of fish populations and certainly cannot provide more in-depth community-related data.

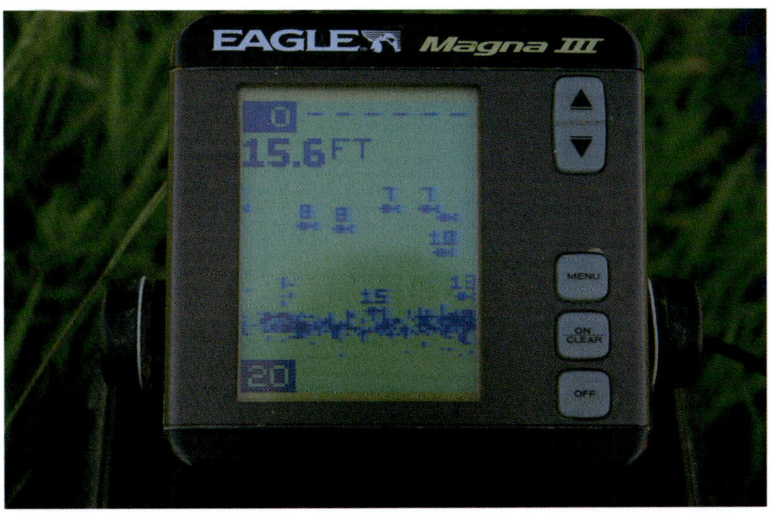

Figure 11.6 Example of an Eagle echo-sounder.

11.3.3 *Further information*

The main advantage of capture techniques over non-capture techniques is that they allow further data to be collected. Once a fish has been caught further information can be obtained including the following:

- length;
- weight;
- scale samples;
- clear identification of the species;
- health samples;
- reproductive state.

The use of this information will be discussed later in this chapter.

11.4 Estimating abundance

11.4.1 *Mark-and-recapture estimates*

The mark-and-recapture method requires that you accept that the following conditions have the same probability of occurring:

- marked and unmarked fish have the same mortality rates;
- marked and unmarked fish are equally vulnerable to capture;
- marks are retained during the sampling period and all marks on recaptured fish are recognised;
- marked fish randomly mix with unmarked fish;
- there is negligible emigration or immigration during the recapture period.

The general process for estimating the size of a fish population using the mark-and-recapture method entails the following:

I. collecting a sample of fish of the target species;
II. giving fish identifying marks, such as a tag or temporary fin clip;
III. releasing fish in good condition back into the same area;
IV. allowing at least 1 day for marked fish to recover and become mixed in the population;
V. collecting another random sample of fish during a subsequent recapture run;
VI. noting the number of marked fish and unmarked fish by species and size;
VII. calculating for each combination of species and size group (to compensate for sampling selectivity) an estimate of abundance using the equation below;
VIII. summing the size group estimates by species to obtain an estimate of the total population within the size range actually sampled.

Box 11.1 Example of a mark–recapture estimate

If 100 fish were marked and released after capture, and a second capture exercise the next day caught 80 fish of which 10 were seen to be marked, then:

$N = (100 + 1)(80 + 1)/(10 + 1) = 744$ fish.

So the population estimate is 744 fish.

Estimation the size of a fish population, N (and variance of N), can be obtained using the following equation:

$$N = (M + 1)(C + 1)/(R + 1)$$

where

C = total number of fish caught in second sample (including recaptures);
M = number of fish caught, marked and released in first sample;
R = number of recaptures in the second sample (fish marked and released in the first sample);
N = estimate of population size.

Box 11.1 gives an example of an estimate made by the mark-and-recapture method.

The mark-and-recapture method is a useful method and generally considered to be more accurate than most. However, it is expensive, time consuming and requires careful handling of the fish to mark them.

There are various tags available for use in fish. Visible implant tags are good examples of a product that is easy to use. An example is a soft Visible Implant Alphanumeric (VI Alpha) tag, which is a small fluorescent tag with an alphanumeric code designed to identify individual specimens. VI Alpha tags are implanted internally just under the skin usually around the head. They remain externally visible for easy identification and show up remarkably clearly under blue LED light. Visible implant elastomer tags are also available, which mark fish but do not have an individual number. These are cheaper and easier to use.

It is not always necessary to use complex tags. It is usually sufficient for mark-and- recapture experiments to mark the fish by simple fin clipping or by injecting a dye.

11.4.2 Depletion estimates

Depletion surveys

Fish capture techniques also allow abundance estimations to be made through the use of depletion surveys. These generally rely on the concept of a constant capture effort being applied to the same population, the size of which declines as the fish are removed. This should cause the catch per unit effort to fall as it gets more difficult to catch the remaining fish.

Extrapolation from a depletion or removal rate is therefore possible with relatively straightforward calculations.

A constant effort, using a selected capture technique, is applied in a defined area, on at least two occasions. The fish caught on the first occasion are retained until after the second capture is complete. The anticipated decline in catch per unit effort is used to calculate (by extrapolation) the initial absolute number of fish present.

The depletion method is satisfactory, although realistically all the data for one site must be collected in a single day. It also works better if the population being estimated is relatively small (say fewer than 2000 individuals).

This method requires that an adequate number of fish be removed on each sampling pass so that measurably fewer fish are available for capture and removal on a subsequent pass. Two types of depletion method are used: two-pass and multiple-pass. Because of differences in sampling selectivity, an estimate, by species and size groups, is normal. For both two-pass and multiple-pass methods, size group estimates and their variances are summed, as with mark-and-recapture methods, to provide estimates of the total population. The following conditions must be met for accurate depletion method estimates:

- Collection effort and conditions, which affect collection efficiency, such as water clarity, must remain constant.
- All fish within a specified sample group must be equally vulnerable to capture during a pass.
- Vulnerability to capture of fish in a specified sample group must remain constant for each pass (e.g., fish do not become more wary of capture).
- Emigration and immigration by fish during the sampling period must be negligible.

Depletion estimates are made using the following the general process:

I. Complete capture operation within a discrete section of water body.
II. Record number of fish removed, by species and size group.
III. Do not return fish to the water body.
IV. Repeat steps 1 and 2.
V. If steps 1 and 2 were completed twice, calculate population estimates using two-pass equations.
VI. If steps 1 and 2 were completed more than twice, calculate population estimates using multiple-pass equations.

Two-pass depletion methods

For two-pass depletion estimates, fish are captured and removed during two capture sessions.

Population estimate N and variance of N are calculated as follows:

$$N = C_1^2/(C_1 - C_2)$$

where

C_1 = number of fish caught in first sample;
C_2 = number of fish caught in second sample.

Box 11.2 Example of a two-pass depletion assessment

Two stop nets are placed across the river 100 metres apart. A single electrofishing pass through this 100-metre stretch captures 223 fish. These are measured and scale samples taken, then the fish are stored in a transport tank. A second electrofishing pass captures 121 fish. The fish are returned to the water. The population estimate is:

$$N = 223^2/(223 - 121) = 488.$$

Variance of $N = (223^2)(121^2)(223 + 121)/(223 - 121)^4 = 2314$.
Standard error of $N = \sqrt{2314} = 48$.
Approximate 95% confidence limits $= N \pm 2\ (48) = N \pm 96$.

So $N = 488$ (lower limit 392, upper limit 584).

Box 11.2 shows an example of a two-pass depletion method.

Two-pass depletion estimates are quite unreliable when fewer than 20% of the population is caught per pass.

Multiple-pass depletion methods

These require three or more catch runs to be completed and rely heavily on consistent catchability of the fish. Although multiple-pass depletion methods are more accurate than two-pass depletion surveys, they require a consistency of effort that is difficult to achieve and for the fish to not show any wariness to the capture method being used. The statistical analysis of the catch data is complex and the requirements of this method are so exacting as to make its use inadvisable.

11.4.3 Population structure

It is normal when capturing fish for a population estimate to take additional samples that can be used to gain further information about the population being surveyed. This additional data may include length, weight, reproductive state, photographs, definitive identifications, scales and health samples. This information can then be used to give a variety of useful outputs.

Length–frequency analysis

Length is easy to measure and there is a broad correlation between length and age.

Once all the fish (or if there are large numbers, a suitable sub-sample) are measured, then a simple graph of the length against the numbers caught can give you some very useful information. An example of a length–frequency histogram is given in Appendix C at the end of the book.

In the example represented by the graph in Appendix C, the first peak relates to one age group (year class) (2-year-olds) at about 50 mm long, then there is another peak at 70 mm long, which could represent a 3-year-old year class of fish. So in this example it is possible, with representative data, to see the distribution of year-class groups of young fish up to three years. However, variations in growth between individual fish of the same age mean that this method of ageing is unreliable, except for the very youngest age groups. The rest of the graph, as expected, shows a fairly steady decline in numbers, as the fish get older. This sort of graph provides very useful information about the population structure and helps to indicate things like poor year classes, which are under-represented in the population. In some cases the decline in numbers can also be used to calculate mortality.

The number of year classes or cohorts present in a sample depends on several factors:

- species longevity: this can range from one year in some 'annual' species to several decades for long-lived species;
- prevailing rate of mortality, which varies with time and place;
- sample size and the selectivity of the sampling technique and pattern of sampling.

As a generalisation, more young fish than old can be expected, but population age structure will also be influenced by year-class strengths. The selectivity of the sampling method will generate differences between population and sample-age structure.

To make this information more useful, it can be validated by ageing some of the fish accurately, as described below.

11.4.4 How to age fish

There are two methods used to determine the ages of fish in a population. These are mark–recapture and use of growth checks on skeletal structures.

Mark–recapture

Fish can be marked, and if the mark is representative for that year class then the age from that year can be recorded and will be known and established when the fish are recaptured. This can provide direct data on growth and mortality over a very precise time interval. Although a whole part of the life history is known, the age at first capture is often not known. Although this method is very accurate it is very time consuming, and tagging methods are often only visible for months rather than for years.

Use of growth checks on skeletal structures

This can provide a precise absolute age (in years at least) for all individuals in a sample. The species captured are identified and measured, and two to four removed, usually from the shoulder region. Otoliths (ear bones), fin rays and gill opercula can all be used to age fish, but their use means the fish must be killed, so scales are preferred. Any scale removed is regenerated by the fish. A few fish species do not have scales suitable for ageing (e.g., eels, tench, pike and perch). If these require ageing, otoliths are the preferred method.

The principle behind ageing fish relies on the fact that many of the body structures, such as scales, have growth rings (circuli) laid down as the structure grows. These rings are calcium based and are probably for support and strength. The growing seasons can be seen on these structures as widely spaced growth rings. When the fish stops growing, for example in winter, the growth rings are not so widely spaced. With the onset of the new growing season the circuli become much more widely spaced again and a mark (annulus), representative of a year, is formed. During the period of no growth it is not unusual for the edge of the scale to be eroded, so when new growth starts it begins on an eroded edge and the new circuli may be overlaid across partly eroded ones. This is a clear indication of an annual mark. The number of annuli is equal to the age of the fish.

The scale in Figure 11.7 shows a 2-year-old roach. It is very clear where the annual marks are. For many species this is true and it is relatively straightforward to tell the ages, particularly of young and fast-growing fish. However, some species such as tench, pike and perch are almost impossible to age from scales.

Furthermore, the older a fish gets the more difficult it is to age. For the first few years of life fish tend to grow quickly, but as they mature they put more of their energy into producing gonads and so grow less quickly. As the annuli get closer together as they get older, they become increasingly difficult to tell apart.

Information gained from ageing fish

The age of fish can give you a lot of other useful information including the following:

- establishing the longevity and individual life history of the fish;
- defining population structure;
- establishing individual and population growth rates;
- determining age and size at maturity.

Knowing the age of each fish and the length at which they were caught gives the average length at age.

Age group notation

In the UK (which is in the temperate northern region) the growth season is regarded as 1 April to 30 September. For fish sampled in this period, growth is still visible by following

Figure 11.7 A roach scale.

the annuli formed the previous winter. This is termed '+ growth'. It is anticipated that the growing season will become extended as the predicted climate change occurs.

Fish that are sampled in the winter months are in a period of little or no growth and the annulus is formed. The edge of the scale is therefore the end of the previous growing season and the age is given in full years ('no + growth').

11.4.5 Estimating health

The health of a fish, and therefore the population, is obviously useful information. Some information on the health of the fish population will be apparent from the result of a fish population analysis. Also, ageing the fish will give information on the growth rate of the fish in the population and age at maturity, again useful indications of the overall health of the population.

Chapter 19 deals with how to obtain health information from live and dead fish, but other parameters can indicate the health of the stock as a whole. For example, if length and weight were both recorded during the initial data collection exercise, the 'condition factor' of the fish can be worked out using a simple formula. This is a mathematical formula for determining the physiological state of a fish. It is calculated by dividing fish weight by length cubed (W/L^3). The heavier a fish for a given length, the higher its condition factor (K). The higher the condition factor is, the healthier the fish is.

However, the overall health of the fish in terms of their disease status is important as well. Knowledge of the pathogens found in a fishery is important for the following reasons:

- in the event of a mortality problem, a 'normal' level of pathogens is known;
- it may allow any potential disease risks to be managed before they become a problem;
- it gives information on the overall health and stability of the fish populations;
- it may inform other fisheries management options, such as adding or removing stock.

11.5 Conclusions

It is easy to generate information from stock assessments, from the estimate of numbers or biomass of different species of fish present, to their growth rate and general health. This is where the clear objectives set out in the planning stage come in. What was the question that needed answering? If the question is clear, then the answer should be obvious once the data are collected.

In most cases the first question a fishery manager wants to know is, 'how many of what species are present in this water body?'. Secondly comes, 'how healthy are the fish in this water body?'. Thirdly comes, 'how healthy is this water body?'. These questions may be asked in a different order but these are the most common.

For most fisheries managers, many of the techniques described may seem daunting or impossible. However, the information gained can be invaluable and so it is worth pursuing some sort of population survey.

For most purposes it is possible to obtain a good deal of information from a general lake survey. Here we would suggest the following:

- a fish survey using a combination of seine netting, electrofishing and fyke netting, depending on the type of water body;
- fish identification: capture as many different species as possible and make sure they are all identified correctly;
- scale samples, particularly from silver fish like roach, rudd and bream;
- invertebrate and plant survey;
- water-quality survey.

12 Stock Manipulation

Stock manipulation is one of the most powerful and commonly used tools available to the fisheries manager. To achieve customer satisfaction, it is common for coarse-fisheries managers to manipulate the stock of fish in favour of the target species for their average angler. Unfortunately, this often takes the form of directly adding more of that target species. As discussed in earlier chapters, the water body can only naturally support a certain biomass of animals at the top of the food chain. If the fisheries manager has no real knowledge of the biomass present in the water body and adds more fish, they could inadvertently tip the fishery out of balance. However, once knowledge is obtained about the levels of stock within the fishery, and the impact it has upon the environmental conditions, many more possibilities present themselves.

All fisheries will have an increased chance of being successful if they offer their customers what they want. This will vary from group to group, but anglers always want to catch particular species and size ranges. For example, only a minority of anglers want to spend the day catching very small perch or roach. In general terms, recreational fisheries management is often seen as an attempt to make sure that there are as many fish present as possible of the target species and of the preferred sizes, and as few as possible of other species and less favoured sizes. What often appears to be an apparently easy way of achieving this is to add lots of the main target species of the preferred sizes and sometimes to counterbalance this with the removal of the non-target species and sizes.

Although stocking individuals of the target species may seem like a reasonable way to achieve the aims of the fishery, several factors must be considered. Any stocking or removal of fish may well disturb the balance of the fishery and have unexpected consequences. It is well worth establishing what outcome the fishery requires from its stock manipulation and afterwards assessing whether this outcome has actually been achieved. It is surprising how often fish are stocked with no benefit at all.

The supplementation of stocks with particular species, or sizes of fish, is always likely to be a useful management tool. This is particularly true of carp, which often do not naturally recruit in still-waters, for several environmental reasons. Not only is it useful, but the paying customer has almost come to expect a certain amount of restocking. However, it is obvious from all that has been written so far that the introduction of fish into a fishery represents the most serious risk to that fishery. Not only is there the risk of introduction of disease, but also of invasive species of fish and other animals and plants. Because of the biosecurity threats associated with stocking, it should be equally obvious that great care must be taken if a decision is taken that new stocks are needed.

There are two common ways a fishery may obtain new stock. It may buy them from a third party or it may produce its own through artificial spawning. Of course, another method is to improve the fishery so that the fish will spawn naturally and the fry will recruit into the next generation. The section on habitat management will outline methods to establish habitats suitable for spawning and fry recruitment.

12.1 Acquiring fish from outside the fishery

12.1.1 Purchasing fish

When purchasing fish it is very important to acknowledge the risk this activity represents to the fishery. Once acknowledged and understood, the risks can be minimised. It is too late to decide it was the wrong idea once the new fish have been introduced!

12.1.2 Sources of fish

There are two ways to obtain fish for restocking: either farmed fish or fish sourced from cropping other fisheries or natural waters. Both of these sources have their good and bad sides, but either way the following questions should be asked about the health of the fish:

- Do the fish have a relevant health check?
- Have the fish undergone any additional health investigations such as testing for koi herpesvirus or spring viraemia of carp testing? You cannot tell if a fish is healthy just by looking at it. Health checks are the best way of ensuring that fish are healthy.

- Was the health check produced by a competent person (a recognised fish health expert)?
- What species have been health checked? A health check only applies to the fish species submitted.

Health checks do not guarantee the fish are healthy, but they do provide the best protection possible.

Ask for details of previous customers you may contact for references of the supplier and the fish. Always talk to other clubs or fisheries for their experiences and recommendations. Shop around and contact various suppliers: very cheap fish are often cheap because they are of poorer quality, or there may be a problem with the fish.

Additional questions are the following:

- What is the history of the fish? For example, are they imported?
- What age are they?
- What is the maximum and minimum size range of the fish you will receive?
- Are the fish the right species and not hybrids or colour variants?
- How will the fish be transported? – (see section 12.1.3)

Be a careful buyer. Buying fish is just like buying anything else. Confirm the cost of the fish, delivery charges and payment requirements. Protect yourself by paying by cheque, obtain a receipt, and keep copies of all of the paperwork. Get written confirmation of your order from the supplier, listing all the important details above.

Visit the source fish farm before purchasing the fish. If the fish are coming from cropping another fishery then ask to be present at the removal.

There is usually legislation covering the introduction of fish into a fishery, so inform yourself of the local regulations. Either complete appropriate paperwork yourself or insist on obtaining copies of all relevant consents and health checks to do with the fish and their introduction. Ensure these consents have been received before the fish arrive at your fishery. This is a legal requirement in the UK and any good supplier will insist this is completed, as should you.

12.1.3 Transport of fish

Introduction

The movement of live fish within the UK is an essential element of still-water fisheries management, as it is with the recreational fishery sector as a whole. The still-water fisheries manager may have occasion to transport fish when purchasing them from an external supplier, when selling fish to other fisheries or when moving fish between fisheries.

The transport of fish is controlled by legislation that applied to movements of all animals, as set out in section 27.4.

There are two basic and distinct transport systems for live fish: the closed system and the open system.

The closed system is a sealed container in which all the requirements for survival are self-contained. The simplest of these is a sealed plastic bag, partly filled with water and oxygen (Fig. 12.1).

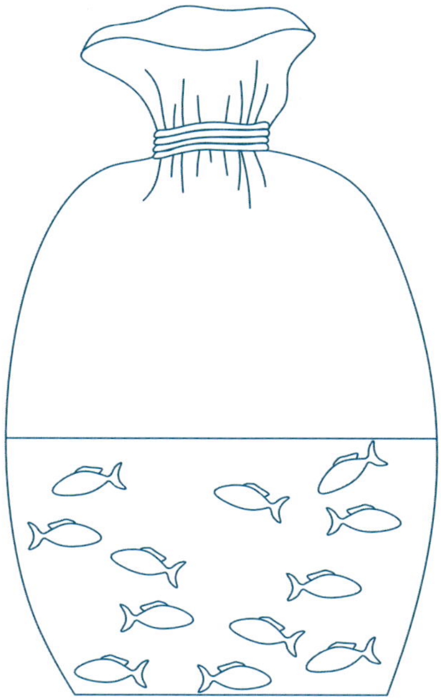

Figure 12.1 Transport of fish in plastic bags.

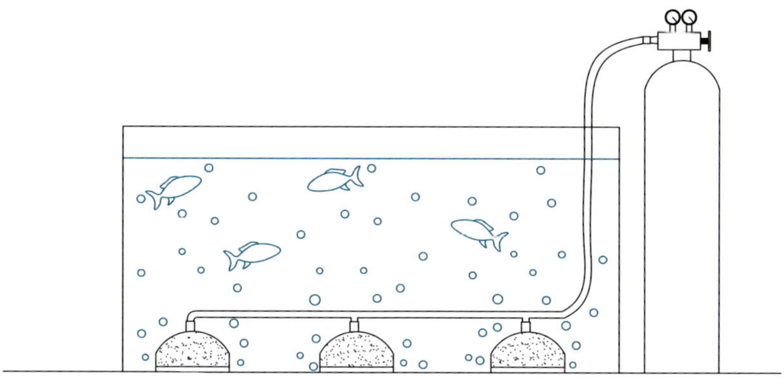

Figure 12.2 Transport tank with aerator.

The open system consists of a water-filled container in which the requirements for survival are supplied continuously from an outside source. The simplest of these is a small tank with an aerator (Fig. 12.2).

Factors involved in fish transportation

There are several factors to consider when moving live fish. Fish must not just survive during transport, but must reach their destination in a condition that ensures their future health and viability for their intended use.

Fish health

Fish must be healthy and in good condition before transportation is considered.

Weak individuals must be removed from the consignment, particularly when the temperature during movement is high. When the fish are unhealthy, even a large reduction of fish density in the transport container may fail to prevent fish losses. Weak fish are prone to mortality, especially when the transport time is extended, so only fish in good condition should be considered for transport.

Fish to be transported should not be fed for least 24 hours before transportation and ideally this will include being 'caught up' for transportation. If the digestive tract of the fish is not totally evacuated, the possible time of transport is reduced by approximately half, though the other environmental parameters (tank conditions) may be the same. Fish with full digestive tracts need more oxygen to fuel the increased metabolism, they are more susceptible to stress and produce waste products which take up much of the oxygen in the water and are toxic to the fish.

When fish larvae are transported, their time of survival without food should be taken into consideration. For example, the transport time of the larvae of herbivorous fishes should not last longer than 20 hours.

Oxygen

The single most important factor when transporting fish is the dissolved oxygen (DO) content of the water, although high levels of DO on its own will not necessarily guarantee the fish will arrive at their destination in good condition. The ability of fish to use the DO depends on a variety of other factors, such as their tolerance to stress, ammonia levels, temperature and pH.

One hundred per cent or greater DO saturation is the standard for fish transport, to maintain safe transport water quality with minimal post-transport mortalities. For warm water fish, the DO should not drop below 6 mg/l.

Fish weight and water temperature directly affect oxygen consumption during the transport process. Heavier fish, and those transported in warmer water, need more oxygen. For example an increase in water temperatures of 10°C (e.g. from 10 to 20°C) will double the oxygen consumption. It is recommended therefore that for each 1°C rise in temperature, the fish load should be reduced by about 11%; conversely, for each 1°C decrease in temperature, the load can be increased by about 11%. It is also worth noting that different fish species have different oxygen requirements. A large fish consumes less oxygen per unit weight than does a small one.

Oxygen consumption also increases with fish excitement and with handling-induced stress. Stress and fish excitement increases oxygen demand three to five times and, for instance, salmonid fry may need several hours to return to the normal level of oxygen metabolism, which in reality is usually after the end of the transport period.

The first hour after putting the fish in the transport system is the most crucial, as the fish are excited so their oxygen requirement increases.

Some conversion coefficients of oxygen demand are indicated by the Formula Research Group Recommendations to the Food and Agriculture Organization of the United Nations (1979). Taking the oxygen demand of carp as 10 mg/l, the converted oxygen demand levels for other fish are given in Table 12.1.

Table 12.1 Oxygen demand coefficients for selected species

Species	Oxygen demand (mg/l)
Trout	28.3
Bream	14.1
Perch	10.1
Pike	14.6
Roach	15.1
Eel	8.3
Tench	8.3

Taken from Formula Research Group Recommendations (1979) as cited: *Transport of Live Fish: A Review*. Fisheries and Agriculture Organization of the United Nations (1986).

In closed systems with pressurised oxygen atmospheres, DO should not be a limiting factor as there should be enough of this gas in the container. Of course if fish density is too high, or the transport time is longer than the contained and stored oxygen can support, then the fish will die.

Should there be mortalities during transport, the live fish will have to compete for oxygen with increased bacterial levels, owing to multiplication of bacteria as the body tissues of the dead fish break down, requiring more oxygen. The multiplication of bacteria may further produce toxic metabolites.

Temperature

Water temperature is also very important. When water temperature is low, the pH remains higher and fish metabolism decreases. The generally applicable zones of optimum temperatures for transported fish have been calculated at 6–8°C for cold-water fishes and 10–12°C for warm-water fishes in summer, 3–5°C for cold-water fishes and 5–6°C for warm-water fishes in spring and autumn, and 1–2°C for all in winter. The one exception to this is the early fry: the early fry of cyprinids should not be transported at temperatures below 15°C, and early fry of salmonids at temperatures higher than 15°C.

In warm weather, water temperature in the transport tanks may rise quite quickly. This would be a significant additional stress and must be prevented. Ice can be used to cool the water, but this should not come into direct contact with the fish. As a guide ratio, 25 kg of ice will cool 1000 litres of water by 2°C. The transport containers must also be well insulated. It is also important to time the journey so the fish are not in transit during the hottest, sunniest part of the day. The transport water should never vary more than 5°C during transport.

pH, carbon dioxide and ammonia

These water quality parameters can be viewed as a function of the length of time that the fish are in transport and the density of fish in the container. The source of water used in the transport process can be tested before the journey. The pH of the water can be used as a control because the proportions of toxic ammonia and carbon dioxide (CO_2) content are direct functions of pH, i.e. when transport time increases, carbon dioxide production through fish respiration decreases the water pH.

Fish transported in tanks are exposed to gradually increasing concentrations of carbon dioxide. Unless aeration is adequate, carbon dioxide levels may exceed 20–30 mg/l. In general, for each 1 ml of oxygen a fish consumes, it produces approximately 0.9 ml of carbon dioxide. If the carbon dioxide level increases rapidly, as may occur with heavy fish loads, fish become distressed. However, elevated concentrations of carbon dioxide can be tolerated if the rate of build up is slow.

Adequate ventilation is a necessity for transport units. Tight covers or lids on the units can result in a build up of carbon dioxide, which will stress the fish. Aeration of the water will reduce concentrations of dissolved carbon dioxide, if there is adequate ventilation.

Ammonia (NH_3) builds up in transport water owing to the protein metabolism of the fish and bacterial action on any excreted waste. Temperature and the time of last feeding are important factors that regulate the excretion of ammonia and other waste. For example, trout held in water at 1°C excrete 66% less ammonia than those held in 11°C water, and fish starved for 63 hours before transport produce half as much ammonia as recently fed fish. As a result, research has recommended that fish larger than 10 cm should be starved at least 48 hours before being caught and transported. Those 20 cm and larger should be starved 72 hours before transportation.

Another important factor is chlorine concentration in water, although – like carbon dioxide – chlorine is also removed from the water by aeration. A concentration of 0.5 mg/l is considered as potentially dangerous, though even lower chlorine levels, (e.g., 0.2 mg/l) may disturb the fish respiration mechanisms considerably.

Stock density and fish activity

Space for the fish in transport must also be considered. Lower stock densities are more desirable. In the Formula Research Group Recommendations (1979), the following ratios between fish weight and the volume of water in the transport tank are given: table carp 1 : 1, stock carp 1 : 1.5, stock pike 1 : 2, herbivorous fishes 1 : 2. This is with good aeration of water, at a temperature of 8–12°C, during short transports (lasting 1–2 hours).

Stock density of the fish in the transport system also depends on the length of transport time; density should go down with the amount of time in transport (Fig. 12.3).

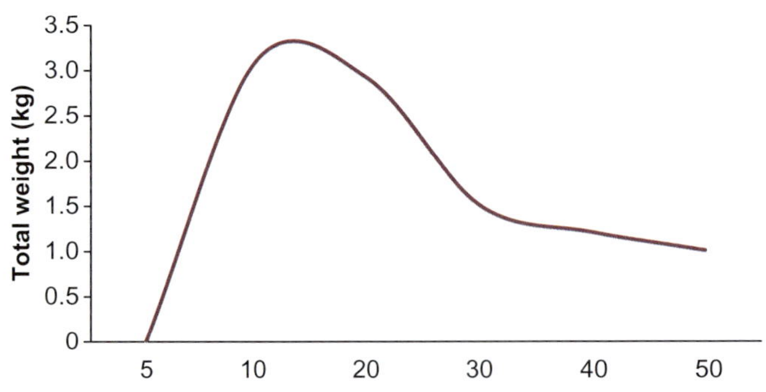

Figure 12.3 Optimum weight of fish transported for various transport times.

The release of fish at the end of transportation can be one of the most critical stages of the entire transport process. The fish are under, at best, some degree of stress in the transport unit. Sudden exposure to water of different characteristics or lower environmental quality will further stress the fish, often beyond what is tolerable without further negative impact. It is recommended therefore that the fish are not just released into water of unknown quality. As a minimum, the fish to be released should be given time to acclimatise to the temperature of the receiving water.

The different characteristics between the transport system and the receiving water can mean that a pH, temperature or gas saturation difference can be significant, and caution is recommended.

Other considerations

The figures quoted above are to be used as guides and are by no means definitive. When moving fish stock it is advisable to consider the transport conditions on a case-by-case basis.

It is also recommended to transport at a fish density at which the time of transport can be prolonged at least 1.5 times to prevent the consequences of a possible delay during transport, for example failure of the transport, failure to stick to schedules and a special allowance for travelling by road on a Friday.

Closed system fish transport

Closed systems for fish transport normally consist of polyethylene bags and other sealed transport units. Closed systems are used primarily for the transport of the early fry, small numbers of brood fish and ornamental varieties such as koi carp. The transport of fry in polyethylene bags with oxygen is widespread and is a very effective method. It substantially reduces the total volume and weight of transport water, making it possible to prolong the transport time, and is economically advantageous.

During transport, the bags with fry are placed in outer cases protecting them against mechanical damage, mainly punching or tearing in contact with the ground or other sharp objects. The case keeps the bags in the desired position, enables easier handling and/or provides thermal insulation of the bags (Fig. 12.4).

Figure 12.5 shows a simple closed system for the transportation of live fish.

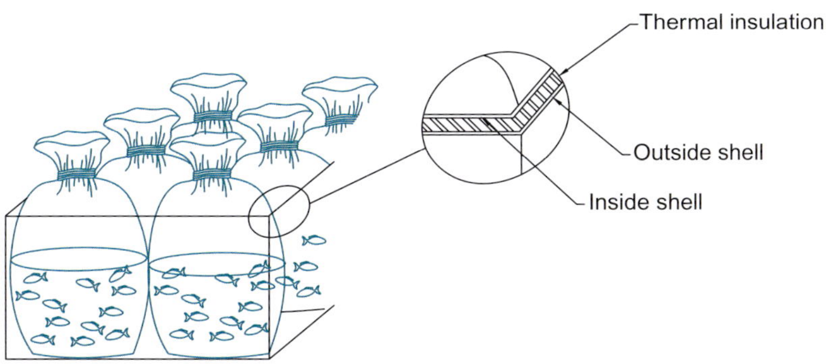

Figure 12.4 Transport case with thermal insulation.

Figure 12.5 Simple closed system for the transport of live fish.

Guidelines for closed transport systems for fish are as follows:

- If water with transported fry is to be cooled, bags of ice should be placed under the fish-transport bags on the bottom of the box.
- The water to be used for fry transport in a bag should comply with all requirements for good water quality; it is considered best to use water of the same quality as that in which the fish were kept before transport, not tap water.
- Before putting the fry in the bag, the procedure of catching, counting and distributing the fry in the bags should have been thoroughly prepared to complete the operation as quickly as possible.
- After transport, or during control on a longer journey, the condition of the fry should be checked before release.

Open system fish transport

The open system is more commonly used when transporting large quantities of fish. It has many technical variations, ranging from small transport fish-cans up to special fish transport trucks with technologically advanced systems. The weight of fish that can be safely transported in any tank depends mostly on the efficiency of the aeration system, duration of the transport, water temperature, fish size and species.

Open systems come in many shapes and sizes, depending on fish type and quantity to be transported. As such, it is impossible to recommend 'best practice' units. Instead, we will focus more on relevant aeration techniques. It is up to the individual fishery professional to decide which systems are appropriate for the fish to be transported and to seek further advice if unsure. When considering the suitability of a system it is also necessary to consider the

Figure 12.6 Four wheel drive with transportation tanks.

vehicle that will house the system; 'off road' deliveries may require a four-wheel drive vehicle (Fig. 12.6).

Aeration systems

A wide variety of methods can be used to aerate water during transport. Because fish are often densely crowded and excited, their stress levels go up, especially after loading into the system. So it is vital to have an aeration system that provides adequate DO rapidly and efficiently.

For most cases it is advisable to maintain oxygen concentrations above 6 mg/l at all times. As an added advantage, high oxygen levels, near air saturation, may lessen the effects of any ammonia build up, especially on long trips.

Aeration systems that blow air direct into the tanks are the most obvious way of introducing oxygen into the transport tank. Heavy aeration, where the surface area of the bubbles creates a large interface as they rise through the water, creates excellent conditions for dissolving oxygen and degassing carbon dioxide. Aeration also creates a lot of turbulence, bringing water to the surface, where carbon dioxide can escape directly to the atmosphere.

Pure oxygen is actually more commonly used and has numerous advantages. Most transporters are equipped with compressed oxygen tanks, but some transporters prefer liquid oxygen because it is cheaper, weighs less and fewer tanks supply the same amount of oxygen. If oxygen cylinders are used then follow all recommended safety precautions. Single-stage pressure regulators are needed for each oxygen tank, and each aeration line to a tank compartment should have a flow meter to adjust supply rates as desired. Various diffusers are used for dispersing oxygen from the bottom of the tank compartment. Aeration lines from the regulator to flow meters and drop lines to the diffusers are usually made of flexible plastic tubing or PVC pipe of three-eigths of an inch internal diameter. Some diffusers are made of PVC frames that fit the bottom of the tank compartment. Always make sure that the hose is suitable to carry compressed oxygen.

Pure oxygen tends to make the fish quite lively, so precautions must be taken to prevent the fish jumping out of the tank at any stage of the journey.

Diffusers

Leaky pipe aerator

The 'leaky pipe' aerator relies on compressors or blowers to supply high or low-pressure air, which enters the water through the diffuser. The diffuser splits the air into small bubbles by passing it through a porous material. The diffuser is placed at a low point in the tank (Fig. 12.7).

Oxygen is transferred from the air bubbles to the water as the bubbles pass through the water column; the smaller the bubble the larger the surface area, thus better oxygen exchange rate. The distance and speed the bubble travels through the water column also impacts on the efficiency of exchange; the greater the contact time the greater the exchange.

Microbubble diffuser

The microbubble diffuser uses oxygen directly. It can be supplied from industrial type oxygen cylinders, liquid oxygen tanks or oxygen generation equipment. This method has significant advantages over air (Fig. 12.8).

The addition of pure oxygen can be used to increase saturation to far greater levels than 100% but problems can be encountered with saturation levels approaching 200%. A benefit of using pure oxygen is that it will force out dissolved nitrogen. This ensures levels remain below the 102% saturation level at which nitrogen gas bubble trauma can occur.

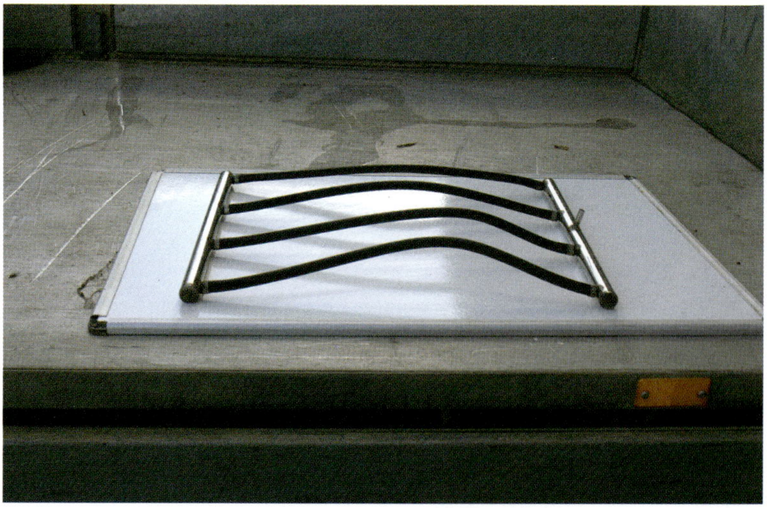

Figure 12.7 Leaky pipe diffuser.

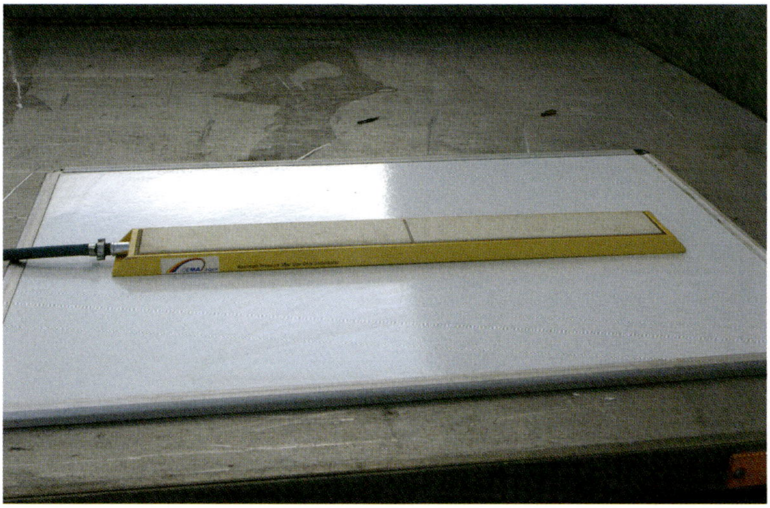

Figure 12.8 Microbubble diffuser.

Leaky pipe (air) versus microbubble (oxygen)

The leaky-pipe diffuser is most suited to supplying high volumes of gas at low pressure. It is most suited to delivering air supplied by a blower. The results indicate that in an emergency these diffusers are capable of putting a substantial amount of oxygen into the water, in a comparatively short time. Economics would dictate that this usage was very occasional and for short periods. In the case of emergencies, therefore, the 'leaky pipe' method may be recommended.

Microbubble diffusers are suited to supplying lower volumes at higher pressure. Their efficiency is much higher than the leaky-pipe (60% compared with 20%), making them ideally suited for small tanks, such as a transport tank or off-system facilities such as hatchery tanks or for boosting oxygen around a system. Because the diffuser works under pressure, its efficiency will not be impaired.

The combination of a microbubble diffuser loading the transport system with DO and a leaky pipe diffuser putting through the coarser air bubble will enable the system to operate with an efficient source of DO and an economical way of reducing carbon dioxide by blowing it off with air, the best of all worlds perhaps.

Agitators

Agitators are sometimes used either alone or in combination with diffusers, depending on the timescale (Fig. 12.9). Agitators can also provide back-up aeration for emergency situations. For short trips, agitators alone may suffice without the need for oxygen.

The combination of agitators with oxygen diffusers works well for fish transport over longer periods. Agitators do not have good mixing characteristics alone, and as such low DO levels may result in the corners of the transportation tank. Oxygen diffusers do not remove carbon dioxide efficiently because of minor agitation at the water surface, but under high-density transportation conditions, they supply oxygen efficiently.

Figure 12.9 An aeration surface agitator.

Note

Many aeration systems suitable for transport systems are 12-volt direct current (DC) units powered by the vehicle's battery and alternator. If this is the case, the vehicle's engine should be kept running during any stops to keep the battery charged. One agitator per compartment is used. Excessive agitation can be harmful to delicate scaled fish, and diffused oxygen is preferred for small fish.

Transportation of mixed species

It is recommended that different species should be transported separately, although it is acknowledged that this is not always practicable. There are circumstances where decisions must be made by the transporter to either transport species together, to reach the destination more quickly or to keep species separated in cages at the departure point for separate transport.

Decisions such as this must be based on the well-being of the fish and, although some species can be transported together with little adverse effect, other species can suffer in mixed transport containers as stress levels increase.

In cases where species must be transported in mixed containers, it should be recognised that certain species do not mix well and that active and dominant species should always be transported in isolation. An example of a species that should not be transported in mixed containers is carp (*Cyprinus carpio*) with silver fish such as roach (*Rutilus rutilus*). Mixed transportation invariably causes stress to the fish and as such, every effort should be made to transport species in isolation.

Fish sedation

Sometimes fish are sedated before transport as a method of reducing stress. However, sedation, unless it is very precise, may lead to significant problems, particularly 'packing', where the fish stop moving normally in the water column and drift to the bottom where they 'pack' on top of one another. The fish at the bottom soon die from lack of oxygen. Furthermore, when the fish arrive at their destination, considerable planning would be required to revive the fish before they were stocked into a fishery. As a result of these issues, sedation is not recommended.

Transportation times

In the UK, Defra has laid out guidelines for 'good practice' for transport times and conditions for moving fish. These guidelines state the following:

- The duration of the transport must be made as short as possible.
- The vehicle is cleaned and disinfected before the journey starts.
- No water escapes from the vehicle.
- All reasonable steps are taken to safeguard the health of the fish.

The transportation of stock is an integral part of the industry. However, where it is possible to avoid movement, or reduce the time in transport, this is considered the desirable option. The longer that fish are in transport, the more stress they are likely to incur and thus the stress-induced risks increase.

From a biosecurity perspective, risks to fish health will become unacceptable after a certain time in transport.

Pre-transport planning

After fish are harvested, carefully remove any debris from the water and give fish a resting period of several hours before they are handled again. If holding vats are available, you can easily grade, clean and sort fish before shipment. Then the fish may be slowly acclimatised to the expected transport water temperature. If fish are to be held in ponds, maintain adequate aeration and water quality:

- Avoid holding large fish or active species in cages.
- Do not overload cages.
- If fish begin to die, diagnose the problem, then treat and return fish to the pond rather than transport them.

Prophylactic treatment with approved chemicals can reduce pathogenic organisms that could cause problems during or after transport. Avoid over-treating with excessive doses or combinations of chemicals. Be prepared to flush any treatment with fresh water if fish show any signs of stress.

The transport unit and equipment, such as dip nets, should be dried and disinfected between loads of fish. This practice reduces the possibility of spreading disease pathogens from one group of fish to another.

To avoid untimely delays, get good directions to your destination. Plan your trip to include any drop-offs or water refreshment. Be certain that all required permits or licences are in order and carried in the vehicle as good practice. Check with the proper regulatory authorities for current information on laws and regulations on transporting and marketing live fish before transportation.

Table 12.2 Regulations for imports into the UK from elsewhere in the European Union

Species	Source	Rules	
		Dead fish	Viscera
All species of salmon and trout	VHS and IHN EU approved zone or VHS and IHN approved farm	No restrictions	No restrictions
	Other	Must be eviscerated before import	Prohibited
Grayling, whitefish, pike (excluding fry), turbot	VHS and IHN EU approved zone or VHS and IHN approved farm	No restrictions	No restrictions
	Other	Must be eviscerated before import	Prohibited
Pike (fry only)	VHS and IHN EU approved zone or VHS and IHN approved farm	No restrictions	No restrictions
	Other	Must be eviscerated before import	Prohibited
Other species of fish	Any	No restrictions	No restrictions

VHS, viral haemorrhagic septicaemia; IHN, Infectious haematopoietic necrosis.

Check your oxygen supply is full and the system is operating correctly. Always carry a list of oxygen suppliers with you, a spare length of hose, hose clips and a regulator and, not least, the correct spanners.

Documentation

Any consignment of fish transported within the UK must be accompanied by documentation that identifies the consignment. This documentation enables the source of the fish to be traced if there is any problem with the consignment. This can also enable the consignment to be linked to a movement document if necessary under EU law (see Table 12.2 for information on rules governing movements of fish into the UK from the EU). This documentation can be provided on the transport container, on a label attached to the system or accompanying the movement document if required. Further details on the legislation surrounding the transport of fish can be found in section 27.4.

12.2 Producing your own fish

To reduce the risk associated with restocking, it is worth considering never bringing fish onto the site. To do this and still have the option of restocking fish, you need to produce your own. For most fisheries this is perfectly feasible but does require some understanding of the techniques, and some pond space. The primary requirement is for a hatchery or for breeding ponds.

12.2.1 Establishing a hatchery

Before beginning to think about the building of a hatchery, various aspects need to be assessed to work out whether suitable conditions are available to establish one.

Water supply

The water supply is obviously critical to the success of any hatchery. However, with this there are major differences in cyprinid culture and other forms of fish farming. For instance, salmonid culture relies on large volume water supply from a variety of external sources; cyprinids require much less water and are therefore often cultured in self-contained recirculation tanks. Here the initial water supply is often tap water or borehole water. This water is then recirculated through the holding tanks and filtered to ensure water quality is maintained. Generally no more than 20% per week of the total volume of the tanks is replaced with fresh water. Furthermore, the whole process, from obtaining eggs to transfer of fry to earth ponds, can be completed in as little as 3 weeks, so the overall use of water is small.

Water quality

It may be possible to produce fry with very little water, but the water does have to be of good quality. Table 12.3 gives the basic water quality requirements for carp compared with rainbow trout and is a good basis from which to start.

The water supply for a fish farm, and particularly for a hatchery, must be within these parameters for successful rearing of fish. Ideally, the ammonia, nitrite, nitrate, biological oxygen demand and a variety of other parameters (lead, copper, etc.) should be as near zero as possible.

The most important parameters are temperature, pH and oxygen content. These are basic requirements. If these are wrong and unable to be corrected by some form of management, then the water source is probably not worth considering.

Temperature

As can be seen from the Table 12.3, the water temperature requirements for carp are quite high. This is true for many cyprinids. Because of the way a cyprinid hatchery works, it is possible to heat the water artificially to get the optimum temperatures required for the particular cyprinid. This does create some problems:

- at higher temperatures pathogens often work quicker and more effectively;
- the fish will need to be acclimatised gradually before they are moved out to ambient temperature growing ponds.

Table 12.3 Water quality criteria for rearing fish

Parameter	Carp	Rainbow Trout
Temperature	Maximum 30°C	Maximum 20°C
	Optimum 22–25°C	Optimum 15–17°C
Oxygen	Minimum 4 mg/l	Minimum 8 mg/l
Carbon dioxide	Maximum 25 mg/l	Maximum 10 mg/l
pH	Minimum 5.5	Minimum 5.5
	Maximum 8	Maximum 8
Nitrite	Maximum 0.2 mg/l	Maximum 0.2 mg/l
Ammonia (NH_3) (total) (if the pH is less than 8)	Maximum 0.05 mg/l	Maximum 0.05 mg/l
Biological oxygen demand (BOD)	Maximum 15 mg/l	Maximum 15 mg/l
Lead	Maximum 0.1 mg/l	Maximum 0.03 mg/l
Copper	Maximum 0.3 mg/l	Maximum 0.1 mg/l

Oxygen

If the water source to be used is tap water or borehole water, then the oxygen levels need to be checked. Additional aeration of the source water is usually a beneficial action (see the section earlier in this chapter for advice on aeration systems).

Other considerations

One of the most important conditions of the water supply is that the proposed source of water has no fish upstream of the point of extraction of the supply. This will mean there is less possibility of contamination from pathogens, wild fish or aquatic predators. Because of the limited amount of water used, it is usually possible to have a fish-free source of water such as tap water or borehole water.

If the system is based on recirculating water then it is more usual for the farm to use tap water or mains water. Mains water must be treated before it is used, as it will contain certain chemicals (e.g. chlorine) that are toxic to fish. It is often beneficial to use borehole water, as this is generally more suitable for fish. Once the systems are full of water then there should be access to a volume of at least 10% of the total water per day in case of emergencies. Furthermore, many systems trickle water through at 1–2% per day, which works well in maintaining water quality.

Water supply systems

Given the information above, the best way of obtaining water would be from the mains supply or from a borehole. The water supply should be through a holding tank. This tank should contain at least 10% of the total volume of all the hatchery tanks. This tank should be above the height of all the hatchery tanks so that any water can be gravity fed into all the other tanks. The header tank should be heavily aerated and held inside the hatchery building to allow the water to acclimatise to the ambient temperature of the building. Consideration should be given to heating the header tank to make sure it quickly gets to the right temperature. Mains water and borehole water is usually cold (5–10°C), regardless of the time of year.

Water quantity

Most cyprinid farms have little or no turnover of water. In the hatchery, fish are maintained in recirculation systems and so the maximum amount of water required is that needed to initially fill the systems.

However, when the fry need to go out into growing ponds there is a requirement for significant amounts of water. Box 12.1 gives an example of calculations of the quantities required for a typical system. There is little turnover in the ponds once they are filled in early spring, and they are not drained until harvest. However, there must be access to enough water to fill all the ponds fairly quickly in spring, 6–8 weeks before stocking, and to compensate for losses due to filtration and evaporation.

During the summer months when the ponds are full, evaporation will be an important factor. Evaporation rates will vary widely, depending on temperature and wind, etc., but they can remove a lot of water. The amount of water that should be considered as being lost by evaporation, and evapotranspiration, can be obtained on a local basis from the regional meteorological service. So, although a cyprinid farm does not require much water compared with a salmonid farm, at certain times of the year it does require large amounts.

Box 12.1 Calculation of water needed in a typical fishery

On Eric's fishery there are four ponds, two medium and two small. All the ponds have an average depth of 1 metre. To work out the water quantity required, do the following calulation:

length × width × depth = volume

$1 m^3 = 1000$ litres

Ponds	Dimensions	Calculation	Volume per pond (litres)	Total volume
Two medium	30 m × 15 m	30 × 15 × 1 × 1000	450 000	900 000
Twp small	5 m × 15 m	5 × 15 × 1 × 1000	75 000	150 000
				Total 1 050 000

So, if we wanted to fill all these ponds in 10 days (10 × 24 hours), it would require a pump rate of 1 050 000 ÷ 240 = 4375 litres per hour.

In conclusion, a cyprinid farm requires water of good quality, available in large quantities when starting the production cycle, with a much smaller amount available throughout the year to compensate for evaporation and any seepage that may occur.

12.2.2 Requirements of a hatchery

The actual building requirements to establish a hatchery are relatively straightforward provided that the basics of husbandry, cleanliness, biosecurity and good water quality are respected.

As discussed above, it is normal for cyprinid hatcheries to be inside a building and to have some sort of temperature control. The building must therefore have good insulation.

The hatchery will require several holding tanks for various aspects of the spawning process. All of these will require filtration systems and will need to be cleaned regularly. The floor of the building will need to be freely draining to the foul sewer.

Much of the rest of the building design comes from the sorts of tanks and the way they are used. Therefore we will now discuss the actual spawning of fish. We will return to the facilities required at the end of the chapter.

Holding units for broodstock

Broodstock management is a key factor in the success of any spawning enterprise.

The broodstock that are selected should be young and vigorous. Egg condition may deteriorate in larger, older fish and they are not so easy to handle, owing to their size. After this, you can select for any characteristics that is important, for example body shape, rapid growth or scale pattern. However, it is recommended that the brood fish come from the same or very similar conditions to those into which the offspring will be introduced.

If the brood fish are to used year on year, then they should be kept in natural conditions through the summer, with plenty of food. You can minimise the stress on the fish by putting them in a pond that is well out of the way. For cyprinids, eggs start to develop in the fish immediately post-spawning. Most cyprinids spawn in spring or early summer, depending on species, and egg development takes all summer. When the winter arrives the eggs should be fully developed, but stop developing through the cold weather. As the temperatures rises in spring these eggs will mature ready for spawning. This stage of maturation is very important and enables the fish to time their spawning so that there will be plenty of food for the fry when they hatch. Other factors are also important, such as day length, oxygen levels and feeding regime. Most cyprinids are shoal spawners, where all members of the same species will be ready to spawn at the same time subject to the same conditions. For example, carp require 1000 degree-days (average daily temperature × number of days) over 10°C to mature their eggs. In England this will usually result in the fish spawning in the wild in the first week in June. This gets slightly later the further north you go.

There are two alternatives for preparing the broodstock fish:

- It is possible to spawn fish slightly earlier than they would in the wild by manipulating the spawning conditions. This is a very simple process of controlling the temperature, lighting and feeding regime to fool the fish into believing its 3–4 weeks later in the year than it really is.
- It is also possible to allow the fish to stay in their ponds until they are nearly ready to spawn. The brood fish are then caught a week or so before they naturally spawn and are brought into the broodstock tanks at that point.

Once in the broodstock tanks any disturbance to the fish should be minimised and the male and female broodstock must be kept in separate tanks.

Egg incubation and hatching systems

Cyprinid eggs are usually incubated in either Zuger jars (see Fig. 12.10) or hatching troughs. Of the two, Zuger jars are by the far the most efficient. However, the eggs must be treated to remove their sticky coating before these can be used. This can harm the eggs and so troughs are often a safer method.

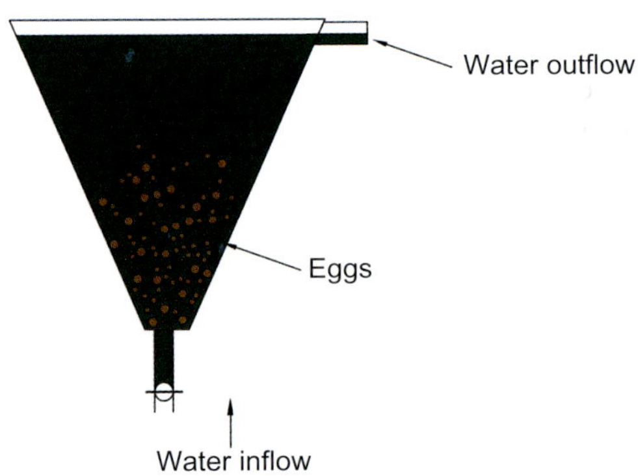

Figure 12.10 A Zugar Jar.

First feeding facilities

The first feeding stage is often critical, and some people prefer to stock fisheries with swim-up fry. However, losses can be very substantial, sometimes total, so taking the fish through first feed in a protected hatchery to prevent these losses is recommended.

The most popular first feeding tanks are those between 1 and 2 metres in diameter. Water depth is usually set at 40–60 cm. The systems are usually based on recirculating water and have a flow of less than 10 litres per minute. These tanks should be heavily aerated and the inlet must be designed in such a way as to reduce flow. The fish are very small and cannot maintain themselves against a flow and will congregate in slack areas.

Rearing of eggs and fry

It is not possible to give information on spawning for all cyprinids. The following is based on the procedure to spawn the carp, *Cyprinus carpio*. For most other cyprinids the procedure is basically the same but with some slight changes to the timings and temperatures.

After 1000 degree-days, the brood fish should be ready to spawn. They will not spawn if the sexes are kept separate, as they require significant additional stimulation to actually spawn. The fish should be closely observed at this time for any signs of spawning behaviour. If you have caught the fish from the wild then they should be ready to spawn as soon as they have recovered from the disturbance of being caught and moved to the broodstock system.

When are the broodstock ready?

The broodstock should display the following signs when ready to spawn:

Females:

- Well-rounded abdomen.
- Fullness extends past pelvic fins to genital pore.
- Genital opening is swollen, red and protruding.

Males:

- Anus is swollen.
- Will release a small amount of milt if squeezed.
- Milt thick and white.
- Spawning tubercles present; skin feels rough to the touch.

Once you have got fish of both sexes ready to spawn, there are two main ways of successfully spawning them: semi-artificial and artificial.

Semi-artificial spawning

Place the fish into a tank with one female to three males. Add some spawning brushes and raise the temperature from 20 to 25°C. Then the fish will almost certainly spawn at first light the following day.

Advantages:

- Very easy and the fish enjoy it!

Disadvantages:

- No control.
- Fish may damage themselves.
- Much lower fertilisation rate.

Artificial spawning

Fully artificial spawning requires the administration of hormones to stimulate the egg and milt release. It is usual to use the ground-up pituitary gland removed from dead carp. This is known as carp pituitary extract (CPE). This works on all cyprinids and quite a few other fish species. You can also use a synthetic hormone, for example Ovaprim.

Injections of carp pituitary extract

The males and females require specific amounts of CPE injected at specific times to allow the eggs and milt to be stripped from them. CPE usually comes as a powder, which needs to be made into a liquid so that the dosage can be injected into the fish.

The dosage required is as follows: female fish receive 6 mg/kg of CPE; male fish receive 3 mg/kg of CPE.

> For the females, the dosage is split into two separate injections, the primer and resolver, whereas the males only have one, the resolver: females 20% (primer) and then 80% (resolver); males 100% (resolver).

The CPE must be prepared and diluted with saline before it can be injected into the fish. The CPE powder is diluted at the rate of 4 mg diluted in 0.2 ml of saline (100 mg diluted in 5 ml of saline). You can use sterile eyewash solution in the following procedure.

(1) Grind down the pituitary into a fine powder.
(2) Mix with required amount of saline in a test tube, 100 mg diluted in 5 ml of saline.
(3) Shake for 5 minutes.
(4) Allow to stand for 20 minutes.
(5) Spin in a centrifuge for 5 minutes at 3000 r.p.m.

Draw off the liquid into 1-ml syringes and use 23- or 25-g needles. Take care not to draw any of the solid material at the bottom.

Timing of injections

The timing of the injection is very important. It is always best for the injections to be timed so that the fish are ready in the morning. So, for example, the following is a good sequence:

I.	Get primer injection at	0 hours	e.g. 9 a.m.
II.	Get resolver injection at	12 hours	9 p.m.
III.	Ovulation and ready to be stripped at	24 hours	9 a.m.

It is possible to make an intraperitoneal (IP) or intramuscular (IM) injection. It is usual to inject IP as there is less seepage and the injection is going direct to the target organ. However, this is not a hard and fast rule, and many will inject IM.

Injecting fish

This is a hectic process so preparation is very important. Make sure everything is ready and prepared. The syringes should be prepared and the amounts required for each fish calculated beforehand. The anaesthetic bath and a recovery tank should be prepared. These should both be heavily aerated.

The fish can be anaesthetised by using benzocaine at 1:40000, 2-phenoxyethanol or clove oil.

Now, use the following procedure:

I. Place the fish in anaesthetic until it is unresponsive but the opercula are still beating steadily.
II. Take the fish out and place on a padded wet surface, on its side. Place a wet towel over its eyes and tail and have someone gently hold it in place.
III. Introduce needle under a scale in a line above pelvic fin, three to five scales up.
IV. Ensure needle has penetrated the peritonea.
V. Inject fluid slowly in a continuous motion.
VI. Withdraw needle quickly and ensure there is no seepage.
VII. Now place fish into recovery bath until it is fully awake. Then place it back into the holding tank.
VIII. Keep a very close eye on the fish to ensure that they recover fully.

Stripping the fish

Once the fish have had their resolver injection, then the eggs should ovulate and be ready for stripping 12 hours later.

Fish may release eggs before this, particularly if they start spawning behaviour with other fish in the tank. Stitching the vent closed can prevent this, although this is a fairly harsh treatment. Alternatively, you can keep them under constant observation. As soon as the females start to show vigorous spawning behaviour they can be stripped straight away. It is common for the fish to be ready before the full 12 hours is complete. They should be stripped as soon as they are ready as they may well release the eggs into the tank and then they are useless.

Females

I. Take the fish out and anaesthetise her.
II. Handle the fish very gently to prevent premature release of the eggs.
III. Holding the fish upside down, dry the body thoroughly. No water should come into contact with the eggs.
IV. Hold the fish horizontally with the vent over a dry plastic bowl. The eggs may start to come out straight away. Very gently squeeze the flanks of the fish. Use the anal fin to guide the eggs into the bowl.
V. Continue squeezing until most of the eggs have been released. Adjusting the position of the fish and wiggling the belly may help to release more eggs.
VI. Eggs should all be of equal size and colour, usually a dark green.
VII. The eggs can be put to one side and covered.

Males

The males are stripped using exactly the same technique, but they will not normally flow freely and so will require squeezing. The milt should be squeezed into a dry Petri dish.

It is usual to mix eggs from one female with milt from two males. You do not have to use all the milt from one male, just enough to cover all the eggs. Motility of the milt can be checked before use.

The eggs and milt should be mixed dry for 20–30 seconds, making sure they are thoroughly, but gently mixed. A large goose feather is good for this.

Now, while continuing to stir the eggs, add a very small quantity, just enough to cover the eggs, of fertilising solution, also known as Woynarovich solution (4 g NaCl + 3 g urea + 1 l distilled water).

This solution will initiate sperm motility and cause the eggs to start swelling. It will also prevent the eggs getting sticky. Eggs should be fertilised in the first 20–30 seconds but sperm motility may last up to 5 minutes.

Use the fertilising solution to wash the eggs four or five times to remove unused sperm.

Now the eggs need to remain in the solution, changing it occasionally, for 60–90 minutes. During this time they need to be stirred very gently to prevent clumping.

After this time the eggs are fertilised and will have swelled to some ten times their original size. They are now becoming sticky and if given a chance they will stick to anything, including each other. If they stick together in clumps the inside ones will die owing to lack of oxygen. Bacterial and fungal infections of the dead eggs will then quickly spread to the live eggs on the outside and kill them too. So there are two choices at this stage for treating the eggs:

- The eggs can be laid out in trays on fibrous matting. This matting should have a complex three-dimensional structure so that the eggs can stick throughout it but be held away from each other. This will allow free movement of water to all the eggs, keeping them well oxygenated.
- Place the eggs in Zuger jars. If added to Zuger jars, then the sticky outer layer of the eggs must be removed. There are various methods to do this.
    - talc;
    - mud or silt;
    - full fat milk;
    - enzyme to dissolve protein layer;
    - tannin.

By far the most popular method is to use a tannin solution. This denatures the protein that makes up the sticky layer. However, great care is needed because if an overdose with the solution can destroy the outer surface of the egg and kill it.

Tannin solution is made up of 5 g of tannin in 10 l of distilled water. Add the tannin solution to the eggs and then wash off immediately (contact for 10 seconds). This may need to be done several times.

All these processes, including the use of tannin solution, change the nature of the outer layer of the egg. This layer contains antibacterial and anti-fungal properties so once treated the eggs are more susceptible to bacterial and fungal infections.

The eggs are now ready to be placed into Zuger jar incubators. Two hundred thousand to three hundred thousand eggs can be placed in each 20-litre jar. The flow should be adjusted so the eggs are in constant motion. They should be regularly monitored and observed. The dead eggs and those that attract fungus become less dense and are mostly carried away with the outflow. However, this is never 100% effective so dead eggs and those with fungus will need to be removed occasionally. The Zuger jars need regular monitoring to check that the flow rate is keeping the eggs in constant motion. The eggs will almost certainly require a treatment for fungus during this period. The only readily available treatment for *Saprolegnia* fungus is Pyceze. Eggs should be treated at least once daily with 20 mg bronopol per litre (1 ml pyceze per 25 litres of water) flowing through the Zuger jar.

Carp will hatch in approximately 60 degree-days. For example, if they are held at 20°C, they will hatch in three days.

First feed

Once the carp eggs hatch, they will survive on their yolk sacs for about 24 hours. During this time they will come to the surface and fill their swim-bladders. When they first hatch, they will not swim freely but remain attached to a solid object such as the tank sides. Once they fill their swim-bladders, they are free swimming and will start looking for food. At this stage they are not fully developed and many of their senses will have to develop further, but their eyesight is good. Therefore, to get them to feed it is important to present them with something that will attract them. This can be done very successfully with live food. This will engage what is known as the 'snap' response. By far the most successful and convenient live food is the brine shrimp, *Artemia* sp.

Brine shrimp

The brine shrimp, *Artemia* sp., is a very unusual invertebrate of salt lakes. To survive long periods when the lakes dry completely, it produces a very resistant egg stage, known as a cyst. On contact with water these dehydrated cysts will hydrate and the hatching mechanism will be triggered. Within 24–48 hours, a live *Artemia* nauplius larva will emerge. These dehydrated cysts are collected and sold in tins or vacuum-packed sachets. Artificially hatching brine shrimp is quite straightforward. The now empty Zuger jars can be reset on a rack with plenty of lighting all around the jars, access to plenty of aeration and an aquarium heater for each jar.

Twenty-four hours before you require the hatched *Artemia* you should do the following:

- Fill the Zuger jar with water at 28 parts per thousand salt (28 g/l).
- Add the aquarium heater, set to keep the water at 28°C.
- Now add the *Artemia* cysts. For one tank of fish, use 20 ml of cysts.
- Add the aeration and stir to sink as many of the cysts as possible. Leave under intense light for 24 hours.
- Remove the aeration and turn the top light off, or cover the top of the Zuger jar for 10 minutes.
- All the hatched *Artemia* should swim to the light and therefore go to the bottom of the Zuger jar.
- All the unhatched eggs and shells should rise to the surface.

- Now you can open the tap at the base of the jar and draw off the first 5 litres of water, which will contain the most of the hatched *Artemia* but none of the unhatched cysts or the cyst shells.
- The remaining water and cysts should be discarded.
- The 5 litres of *Artemia* can now be added straight to the hatchery tanks or sieved through a muslin cloth to separate the hatched larvae from the salt water. This is recommended if continuous feeding is to occur over a significant period. Too much salt water added to the fry tanks may alter the salinity too much.

Although *Artemia* is very good for getting the fish to start feeding, it is not nutritionally complete and your fish will not grow and survive on it for long. So at this point you need to also start feeding with a very finely powdered diet. This type of diet is readily available and is usually known as a fry starter diet.

After approximately 2 weeks the fish can be stocked out, or moved into a prepared pond at a stocking rate of 200 per cubic metre to on-grow for a while longer before being stocked out. It is difficult to feed the fish enough in a recirculation system to get them through the next month as they grow to around 2.5 cm (1 inch). A well-prepared pond will provide the perfect environment.

Hatchery cleaning

In the early stages in the life cycle of a fish, it is at its most vulnerable. For the first 5–10 days it has no immune system. The skin mucus and gills are not fully developed, neither are many of the internal organs. This means that it is important not only to avoid contact with any pathogenic organism, but also that the fish are vulnerable to damage caused by excessive numbers of any bacteria or fungi associated with uneaten food and fish waste.

Because the fish are very small and undeveloped, they will have difficulty hunting for food, but they will require small amounts of food very frequently. Feeding regimes therefore need to include feeding lots of food at regular intervals; much of this will remain uneaten. In addition to this, the screens at the outlet of the fish tanks need to be very fine to prevent fish escaping. This will result in a build up of dead fish, uneaten food and fish waste in the tank. This must be removed very regularly. There are a variety of ways of achieving this but some patience and a siphon tube is the best method.

There is no such thing as natural mortality, just death by neglect. The hatchery must be kept perfectly clean throughout the time that fry are there.

Cyprinids are usually kept inside for around 2 weeks after they hatch. You may wish to stock these 2-week-old fry straight into the receiving water as seed fry. However, to give the fry a bit of an advantage, they can be on-grown. This can be achieved by maintaining the stock within the recirculation tanks and weaning them onto dry food. However, for restocking fish it is vastly preferable to introduce them into prepared earth ponds to on-grow.

12.2.3 On-growing in earth ponds

Two-week-old fry are relatively robust and will grow very quickly if supplied with a good food source. However, the fry are not good at hunting for food and are very vulnerable to predators. Even insects like water boatmen will eat them at this early stage. Therefore the on-growing ponds must have a lot of food present and be free of predators. The favoured first food of fry is rotifers. Once they start to grow they will quickly move up to larger zooplankton like daphnia and cyclops.

The pond should be shallow, going from 0.45 m (18 inches) to about 1.2 m (4 feet) at the deepest part. It should be empty through the winter so the bottom of the pond can dry out completely and no aquatic animals can inhabit it. About 4 weeks before the stocking of the fry, the pond should be fertilised; dry chicken manure is good. The pond should then be filled with freshwater, preferably from a borehole. This clean freshwater will have little or no life in it but it will have plenty of nutrients. It should get warm quickly because of the shallow water and with plenty of sunlight it will quickly develop a plankton bloom. This plankton bloom will provide an enormous quantity of food for any zooplankton, which will quickly multiply in the absence of any predators. The zooplankton bloom is usually dominated to start with by the very fast growing and reproducing rotifers. As this bloom reaches its peak is the best time to introduce the fish fry. There is an abundance of food, but none of the larger invertebrates have had sufficient time to move in or grow large enough to eat the fry. The fry will now start to clear the rotifers, allowing the phytoplankton to repopulate and attract the larger zooplankton such as daphnia.

It can be very difficult, but achieving this delicate balance is very important. Obviously the exact timing can be changed by prevailing weather conditions so it pays to be a bit flexible.

Given these conditions, the fry can grow extremely quickly and may consume all the natural food. A stock of supplementary dry food should be available if the pond goes completely clear and the natural food source dries up.

13 Invertebrate Survey

The invertebrate populations present in any water body can tell the manager a great deal about the quality of the habitat, not only at the present time but, because many larger invertebrates are very long lived, they can tell us something of the history of the still-water. So the invertebrate populations in a water body are particularly significant as they can be used as static, long-term indicators of pollution and other problems, such as organic enrichment. Invertebrates are relatively easy to sample, they can be long lived and they are variously sensitive to pollution problems, depending on species. They are the most widely used source of information on the quality of water bodies.

Sampling is generally straightforward and there are several methods, depending on the type of water. In still-waters, sweep sampling – moving through the water, a hand-net swept backwards and forward in front of the operator, taking care to include any weedy areas – is the main method. This process should be done close to the marginal edge and for around four minutes duration.

Benthic sampling is the use of grabs or nets to extract areas of the benthic substrate. This is then washed through a sieve to get rid the mud and silt.

The samples should be assessed while the invertebrates are still alive, as soon as possible after collection. All the invertebrates seen should be identified, usually to family level. An approximate count of the individuals in each family should be made to give an idea of the abundance and to determine the dominant family and community structure.

Table 13.1 Bands developed to give an outline of the biological condition of a water body, including water quality and habitat

<10	Unsatisfactory
10–25	Poor
26–50	Moderate
51–100	Good
101–149	Very good
>150	Excellent

These results can then be put through various scoring systems to allow comparison with other sites, and to allow the data to be interpreted and presented in an accessible way. The most widely used scoring system is the British Monitoring Working Party score, first developed in the 1970s. Each family of invertebrates is given a score (up to 10) based on its tolerance to pollution, particularly organic pollution. The families most tolerant of organic pollution score low, 1 or 2 (for example rat-tailed maggots); the most susceptible species score 10 (for example most species of cased caddis-fly larvae). The scores of all the families are added together to give a total British Monitoring Working Party score. The bands in Table 13.1 were developed to give an idea of the biological condition of a water body, including water quality and habitat.

14 Control of Water Quality

14.1 Water quality survey

It is extremely important to survey the water quality, but it is even more important to monitor the water quality regularly and frequently. This is because the biggest single cause of fish mortalities in still-water fisheries is poor water quality. Even where mortality, poor fish condition or poor catch rate can be blamed on some other cause, such as disease, there is often a water quality problem underlying the more obvious problem. Monitoring of water quality is vital to give an early indication of a problem, so that appropriate measures can be taken to prevent it getting worse and harming the aquatic life. The various physical and chemical factors influencing water quality are described in detail in Chapter 2.

14.1.1 Obtaining results

There are a host of techniques for recording water quality parameters, ranging from cheap and easy to complicated and expensive. Unsurprisingly these also range from inaccurate to highly accurate. For most fisheries the cheap, easy and slightly inaccurate models are perfectly adequate, as long as good records are kept so that it is easy to compare results across a range of time. Occasional samples can be sent for laboratory testing to give some parallel testing to confirm the results of the cheaper test kits.

For dissolved oxygen (DO) measurements, the only option is a hand-held electronic meter. These are expensive (costing about £200 for a simple hand-held meter), to up to £400 for more elaborate ones (2008 prices), but they are accurate and easy to use for routine monitoring. Recording DO fluctuations, during the summer especially, is such an important part of the management of a fishery that the cost of a good quality DO meter is well worth it.

A hand-held electronic pH meter, costing about £50 (2008 prices) will also include a good quality thermometer. Again, they are generally simple and easy to use, making them ideal for routine monitoring.

DO, temperature and pH should really be recorded every day, preferably at the same time of day, at least in the spring and summer months.

Measurements of ammonia, nitrite, nitrate and phosphate dissolved in the water can be achieved using test strip paper. These work by being held in the water for a proscribed amount of time. The test paper will change colour according to how much of that chemical is present, and then it is simply a matter of reading off a colour chart. These are useful tests but not very accurate or precise so the results need to be treated with caution. Testing once a week, however, will give you useful data and will help to identify any changes in the levels of these particular chemicals. Twice yearly (late spring and late summer) laboratory tests for these chemicals can also help to confirm how accurate the readings are.

Another parameter that may need recording is the total suspended solids or turbidity. Simply, this is the cloudiness of the water and can have a major impact on the ability of plants to grow in the water body; sudden increases can indicate a pollution incident. The only viable way of measuring turbidity routinely is using a Secchi disk. This is a disk of material with an alternate black and white pattern. This is lowered into the water (preferably between 10 a.m. and 2 p.m.) off the shady side of a boat. The depths at which it disappears from view on the way down and reappears on its way back up are recorded. The Secchi depth is the average of the two values.

The expected values for the main physical and chemical factors influencing water quality are given in Box 14.1.

14.1.2 Recording results

It is very important to record all the results you get so recording of all the water quality results in a diary is vital. In many cases the absolute value is of less importance than trends in values over time. One of the most useful aspects of monitoring water quality is describing and understanding the changes that occur during the seasons. This will help enormously in understanding how the lake is working. DO, temperature and pH should be recorded every day, particularly in the spring and summer. All the other parameters should be recorded once a week.

Box 14.1 Critical levels of various physical and chemical factors for maintaining the water quality of a still-water fishery

Parameter	Optimum levels	Dangerous levels
Dissolved oxygen	Above 8 mg/l	Below 6 mg/l
pH	Between 6.5 and 8.5	Below 5.5; above 9.5
Temperature	0–28°C	Above 28°C
Ammonia (total)	0 mg/l	Above 0.2 mg/l
Nitrite	0 mg/l	Above 0.2 mg/l
Nitrate	Less than 20 mg/l	More than 50 mg/l
Phosphate	0.1 mg/l	More than 0.5 mg/l

N.B.: a sudden change in any of the parameters is particularly dangerous. Routine monitoring is important to identify any sudden changes, particularly in DO, pH and ammonia.

14.2 Taking control

Once the fishery has been assessed and the monitoring programme for the water quality has been established, the fisheries manager has a good idea of what the fishery is like. Now we need to understand what can, and needs to be controlled and how we go about enforcing that control.

The control of water quality is a key area. It is often said that a fish keeper's real job is to keep the water quality good and the fish will look after themselves. Whilst this is not totally true in a fishery it is certainly the cornerstone of good fisheries management. In this section we will consider the two main areas of control: the control of DO and carbon dioxide through the use of aeration systems and the control of toxic substances such as ammonia and nitrites. It is important to recognise, however, that many of the habitat management techniques described later in the book will make a huge contribution to the overall quality of the water.

14.2.1 Aeration

Given that fish, in common with most aquatic organisms, are totally dependent on DO, it is perhaps appropriate to deal with this first. The circumstances that cause low levels of DO have been discussed elsewhere. It is clear that, although low DO is often a direct result of perfectly natural functions, where this impacts on a still-water the fisheries manager must take action to protect against it.

Monitoring

Before any action is taken a comprehensive monitoring programme should be in place as part of the overall management plan for the water body, as described in the previous chapter. Monitoring the DO content of the water body on a regular basis (ideally early every morning, over several months) will mean informed strategies can be put in place should there be a need for aeration. In the absence of a monitoring programme false assumptions may be made which do not represent the whole picture. Early morning DO levels are, for example, expected to be lower in the summer months, so a few measurements taken at this time of year can falsely represent conditions throughout the year. Information gathered during monitoring will guide the choice of an appropriate technical solution to the problems in any particular still-water, the type of mechanism to be adopted and the duration of its use.

The appropriate actions can be divided into two approaches; the first is to increase the amount of oxygen available to the system. Areas to consider might include:

- increasing the amount of do available by injecting air;
- managing the effects of stratification;
- establishing plants to enrich DO by photosynthesis.

The second approach is to consider reducing the biochemical demand for oxygen. Areas to consider might include:

- reducing the fish stock;
- reducing the biological oxygen demand;

The role of aeration

The aeration of freshwater can be crucial in keeping the DO content of the water at a level where it can sustain aquatic life under natural or management induced (over stocking) low oxygen conditions. If there is not enough DO in the water fish will ultimately suffer oxygen stress (the physiological effect of lack of oxygen) and in extreme cases the absence of DO in the required amounts can lead to mortalities.

Water can only hold a finite amount of oxygen, given a set of environmental parameters unique to a given water body at any specific moment in time. The amount of oxygen that can be held in solution (Dissolved Oxygen) is determined by atmospheric pressure, temperature and salinity (see Chapter 2). In a natural setting, oxygen is added to water by atmospheric diffusion at the surface, by wind circulation (augmented surface diffusion) and by photosynthesis (oxygen produced by plants, particularly phytoplankton). Where natural processes are insufficient to keep DO levels high, artificial means to augment oxygen levels will need to be considered.

Artificial aeration can aid in increasing usable habitat, preventing fish kills, and improving water quality. In some cases, nuisance algal blooms can be reduced, or a shift to less objectionable algal species can occur. However, aeration can be misused. It is not a 'cure-all' and it is important to understand what aeration can and cannot do for a water body, to avoid unexpected or unwanted results.

Aeration increases a still-water's DO level by forcefully circulating the water to expose more of its surface to the atmosphere or to expose the water column to diffused air or oxygen. Proper choice and design of an artificial circulation system depends on both the management objectives and the physical characteristics of the specific water body.

Aeration/circulation systems

Different aerators are used for different purposes and in this section we outline some of the more common forms of aeration along with their relative benefits and drawbacks.

Display 'feature' aerators

A common form of aeration in 'commercial' still-waters is to use fountains, waterfalls or other water features. This solution offers obvious aesthetic benefits while aiding in the circulation of water and increasing atmospheric diffusion at the surface. A simple fountain can, for example, reduce the possibility of algal blooms by keeping the surface of the still-water agitated. The use of display aerators are, however, somewhat restricted to ornamental ponds and to fishing ponds and lakes where the still-water is being managed for recreational angling.

Fountains or jets are primarily used to produce water surface aeration to the water. They agitate the surface and help control algae dispersion in the water column; they do not aerate the entire water column. This type of aeration looks attractive but as research indicates, is one of the most ineffective ways to keep freshwaters healthy. Oxygen does not get to the bottom where it is needed to organically break down the hydro-soil.

Display aerators are also commonly powered by electricity, although there is scope for harnessing wind power. This is due to their intrinsically ornamental purpose, which cannot be guaranteed through wind power alone.

Surface agitator (paddle wheel) aerators

Paddle wheels offer a simple solution to aeration problems and can be used in several situations. The concept is simple and works in much the same way as that used to operate paddle driven boats. Due to the design, paddle wheels also offer the advantage of increasing circulation in a still-water, particularly if stratification is a problem. They are therefore more commonly used in larger ponds.

Electric paddle wheel aerators are usually mounted on floats and anchored to the pond bank (Fig. 14.1). Large-diameter paddle wheels obviously transfer more oxygen than smaller diameter aerators, and flat paddles are less effective than other designs. For a given design, oxygen transfer can be increased by increasing paddle depth and hub rotation speed.

Figure 14.1 Paddle wheel aerator.

Increasing the diameter, paddle depth and speed can increase the power required for operation, but also the rate at which oxygen is introduced. Paddle wheels are not very energy efficient and are viewed as more of an emergency measure to increase DO content rapidly. They are, however, more flexible than other aerator designs and, because they are portable, they can be moved from pond to pond as needed and placed anywhere along the bank. Paddle wheel aerators are also available powered by butane gas; these units have the ability to be more mobile.

Wind-driven mixers

The wind can be harnessed in several ways to aerate still-waters driving both display aerators and deep circulation mixers. While display aeration techniques are used primarily to improve the aesthetic quality of still-waters, deep circulation mixers provide far more benefits and are used more appropriately in fishery management. A variety of wind driven mixers are available on the market (Fig. 14.2).

The principle of mixing the water column differs somewhat from other forms of artificial aeration such as the paddle wheel.

The system, which works on the principles shown in Figures 15.1 and 15.2, enhances both the water quality and the hydro-soil of the still-water, this decreases the biochemical oxygen demand (BOD), thus leaving more oxygen available for the living and breathing components of the aquatic environment, uppermost in the managers mind being the fish.

Deep mixing using equipment such as this can aid in reducing deoxygenation and algal blooms, including blue-green algae.

Deep circulation can greatly accelerate the natural aeration process, thereby significantly increasing the oxygen transfer rate. By bringing oxygen-depleted water to the surface, it can absorb oxygen from the atmosphere and from wind and wave action. This oxygen-rich water

Figure 14.2 Wind mixer.

is then drawn back down to the lower strata where it can most effectively break down waste products. Without 'deep' circulation a body of water can thermally stratify and, even in shallow water bodies that do not form a thermocline, the hydro-soil or bottom mud can become depleted of oxygen (anaerobic).

Without good circulation the upper warmer, lighter, oxygen-rich zone can separate from the lower colder, heavier, oxygen-poor zone. The lower or bottom zone contains foul-smelling, incompletely oxidised animal and plant matter, which gives off toxic gases, mostly hydrogen sulphide (H_2S) and methane (CH4), caused by anaerobic (without oxygen) bacterial action. A lack of oxygen has the effect of stopping many life forms, including facultative bacteria, from breaking down waste products into substances that are naturally absorbed into the aquatic environment as part of the natural cycle.

With deep circulation, positive changes in the body of water occur. Aerobic bacteria and other lower forms of oxygen-using aquatic life use up plant nutrients and waste materials, thus helping to clarify the water.

Air diffusers

The concept of using diffused air is a simple one, but can be comparatively costly and should be used in accordance with the management objectives of the still-water. The process involves a 'compressor' or 'blower' on shore, which delivers air through lines connected to a perforated 'leaky' pipe, or other simple diffusers, placed on or near the bottom of the water body.

The 'leaky' pipes or diffusers should be placed on the bottom to allow air bubbles to rise through the water column for maximum contact time with the water. The rising column of air bubbles also causes the cold bottom water of the hypolimnion layer to rise, pulling this water into the epilimnion (warm, surface water layer). It should be noted that thermoclines are generally only stable in water that is deeper than 10 metres, but chemical stratification can occur that impedes the diffusion of oxygen throughout the water column.

When the colder, hypolimnetic water reaches the lake surface it flows across the surface and eventually sinks, mixing with the warmer epilimnetic water. If the system is adequately powered and enough air is injected, this process continues and the metalimnion or thermocline (the transition zone between the epilimnion and hypolimnion) is broken down. Eventually, the entire lake becomes of nearly equal temperature with oxygen distributed throughout.

Most oxygenation occurs through the water's contact with the atmosphere (atmospheric diffusion) and relatively little of the increase in DO occurs through direct diffusion from the bubbles. This aeration technique is sometimes referred to as the airlift method of circulation, because bottom waters are 'lifted' to the lake surface through the action of the injected air.

The technique has many applications and can be used for fish transportation alongside compressed oxygen.

Effects of aeration/circulation

Aeration, and the increase in circulation, can have several effects on a water body, not all of them positive. In larger, deeper water bodies the aim is often to de-stratify the water column and, whilst this is desirable for most species, there can be negative implications.

The most common result of destratification is an improvement in DO levels. Destratification is generally considered beneficial for warm water fish. Fish require adequate DO levels and cannot survive in an oxygen-deficient hypolimnion. Fish preferring warmer water, such as carp (*Cyprinus carpio*), require a minimum DO concentration of 5 mg/l, whereas those preferring colder water, such as trout (*Salmo trutta*,) need 6–7 mg/l. Destratification allows warmer water fish to inhabit the entire lake, and enhances conditions for fish food organisms as well. However, because destratification warms the deep waters, some colder water fish species may suffer adverse effects, or be prevented from flourishing in the water body.

Another common effect of destratification is an improvement in water quality. Under anoxic conditions the hydro-soil may release metals (e.g. iron, manganese) and gases (hydrogen sulphide and methane). When the anoxic hypolimnion is eliminated, these problems are eliminated or greatly reduced.

The effects on phytoplankton (algae) are less predictable. Aeration and circulation may reduce algae through one or more processes:

- Algal cells will be mixed and enter deeper, darker areas, thereby decreasing the cells' time in sunlight and reducing their growth rate.
- Some algal species that tend to sink quickly and need mixing currents to remain suspended (e.g. diatoms) may be favoured over more buoyant species such as the more noxious blue/greens.
- Changes in the water chemistry (pH, carbon dioxide, alkalinity), brought about by higher DO levels, can lead to a shift from blue-green to perhaps more favoured green algae or diatoms.
- Mixing of algae-eating zooplankton into deeper, darker waters reduces their chances of being eaten by sight-feeding fish; hence, if more zooplankton survive, their consumption of algal cells also may increase.

Although algal blooms have been reduced in some lake destratification/circulation projects, in other lakes phytoplankton populations have not changed or have actually increased. In particular, it is less likely that complete circulation would result in any of the above benefits in shallow still-waters because algae are less likely to become light-limited there, and any changes in water chemistry would not be as pronounced.

Aeration also has the potential to reduce phosphorus concentrations in some water bodies. During summer stratification, when the hypolimnion is oxygen-poor, phosphorus becomes more soluble (dissolvable) and is released from the bottom sediments into the hypolimnion. This allows greater amounts of phosphorus to 'escape' into the epilimnion, even in stratified lakes which at times may partly mix. These increased phosphorus levels in the lake's surface waters can stimulate algal blooms.

To aerate or not to aerate?

The advantages of aerators becomes more apparent in larger water bodies where stratification may be an issue; in situations such as this, the circulation that an artificial device provides can be just as important as increasing the DO content.

If artificial aeration is considered the best way forward, then it is essential to seek professional advice when considering the installation and design of a properly sized aeration system. The first question to consider is whether the still-water can really benefit from a destratification/circulation (or other) aeration system. Other important questions also need answering before installing such a system, such as whether the system will operate all year or would summer and/or winter operation be most effective. Seek the advice of unbiased fishery professionals and do not limit the advice to just the one individual or company proposing to sell you an aeration system!

By examining the characteristics of the still-water with the goals of the management plan, you can then better determine whether artificial aeration is needed, and if so what type of system might meet your objectives.

Establishing plants to enrich dissolved oxygen by photosynthesis

Increasing DO by establishing more plants (Nature's way) should be the easiest and most obvious way to ensure that DO levels in your fishery are always above the minimal requirements. That is that there should always be enough oxygen for all the living components. Of course establishing plants is not always that easy and is addressed in Chapter 15. As with all techniques that use a living component, it will always be a matter of balance, and the fisheries manager needs to be aware that plants use oxygen as well as produce it.

The ecological approach should always be the first used to increase DO content within still-waters, before artificial solutions are sought. It is recognised that this is not always possible and problems such as stratification can still occur, prompting the need for mechanical remediation measures. Where possible, these measures should be as ecologically friendly as possible and as such the use of alternative energy sources such as wind power should always be considered more desirable.

Aeration of still-waters can potentially serve several valuable functions, promoting good environmental conditions. It must not, however, be seen as a 'cure all'. Increasing the DO content within a water body is indeed beneficial but to look at this parameter in isolation is to miss the bigger picture.

Reducing the fish stock

This is often a very difficult area for the fisheries manager to persuade his superiors to consider. It is based on the principal that by removing some of the fish the same amount of DO becomes available to fewer fish and other aquatic organisms. The decision should be based on a fish survey, combined with an analysis of water quality. If this indicates that the still-water in question is overstocked, then it certainly will be having an impact upon DO.

Reducing the biochemical oxygen demand

Oxygen is removed from the water by other organisms than fish and invertebrates. For example, bacteria in waters can lower DO. Excessive nutrients and dead organic matter can lead to a major explosion of bacterial numbers and result in the available oxygen being removed from the water very quickly. Where this occurs the quality of the inflowing water should be examined for excessive nutrient concentration and the management regime of the still-water should also be monitored for excessive build-up of nutrients. The bottom sediments (hydro-soil) trap a large amount of organic matter that is slowly degraded by bacteria and fungi. Healthy functioning hydro-soil will have a reduced call upon DO as it will decompose organic material through several efficiently functioning biological pathways.

Although bacteria occur in every type of habitat, and on most living and dead matter, in still-waters the bacteria that most influence oxygen concentration are those responsible for the process of decomposition of the hydro-soil.

In warm conditions, bacteria can quickly multiply, but they are much less active in cold conditions. In warm conditions, as bacteria increase, the amount of oxygen they consume rises. If sufficient food is available, bacteria can quickly reproduce to the point where they use large amounts of the available DO, leaving a reduced amount for the other aquatic plants and animals, including fish. In some instances a specific event, such as a fishing match, may introduce large amounts of food into the system in the form of bait and loose feed. The bacteria multiply and use up more DO, reducing it to levels that may have significant impacts upon other aquatic life.

If the amount of organic matter entering a water body is greater than the amount that can be broken down, it will accumulate in the lake. A healthy hydro-soil is essential to ensure that the bacteria function efficiently, breaking down the maximum amount of material per milligram of DO consumed.

14.2.2 Ammonia, nitrites and other pollutants

Water quality may deteriorate for several reasons, but it is often the levels of nitrites or ammonia that gives the first indication of a problem. There are usually two possible reasons for elevated levels of these toxic chemicals, and it is by understanding these that the manager can to deal with the situation effectively.

First, these chemicals may be on the rise because the nitrogen cycle (see page 2) has broken down. As described earlier in this book, part of the nitrogen cycle breaks down ammonia, the toxic but normal waste product of animal metabolism, to nitrite as the intermediate stage to nitrate. This process only occurs where certain types of bacteria are present to complete this chemical change. A breakdown in this process therefore means that either there are insufficient bacteria present, or that there are too many animals (usually fish) producing more ammonia than the bacteria present can cope with. There are two obvious solutions to this problem: the manager either has to reduce the fish stocks to a more sustainable level or increase the number of available bacteria to process the increasing levels of ammonia. Manipulation of fish stocks will be dealt with in Chapter 12. The bacteria prefer habitats with a firm substrate to grow on and a slow flow of water. This ideal habitat is provided in a still-water predominately on the surface of large emergent and submerged plants. Improvements to the marginal vegetation and the provision of weedy shallow areas helps enormously in providing sufficient bacteria to process the quantities of ammonia that can be produced by the fish in fisheries.

The second possible cause of a rise in ammonia is a sudden influx of water contaminated with ammonia or high in organic matter. This will lead to a sudden increase in decomposition activity by the various invertebrates, bacteria and fungi that breakdown organic material. Once this has occurred it is difficult to prevent severe problems, but fisheries can attempt to protect themselves from this sort of pollution incident. Any fishery with flowing water inlets should at the very least make themselves aware of what is upstream, be it just farm land or even industrial estates. In any case, the provision of a silt trap (section 23.2.1) to remove at least some of the suspended solids should be considered. Reed beds provide one of the most effective ways of reducing most toxic pollutants that are likely to come into a fishery. The creation of reed-beds is dealt with in detail in section 23.2.3.

15 Control of Aquatic Plants

15.1 Introduction

The chapter on aquatic plants (Chapter 4) showed that plants have evolved to be highly adapted to take advantage of both general and niche habitats. This can pose problems for the fisheries manager in controlling or reducing the growth of certain nuisance species. The plant's adaptability to varying environmental factors often means that there are a range of species that can become a 'nuisance' and that these vary from fishery to fishery; what is a problem in one may not be in another. It is when a plant becomes a nuisance, or impacts upon the management strategy, that it is labelled a 'weed'. The term 'weed' will be used here to mean a nuisance aquatic plant that potentially may require control.

It is essential for the manager to have a working knowledge of the biology of the target species, so that the most effective method of control can be applied. It is widely accepted that aquatic plants can be controlled by four methods, which are:

- biological;
- chemical;
- environmental;
- mechanical.

However, before taking action to eradicate any aquatic plant, its positive benefits to the fishery must be considered alongside any negative impacts. This is to ensure that the prescriptive response is proportional and will not upset the dynamic balance of the waters ecology. Thus the importance of aquatic plants should be identified and understood, especially the advantages of submerged plants, which are those often targeted by managers as a response to complaints from anglers.

It must be emphasized that total weed eradication will be unnecessary in 99.9% of cases, and would have a significant detrimental impact upon the ecology of the aquatic system. Control should therefore be limited to reducing growth and spread to within acceptable limits. Before a weed control programme is incorporated into a management plan, the desired result should be carefully considered and the plan should be tailored to deliver that desired outcome.

Before deciding upon the most appropriate control method, the factors that are causing the excessive growth should also be considered.

15.2 Biological control

Biological control is undertaken by introducing, or using, living organisms to control undesirable organisms. This can be direct or indirect. It is an attractive method in certain aquatic situations because:

- it avoids the use of chemicals (herbicides);
- long-term control may be possible, providing the introduced control organism can reproduce;
- once an introduction is made, it need not be labour intensive.

Biological control is rarely the complete answer. The control of submerged weed can be undertaken by using fish as the control method. Only one species, the grass carp (*Ctenopharyngodon idella*), is suitable for consideration for use in the UK. It is also possible to control weed growth by increasing the biomass of some species to such an extent that they effectively overgraze the animal and vegetable food source.

Using fish in this way can harness a secondary effect of the overgrazing that occurs when fish root for food ever deeper in the sediment; which is that silt is moved into suspension and as a result the water becomes more turbid. This reduces submerged plant growth, as the amount of light entering the water column, which is a vital component of plant growth, becomes less effective at promoting growth. This is, of course, an environmental consequence of a biological control.

15.2.1 Grass carp

Grass carp (*C. idella*) belongs to the cyprinid family; it is generally regarded as a true herbivore, although it is capable of adjusting its diet to that of an omnivore if its source of plant food becomes depleted.

It can grow to over 15 kg (33 pounds) in the UK, although a 4.5 kg fish is considered large, with an occasional 9.0 kg fish being reported. Often these growth rates would be on a diet of aquatic plants that in angling waters may be supplemented significantly with angler's bait. For various reasons the grass carp is not ideally suited for its role as a herbivore. Consequently it only digests 50% of the vegetation it consumes. It should also be noted that an indirect effect of stocking grass carp is nutrient enrichment of the water from their excrement. This enrichment may have the effect of creating algal blooms, which, although helpful in restricting light penetration to the weed, can also cause additional problems.

Experience suggests that optimum feeding occurs when water temperatures are maintained between 20–28°C, although effective grazing will begin at temperatures greater than 16°C.

It follows that typical UK weather patterns only allow effective feeding to occur over a relatively short period. This may well become extended in the future as a consequence of climate change. The degree of control is directly equivalent to the density (biomass) of grass carp stocked in water. There is no one stocking density that will guarantee results; however, stocking grass carp at 500 kg per hectare (500 pounds per acre) of fish that are in excess of 500 g (1 pound) could be expected to have a significant impact upon weed. Grass carp attract a high price in the fish supply market, so control by this means can be costly. As the fish grow, overly dense stocking levels can also result in an excessive degree of weed removal, where the submerged and emergent plants are grazed to the point of complete eradication.

An alternative strategy would be to stock smaller fish of about 0.1 kg (4 ounces) at 200 kg per hectare (200 pounds per acre); allowing the biomass to increase over a two- or three-year period to provide effective control. This method is less costly, but slower in effect. Both strategies may, of course, lead to a harvestable surplus once the weed has been significantly reduced.

A disadvantage of introducing small grass carp directly into a still-water is their susceptibility to predation. The cormorant (*Phalacrocorax carbo*) regularly takes fish of 0.5 kg (1 pound), and it is not unknown for fish as large as 1 kg (2 pounds) to be taken. The naïve nature of the grass carp to this predator can lead to high losses.

Submerged plants, such as Canadian pondweed (*Elodea canadensis*), water milfoil (*Myriophyllum spicatum*) and starwort (*Callitriche stagnalis*), are readily eaten by grass carp. Grass carp will eat filamentous algae (blanket weed), but only after the more palatable species of submerged weed and other food sources, such as angler's bait, have been consumed.

It should be noted that fish removal from larger ponds and lakes might pose considerable logistical problems, as grass carp are very adept at jumping the float line on a seine net. Grass carp will not breed in UK climatic conditions, so the stock will not be self-sustaining.

Grass carp are considered to be an 'exotic' or 'non-native' species by the UK Government. Permission to stock these fish is granted in suitable circumstances under the Import of Live Fish Act, the Wildlife and Countryside Act and Section 30 of the Salmon & Freshwater Fisheries Act (1975). Regardless of consent, the fisheries manager should also be aware of the ecological reasons for this precautionary approach to stocking species listed under the Import of Live Fish Act. Not only is there a potential risk that these non-native species can assert a pressure on native species, outcompeting them for their niche, but a significant risk also lies in the fact that the stocking of non-native species brings with it the associated risk of the introduction of novel parasites. The impact upon our native fish stocks can be significant as, unlike the non-native fish, our native species will have no natural immunity that comes from co-evolution and adaptation.

15.2.2 Control over plants by overgrazing using carp

Common carp (*Cyprinus carpio*), including mirror, leather and wild carp varieties, also belong to the cyprinid family. They grow to an excess of 25 kg and now frequently occur at this weight in the UK. Although carp will graze on submerged plants, it is not a herbivore and has a truly omnivorous diet. It spends much of its energy grubbing in the hydro-soil for small organisms, including worms and other invertebrates. In angling waters its favourite food is angler's bait, preferably those morsels not attached to a hook!

Although the carp consumes plant material by grazing new shoots, it is the grubbing action of feeding that has an indirect 'environmental' effect on weed control. Carp feed very actively as the water warms up in spring; this high rate of feeding continues throughout the summer. The process of feeding in the hydro-soil may result in large volumes of mud or silt particles becoming suspended. The initial result is a loss of water clarity. Light's ability to penetrate water decreases with decreased clarity and this will considerably reduce the growth potential of submerged plants.

A further consequence of this feeding activity may be that an increased amount of stored nutrients will be released from the bottom mud (hydro-soil). The nutrients released will have a similar effect to those discussed for grass carp. Unicellular algae will readily take up nutrients, and as these organisms can reproduce very quickly the effect is to produce an 'algal bloom'. The water will turn green in colour and may further reduce the penetration of light to the submerged weed.

The degree of control is directly equivalent to the density (biomass) of carp stocked into a water body. As with all fishery matters no two waters are the same, and there will be a certain amount of trial and error required (with caution).

Optimum feeding occurs when water temperatures are maintained between 18 and 25°C, although effective grazing will begin at temperatures greater than 12°C. It therefore follows that typical UK weather patterns will allow effective feeding to occur from late March to late October. This period is likely to be extended as the effects of climate change become more pronounced.

This biological switch on a fishery's ecology requires a high level of biomass, which may be equivalent to environmental vandalism, so although this is a tool for the toolbox, the manager should think very carefully before going down this route to control weed.

Biomass of carp or bream at 1,000 kg per hectare (1000 pounds per acre) of fish could be expected to provide significant control. This rate of stocking would be expensive as carp and bream attract a buoyant price in the fish supply market.

Biomass levels of 400 kg per hectare will impact on plant growth in a silty bottomed still-water as the fish stir bottom sediments and cause enough turbidity to prevent submerged weed growth. Although the biomass is lower, there will still be a considerable impact upon the wider ecology of the pond.

Compared with other freshwater fish, for example bream, carp are very hardy and physically strong. Dependent on their genetic strain and environmental conditions, their life expectancy varies between 20 and 50+ years.

A disadvantage of introducing small carp or bream directly into a large water body is their susceptibility to predation, especially by the cormorant. As discussed carp are very strong and display good survival rates in cormorant predated water when stocked at above 0.5 kg (1 pound) in weight.

Owing to their popularity with anglers as a sport fish, there is a ready market for their disposal should their effect or their biomass become excessive. It should, however, be noted that fish removal from large ponds and lakes can pose logistical problems. Carp and bream will breed successfully in UK climatic conditions, so the stock may be self-sustaining, although in heavily stocked waters recruitment is often poor.

Carp are considered to be a 'native' species by the Environment Agency in that they have been in the UK for several hundred years.

There is an ever-increasing assumption, however, against allowing stocking into fisheries that are connected to the 'wild' as escape of the stock is a real threat to the wider environment (See Chapter 12). Furthermore, stocking at these very high levels may be stopped because of ecological considerations. Permission to stock must be obtained from the Environment Agency under Section 30 of the Salmon and Freshwater Fisheries Act (1975).

15.2.3 Biological control of algae using barley straw

Biological control over the growth of filamentous algae (blanket weed) and blue-green algae (cyanobacteria) can be achieved by using 'cereal' straw, in particular that of barley. The effect is not yet fully understood, although the mode of action is known to include the following biological effects.

The straw can act as a medium for bacteria. The bacteria help the decomposition of the straw and favour organisms that compete with the algae for the available nutrients.

The straw is also thought to produce an algicide (a chemical that kills algal cells). However, even this description is not strictly accurate as it well may be an inhibiting chemical (affecting growth and/or reproduction of cells) as opposed to an algicide. Whatever the mode of action, properly applied barley straw can be very effective in the treatment of algae and is effective in perhaps six out of ten applications; the reason for success or failure is equally difficult to quantify.

Barley straw offers the following advantages over other methods:

- it is inexpensive (dependent on cost of labour);
- it is environmentally acceptable (providing it is removed when spent);
- it is effective in both large and small water bodies.

As with all matters pertaining to the management of the aquatic environment, a basic understanding of the principles will help those who are responsible for the implementation of management strategies. Some important principles for applying barley straw as a management tool are shown in Box 15.1.

When barley straw is placed in water near the surface (in the presence of oxygen), the decomposition process begins immediately. It is at this stage that chemicals that have an inhibiting effect on the algae are produced and released. The process of decomposition is microbial; it is also temperature dependent.

Box 15.1 Rules for application of straw for algal control

The application of straw is relatively simple given a basic understanding of the process. Experience has shown that some simple rules will optimise its effect and avoid its misuse.

- Barley straw should be used in preference to other straws such as wheat.
- Straw will not kill existing algae.
- It should be used early in the season to inhibit algal growth.
- The decomposition process requires a good supply of oxygen.
- It should not be applied in the form of bales, but loosely packed nets.
- The higher the turbidity of the water, the more straw that is required.
- The correct quantity of straw should be used.

It may take 6–8 weeks for straw to become active when water temperatures are below 10°C, but much quicker, perhaps days when the temperature is above 20°C. The straw is likely to remain active for three months after the process begins. To ensure this, it is important that the straw is not applied in bales, as most of the tightly packed strands will not have the benefit of the movement of well-oxygenated water through to them. It should be split and placed in loose nets. The type of packaging used commercially for Christmas trees is very suitable, especially for larger applications.

The application of straw is relatively simple, given a basic understanding of the process. Experience has shown that some simple rules will optimize its effect and avoid its misuse.

As discussed above, the application of straw into still-waters is best done by breaking bales into loose straw and repackaging it using commercial Christmas tree packaging net, and packaging machinery. The straw nets are made up into 'sausages' that can be as large as 20 m long and would typically contain 40 kg of straw.

The net should incorporate some form of flotation, as the straw will tend to sink as it becomes waterlogged. The sausages when first constructed are very buoyant; this helps considerably with the siting of the straw sausages. Mooring the straw sausages from one end only will allow them to move with the wind and should be considered when the surface area allows.

In still-waters that have multiple uses, consideration will have to be given to anglers and other users of the amenity, such as boaters. In this case, practical consideration may determine the placement and number of sausages eventually sited.

All straw sausages that are spent should be removed to avoid deoxygenation and nutrient enrichment of the still-water as the decomposing process continues without the initial benefit of inhibiting the algae.

Dosing with barley straw

The amount of straw required is dependent upon the surface area: the volume of water is ignored for the purpose of the calculation. The dose rate can vary between 5 g/m^2 in a lightly colonized or clear lake to 50 g/m^2 in a heavily colonized or turbid lake. To obtain the weight of straw required, the dose rate is multiplied by the surface area. As an example, a lake of

3.5 acres has a surface area of 14 175 m^2; if a dose rate of 40 g/m^2 was decided upon, this would give the following calculation:

$$14\,175 \times 40 = 567\,000 \text{ g} \div 1000 = 567 \text{ kg}$$

A small rectangular bale of straw weighs about 20 kg.

$$567 \div 20 = 28.35 \text{ (so 28 bales are required)}.$$

A large straw sausage (Christmas tree packaging) of say 20 m in length takes approximately 40 kg of straw. The number of sausages to be placed in the lake would be:

$$567 \div 40 = 14.18, \text{ therefore 14 sausages are required.}$$

If smaller sausages were felt to be more appropriate, say 5 m long, we know from the figures above that they would require 10 kg of straw to fill each, therefore:

$$567 \div 10 = 56.7, \text{ so 57 sausages are required.}$$

The area of water that will be treated by each net at the prescribed dose rate is calculated by:

$$10 \text{ kg} \div 40 \text{ g/m}^2 = 250 \text{ m}^2$$

Using πr^2 = area of a circle, the diameter of a circle of water that will be treated can be calculated:

$$\pi r^2 = 250$$
$$r = \sqrt{250 \div 3.142}$$
$$r = 5.03$$
$$r \times 2 = 10.06 \text{ m (the diameter of the circle).}$$

The diameter of a circle of water that will be treated by one 5 m net of straw is an ideal; the placement of nets does not need to be exact. For example, it may be required to leave the areas within 40 m of an angling swim station on the bank clear to allow uninterrupted angling.

Experience has shown that figures close to 40 g/m^2 are closer to an average effective dose rate.

The cost of barley straw is not given, in the knowledge that historically the cost of straw is market dependent and can fluctuate greatly on an annual basis. The machine to aid the placement of the straw into the sausage nets is commercially available; there are several manufacturers that supply the basic machine, which costs a few hundred pounds.

It should be noted that an additional effect of straw treatment may be to inhibit the growth of green unicellular algae. This would have the effect of making the water less turbid, which would increase the light entering the water column, which in turn may have the effect of causing a biological switch, affecting the dynamic balance of the ecology of the fishery. In some circumstances this may encourage the establishment of a noticeable increase in the size of blue-green or filamentous algal communities for example. The manager, as always, should be cautious before undertaking any actions, particularly those that may have an effect that is difficult to reverse.

The eradication of algae effectively destroys the base of the food web, so control should be proportional and selective to control only the problem species.

Barley straw concentrate

Barley straw concentrate is simply a liquid that has been distilled from barley straw. It is processed to breakdown the facultative component of the straw, allowing it to be decanted from a vat and sold in plastic containers for direct application to the fishery. It is commercially available. This can be a cost-effective alternative to straw sausages, as it requires far less labour, although application is not without effort. Like barley straw, it works in some waters and in some it does not; generally it does not work where straw did not.

15.2.4 Control of Azolla using weevils

Azolla, also commonly known as 'fairy moss', can be controlled by the introduction of a weevil that grazes on it. Work is still experimental and experience has shown that the population can have difficulty in overwintering. When a robust population is established it can exert stunning control over the plant. The risk with all biological control using living species is that the organism can adapt and evolve over time to have other significant environmental impacts not previously considered.

15.3 Chemical control

The chemical control of weed is performed with an ever-decreasing range of approved herbicides. The number of chemicals available to the fishery manager is always under pressure as environmental standards and the protection afforded the environment increases. The still-water fisheries manager effectively now has the choice of one herbicide, not really a choice at all. It is probably unavoidable that, sooner rather than later, the use of chemicals to control nuisance plants will become even more difficult, as the remaining aquatic herbicide approved is removed and consent for its use is tightened. The use of such chemicals (herbicides) in aquatic situations is strictly controlled; permission must be sought from the Environment Agency for their use.

As discussed, the eradication of vast amounts of vegetation, particularly submerged species, may lead to deoxygenation problems as subsequent die-off and decomposition takes place. It is for this reason that treatments are normally performed in spring, when growth has just started, plant respiration is not at its greatest and foliage is at a minimum. Climate change is inflicting pressure to treat nuisance plant growth both earlier and later in the year. Always remove treated plant material to ensure that it is not left to decompose in the water;

where this is not practical, then only treat 20% of the area at one time, allowing the plant material to fully decompose before performing more treatment.

The use of aquatic herbicides is subject to the requirements set out in the 'Guidelines for the use of herbicides on weeds in or near watercourses and lakes' (available free from Defra).

Unfortunately the last of the suitable herbicides to treat these types of weed has been removed from the list. The remaining herbicide is suitable for emergent and certain floating plants.

15.3.1 Glyphosate

The chemical glyphosate is also known by its trade name: Roundup Pro Biactive.

Glyphosate is a very effective, approved aquatic herbicide that gives consistent results when treating susceptible species. A general description of the product is given below.

The active ingredient in Roundup Pro Biactive is glyphosate. For any herbicide to function, it must first break through the waxy surface of the leaf and enter the plant cells. This is achieved by inclusion of an adjuvant, or surfactant, which breaks down the waxy cuticle. Roundup Pro Biactive contains two adjuvants that work in combination, making it more effective. Glyphosate accumulates in actively growing parts of a plant. Glyphosate controls weeds by inhibiting a plant's enzymes. These particular enzymes are present in plants, but not in humans, animals, birds or fish. It can provide good long-term weed control with use over a range of applications and weather conditions.

Table 15.1 is taken from the manufacturer's instructions.

Table 15.1 Roundup Pro Biactive: recommended applications

A. By weed type

Target weed	Amount of Roundup Pro Biactive
Emergent plants: reeds, grasses, rushes	50 ml per 100 m^2 to be treated, mixed in 2.0–4.0 litres of water volume.
Floating plants: water lilies	60 ml per 100 m^2 to be treated, mixed in 4.0 litres of water volume.
Cut tree stumps	
Deciduous trees	10% in a solution of water.
Coniferous trees	20% in a solution of water.
Chemical thinning	2 ml of neat chemical per 2 cm of diameter.

B. By equipment

Description	Application rate	Water volume (l)
Enclosed still-waters and land immediately adjacent to aquatic areas.		
Hydraulic sprayer: emergent weeds		
Boom sprayer	5.0 litres per hectare	200–400
Knapsack sprayer	50 ml per 100 m^2	2.0–4.0
Hydraulic sprayer: floating weeds		
Boom sprayer	6.0 litres per hectare	100–200
Knapsack sprayer	60 ml per 100 m^2	4.0

Table 15.2 Susceptible species to roundup

Susceptibility	Species
High	Arrowhead (Em) Water-cress (Em) Water dock (Em) Yellow water-lily (Fl) Amphibious bistort (Fl) Branched bur-reed (Em) White water-lily (Fl) Flowering rush (Em) Bulrush (Em) Club rush (Em) Marsh marigold (Em) Reed canary-grass (Em) Soft rush (Em) Sedges (Em) Common reed (Em) Yellow iris (Em)
Moderately susceptible	Duckweed (Fl) Water horsetail (Em) Water-plantain (Em) Water-dropworts (Em) Lesser water parsnip (Em) Broad-leaved pondweed (Fl)
Low or resistant	All submerged weeds and algae

Em, emergent; Fl, floating; Sub, submerged.

Partial or area-selective treatment is possible with glysophate. However, in practice the herbicide does have a tendency to 'drift' in the wind onto areas adjacent to those being treated. For this reason, it is recommended that areas that should not be contaminated are avoided by only spraying in calm conditions.

The fisheries manager should also be aware of the use of glyphosate to treat trees that have been managed (felled or reduced), to inhibit regrowth; or indeed it can be used to thin trees chemically.

A list of susceptible species is given in Table 15.2.

15.3.2 Chemical control of algae

The chemical control of algae is possible, but generally considered to be less than reliable, and very expensive.

In the USA and mainland Europe, filamentous algae are often controlled by the addition of copper sulphate. Unfortunately, this is not approved for use in the UK. There is no longer a herbicide on the approved list that can successfully kill species of filamentous algae.

It is undeniable that the use of herbicides to control weed has many worthwhile applications. However, it must be stressed that unless the root cause is also addressed, it may well prove that the control of one problematic plant merely opens up a niche for another.

15.4 Environmental control

Plants need three basic requirements to grow. These are light, nutrients and oxygen. The fisheries manager who can reduce any one or more of these factors will be able to inhibit the growth of the invasive plant to some degree. It is not always a straightforward proposition, as any prescription to treat nuisance plant growth by manipulating environmental factors may affect another significant part of the ecosystem.

Tree planting on marginal areas of a still-water can create areas of shade that limits plant growth locally, although this technique is perhaps more suited to rivers than larger still-waters.

Light can also be excluded by inserting a sheet of black polythene or geotextile over areas of submerged weed growth. This may not be a suitable method when the weed covers a large area or where there is significant flow.

The control of nutrients is a technique that has gained in popularity over recent years and there are many well-documented successes. Most have been in response to human degradation of a valuable water resource and ecosystem. The response of managing an entire catchment to reduce nutrient loading is likely to be prohibitively expensive. It will require in-depth scientific analysis of the water body and its surrounding catchment (the area where its water comes from).

The understanding of where excessive or high nutrient loads may come from can help to make common-sense judgements, without the expense of numbers of soil samples from around the entire catchment. Usually nutrient management at basin level goes far beyond the responsibilities of managers of individual fisheries and forms part of river basin catchment planning involving many interested parties. However, the fisheries manager should be able to highlight areas of concern and participate in the formulation of river-basin plans. This means the manager should be aware of any major problems in their catchment and of the approaches to resolving them. These are often likely to be run-off based and can be managed by various techniques, which include the following:

- building a bypass channel;
- change of agricultural practice (land use);
- locking-up of the limiting nutrients;
- planting of buffer strips;
- establishing aquatic plants, especially reed-beds and plants growing hydroponically.

Environmental control can also take the form of 'deep mixing'. Deep mixing is performed by pulling water from the bottom of the water column and introducing it back at the top just below the surface. This is shown in Figure 15.1.

The movement of the water requires that the water is pumped, which can be powered by electricity or (as shown in Fig. 15.2) by wind. Deep mixing has the effect of keeping algae distributed throughout the water column, effectively disrupting their life cycle.

15.5 Control of algae using ultrasound

A relatively recent development has been the use of ultrasound to control algae, particularly varieties such as filamentous and blue-green algae. Both can and do have significant impacts

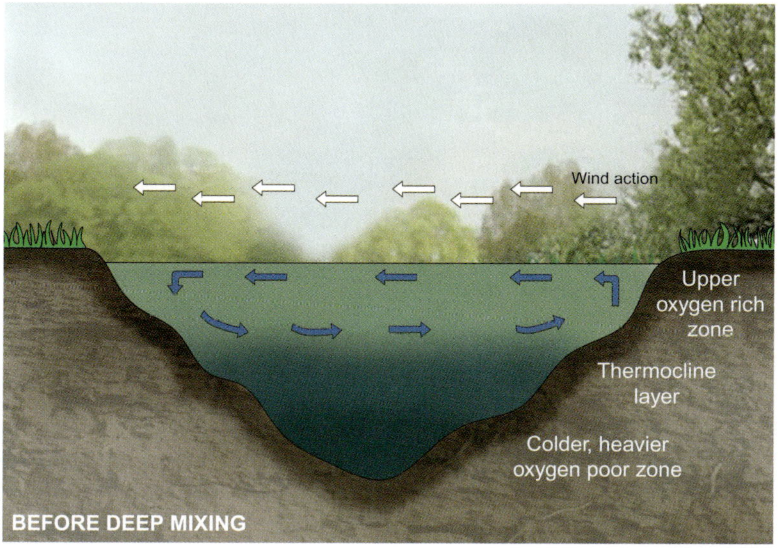

Figure 15.1 The principle of deep mixing.

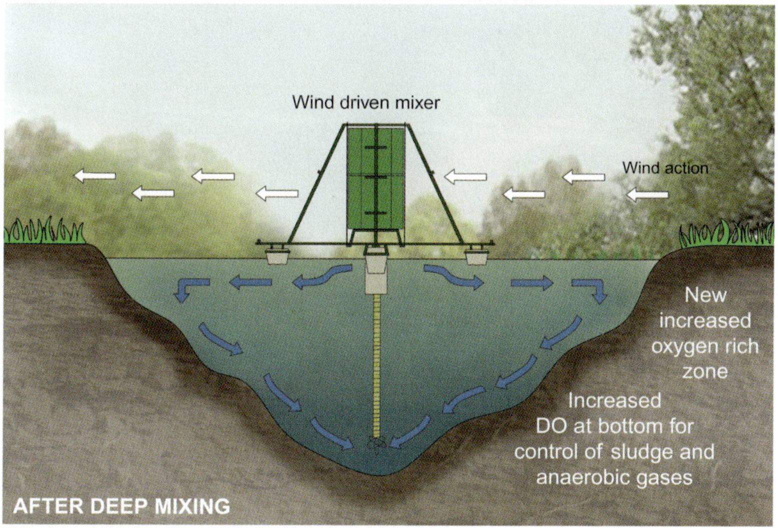

Figure 15.2 The effects of deep mixing.

on still-water fisheries. Before considering ultrasonic control, it may be worthwhile to consider the nuisance that these two types of algae cause.

15.5.1 Filamentous algae (blanket weed)

The nuisance from filamentous algae, commonly known as blanket weed, is largely from an amenity point of view. The filamentous algae can block large areas of what would otherwise

Figure 15.3 Blanket weed (filamentous algae).

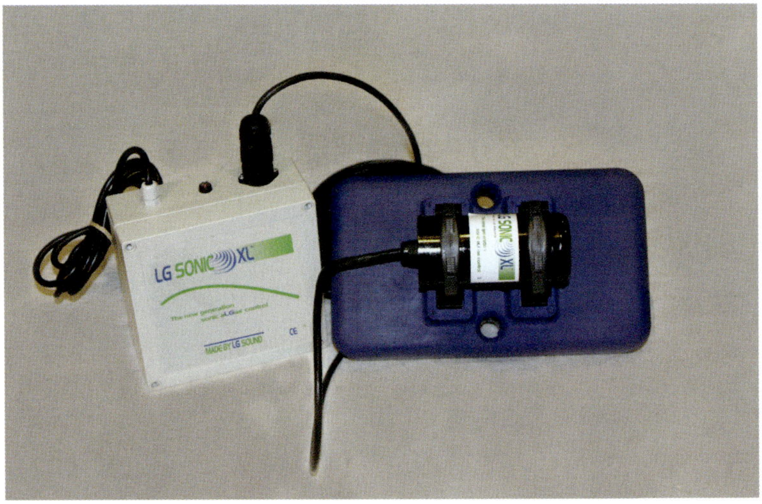

Figure 15.4 Ultrasonic equipment.

be 'open' water, and can tangle fishing gear (Fig. 15.3). It shades out light from the water below. Dieback and subsequent decomposition add to the nutrient reserves, increasing eutrophication. Situations appear to be increasing where filamentous algae become established and then dominate algal communities in still-waters.

15.5.2 Blue-green algae and the need to ensure public health

The potential risk from blue-green algae to human and animal health is dealt with in section 3.2.1.

Using advanced ultrasonic technology, a submerged transducer (Fig. 15.4) transmits ultrasonic waveforms specifically to target the algae cell structure. A complex pattern of ultrasonic vibrations through the water cause the cell vacuole to rupture. Whether floating or rooted, the single cell is destroyed, rendering the algae dead.

The use of ultrasonic waveforms is an environmentally friendly method of effective control. It can negate the nuisance of blue-green (cyanobacteria), filamentous algae (blanket weed) species and other nuisance algae. Ultrasonics is safe and is believed to present no negative impacts to health or the environment.

15.6 Mechanical control

There are many types of tool, both manual and automated (engine powered), that have been designed to cut weed. Cutting and removal of weed by pulling is obviously very effective: what you see on the bank is what you have removed. Efficiency and effect can vary greatly; for example, it can be very effectively and efficiently done with a professional weed-cutting boat, whereas cutting weed by hand can be very time consuming and far from efficient. The deciding factor may often be the scale of the task. It would not be appropriate to use a boat to cut small areas in front of angling stations perhaps, whereas it may be the only mechanical solution in wide, deep still-waters.

15.6.1 Manual weed cutting

In its simplest form, weed cutting by hand is done with a Turke scythe (Fig. 15.5) which in experienced hands is a very effective tool. It is perhaps better suited as a tool of the river keeper and to some extent depends on a walking action as part of the efficient rhythm of use.

Another form of manual cutting is with a chain scythe (Fig. 15.6); this method requires two people to operate it. Again, it can be effective in experienced hands. Of course, the more effective the cutting, the more cut weed needs to be removed. It may be best to allow as much time to remove the weed with a seine net as that allowed to cut the weed.

A disadvantage of cutting, which is sometimes overlooked, is that many aquatic submerged plants are easily propagated by fragmentation and, of course, cutting does exactly this to the plants, often making the infestation worse.

15.6.2 Weed pulling

In principle, raking as a method of removal as opposed to 'raking up' after cutting is a more efficient technique than cutting because, if it is done well, it will remove the root system, greatly inhibiting regrowth.

Electric scythe

This 12-volt DC, battery powered scythe, which is boat mounted (Fig. 15.7), is a useful tool for small waters where the fishery manager wishes to control small areas of submerged weed mechanically.

Figure 15.5 A Turke scythe.

Figure 15.6 A chain scythe.

Figure 15.7 A boat-mounted 12-volt reciprocal cutter.

Figure 15.8 A weed boat.

15.6.3 Weed boat

If mechanical control is the chosen method and the size of the water warrants it, then investment in a weed boat should be considered. Planning the use of a weed boat should take into account that cutting may need to be done on an increasingly regular basis, working on the principle of 'the more you cut it, the more it grows'. Modern weed boats not only cut the weed, they also harvest it (Fig. 15.8).

16 Control of Erosion

16.1 Introduction

Erosion is the displacement of solids (sediment, soil, rock and other particles), usually by natural forces (such as, water, wind or ice) by downward, or down-slope, movement in response to gravity. It is an intrinsic natural process but in many cases it is increased by human activity. This is particularly true when considering the inputs of sediment to the aquatic environment. Freshwater systems that are managed as an amenity will generally experience a greater rate of erosion than under natural conditions. It is important to understand that erosion rates are exacerbated where vegetation cover has been reduced. This disturbs both the soil structure and plant roots that would otherwise hold the soil in place, so we need to understand erosion control alongside the vegetation control talked about in the previous chapter. However, improved management practices can limit erosion, using bank stabilisation techniques such as bioengineering and turf reinforcement.

A certain amount of erosion is natural and healthy for the ecosystem. Excessive erosion, however, does cause problems, such as receiving water sedimentation, ecosystem damage and outright loss of soil.

Anthropogenic processes have undoubtedly increased sediment loadings in water bodies, often to the detriment of the indigenous flora and fauna. Changes in agricultural practice (winter planting of crops, ploughing to the very top of the bank, overgrazing by stock), coniferous forestry (increased drainage, construction of forestry roads) and increases in the areas of 'hard' landscape (property developments, roads) have all contributed to an increase in sediment inputs into aquatic systems.

Impacts of increased sediments through erosion are manifold. In all systems, they have the potential to smother benthic invertebrates, reducing their number and diversity. Increased erosion rates may also cause significant changes in macrophyte flora, favouring silt-tolerant species such as starwort (*Callitriche* sp.) at the expense of species such as hornwort (*Ceratophyllum demersum*) and milfoil (*Ceratophyllum demersum*).

This chapter will look at some of the most common causes of sediment introduction and bank erosion in still-waters and offer methodologies to control the displacement of solids and reduce erosion rates. These methodologies will focus on 'soft' bioengineering techniques, as opposed to 'hard' revetment methodologies which, although in many cases are effective, have proved to be unsustainable in an ecological context. Where the overarching aim is to improve the management of fisheries, consideration must be given to the wider ecology of the site, and erosion and sediment control are essential elements of the management process.

16.2 Causes of erosion and sediment introduction

The causes of erosion in still-waters can be broadly split into two distinct categories, natural and human. Natural causes of erosion, although often acknowledged as serving valuable functions, still have the ability to negatively impact on the aquatic environment. Human causes of erosion can largely be seen as increasing nature's susceptibility to erosion through poor planning or management.

16.2.1 Natural causes

The rate of erosion depends on many factors, including the amount and intensity of precipitation, the texture of the soil, the gradient of the slope, ground cover from vegetation, rocks and land use. The first factor, precipitation or rain, is the agent for erosion, but the degree of erosion is governed by the other factors. Wave action is a further important natural cause of bank erosion.

In general, given the same kind of vegetative cover, one expects areas with high-intensity precipitation, sandy or silty soils and steep slopes to be the most erosive. Soils with a greater proportion of clay, which receive less intense precipitation and are on gentle slopes, tend to erode less.

Depending on the topography of the area, sediments can in some instances be traced back over large distances. These are usually fine particles that get carried in suspension as overland flow after precipitation events. The fine sediment eroded from the soil is washed down slopes, where it settles at the lowest point in the landscape. Lakes and ponds are often positioned in low topographical areas in the landscape, being fed either directly by rivers or streams, by overland flow in the form of run-off or from groundwater. This makes lakes and ponds particularly susceptible to this form of sedimentation. In some instances this does not directly contribute to bank erosion, it impacts on the aquatic environment by contributing sediment to the system. However, more often than not, high levels of overland flow will directly erode banks as the water finds the quickest path to the lowest point. Areas of weak substrate will naturally be carried with it, forming rills and gullies which feed sediment directly into the lake.

16.2.2 Human causes

Although human structures, development and interference can cause direct erosion at specific sites, more often than not humans simply increase the susceptibility of the landscape to erosion. This can be through poor planning and management of the wider environment which, in a still-water context, is often the result of the removal of vegetation, agricultural practices or the modification of watercourses.

In the past, little consideration was given to the setting of man-made or artificial lakes in the landscape or to the local flow of water throughout the hydrological cycle. Poorly planned artificial still-waters can suffer from a whole host of problems including excessive sediment inputs and erosion. It is now understood that an appreciation of the geology of the area is essential so that management strategies can be put in place (where necessary) to control potential erosion issues.

Thought must also be given to the profile of the bank; incorporating the positioning of the angling stations, marginal vegetation and access. Ideally, the profile of the bank should be gently sloping to reduce the risk of erosion more commonly associated with steep-sided banks. Marginal plants also play a part in stabilising the bank, and banks should be profiled to accommodate this vegetation.

A shallow shelf also has the advantages that it promotes marginal aquatic plant growth and acts as a trap for sediments carried into the system by run-off. Both of these aspects considerably reduce erosion caused by wave action as they act as a damper to reduce the power of the waves.

Existing fisheries often suffer from bankside erosion due to members of the public trampling the sides of the banks while angling, especially when landing fish, which reduces vegetation cover and makes the banks more susceptible to erosion. This is a classic example of where erosion can be directly attributed to human interference and, although the solution is simple, controlling human activity is not. By providing designated angling areas (angling stations) around the perimeter of the lake that are easily accessible from a perimeter path and set back from the banks of the lake, trampling of the margins can be easily reduced, although seldom eradicated.

The impact of waterfowl on banks should not be underestimated, especially species of geese such as the Canada goose, which disturb the banks when they enter and leave the water, as well as fouling the surrounding bankside.

16.3 Control methods

Controlling erosion and sediment input into a still-water largely depends on site characteristics. Where a still-water is part of a flowing system, with inflowing tributaries and discharge channels, control over inputs is a problem because upstream bank erosion and sediment transport can effectively turn still-waters into large silt traps. Although the control measures outlined here are good at reducing erosion and controlling sediment inputs, it is worth acknowledging that there is always a potential for problems where inputs of sediment cannot be controlled 'at source'.

The bioengineering methodologies outlined in this chapter are based upon providing green solutions to modern problems. However, it is sometimes necessary and indeed advantageous to combine 'soft' bioengineering techniques with more accepted 'hard revetment' measures. The main considerations when applying bank stabilisation and erosion control techniques must be the natural course of the waterway, the source of sediment or erosion and how that erosion can be controlled and managed in a sustainable way, although there are also other factors that must be considered.

16.3.1 'At source' erosion control

By far the most desirable solution to erosion is to control it at source, before sediment is transported to the aquatic environment. Such solutions involve changes to, and management of, the terrestrial environment immediately surrounding the still-water. The terms 'habitat enhancement' and 'habitat restoration' are often used in fisheries management, primarily when discussing the management of the bankside. The importance of habitat type should not, however, be confined to this bankside zone and instead should encompass all riparian and terrestrial habitats in close proximity to the still-water.

By including the local land management in the fishery manager's remit, the inputs to the aquatic environment can be understood and managed, providing a more holistic approach. It is true that the fishery manager may not always have an input in local land-use decisions, but where a fishery is in a relatively secluded location, or there is scope to influence local land use, then the immediate terrestrial environment must be given due consideration.

Bank erosion rates, overland run-off and sediment inputs into the aquatic system can be dramatically reduced through changes in the immediate environment surrounding the still-water. Significant benefits in reducing run-off of sediment-laden surface water can accrue through changes in habitat and local forestry/agricultural practices including the following:

- allowing permanent and undisturbed vegetation cover;
- where possible removing, impounding or attenuating drainage systems to hold back run-off;
- the addition of silt traps to all drainage systems that run into the still-water;
- controlling run-off from local tracks and other roads;
- reducing agricultural stocking rates.

In many cases, these changes will be beyond the immediate influence of fisheries managers. However, beneficial changes to local land use should be identified in the fishery management plan. Managers can then seek to modify damaging land use practices through the mechanism of government environmental protection initiatives such as the River Basin Management Plans which operate under the Water Framework Directive in England.

16.3.2 Bank erosion control

Marginal/bank side planting

The benefits of marginal vegetation have long been advocated as an erosion control measure, as well as providing valuable habitat above and below the waterline and removing excess nutrients from the water. Plants help to stabilise unconsolidated soils because the roots effectively holding the soil together, thus reducing erosion rates. Vegetation over the entire site is beneficial for slowing water velocities and encouraging infiltration into the soil rather than overland run-off. This is never more pertinent than in the riparian/marginal zone, where an abundance of emergent plants will reduce bank erosion and trap sediment inputs from the run-off.

Marginal plants not only consolidate loose bank material, but also form a natural revetment to protect the bank from wind and wave action that can undercut the bank in some circumstances.

Planting vegetation is the least expensive method to stabilise the banks of still-waters. If any vegetation already exists along the bank, the best solution may be simply to add to the already existing vegetation by planting similar plants. Grasses produce an extensive root system to stabilise the bank. Perennial rather than annual grasses should be used to provide year round protection. Emergent aquatic plants can be used to protect vegetation from wave or current action. The general succession of a healthy, vegetation stabilised bank has grasses and rushes at the water line, proceeding to woody, emergent, flood-tolerant shrubs and then to flood-tolerant, moist soil trees.

The main problem with this is that it is very difficult to establish plants on a bank subject to erosion, because they may be washed away by the continuing erosion pressures. Furthermore, new plantings are very attractive to wild fowl and often disappear almost overnight. Put simply, if native plants could establish there they would; if they have not, then they probably need some protection to get established.

Coir rolls

There is often an inherent difficulty is establishing aquatic plants in a margin that is prone to erosion, because erosion has already occurred and therefore the soil has no structure and is often only a vertical face with no suitable shelf for planting.

Coir rolls provide an excellent method of establishing plants to margins susceptible to erosion as they provide an immediate habitat and protect the young plants from erosion until they become established.

The principle is a simple one: to protect the existing bank from erosion by effectively building a revetment (retained vertical barrier) in front of it. It is also possible to expand the technique to reclaim previously lost bank, by backfilling between a coir revetment built out from the existing bank (Fig. 16.1). The substrate used for the backfilling can be silt excavated from the still-water, so excess sediment is also being removed from the aquatic environment, which is often beneficial.

Indeed, this can be developed by increasing the infill (the area between the new revetment and the existing bank) to 2–3 metres to remove significant amounts of silt.

A suitable geotextile (permeable synthetic membrane) should be positioned behind the coir revetment before backfilling takes place, to reduce sediment moving through the revetment in times of high run-off. Any wave energy is dissipated by the coir roll, so no

Figure 16.1 Backfill behind a coir fibre roll revetment.

undercutting of the bank can occur and bank erosion is reduced as any loose, unconsolidated material from overland run-off builds up behind the coir roll. Where necessary, the coir roll can be positioned on faggot bundles to give the desired elevation in the water (Fig. 16.2].

The coir rolls soon become waterlogged and immobile and where they have been preplanted are virtually impossible to detect after a matter of months. Ideally, the area behind the coir revetment will be planted with appropriate marginal plants, further acting to combat erosion as described in the previous section (Fig. 16.3).

The main advantage of this method is that the bank will quickly become naturalised. It negates the need for 'hard' engineering solutions; the method is primarily used where marginal planting in itself would not control the erosion, or where the establishment of marginal plants is difficult.

Willow spiling

Willow spiling as an erosion control measure is more commonly associated with flowing water and is widely used on rivers and streams. The method provides an environmentally sound solution to erosion problems by intertwining live willow strands around stakes (Fig. 16.4). This ancient technique for building walls and fences works equally well as an erosion measure, and the live willow will quickly start to grow, providing a green solution. As with the coir roll method, the area behind the willow spiling revetment can be backfilled with a suitable substrate, most commonly reclaimed silt from the water body.

This traditional technique provides a living revetment, with the installation usually being done between November and March when the willow is dormant. When the growing season starts, the willow produces leaf and provides an aesthetically pleasing natural solution to erosion control.

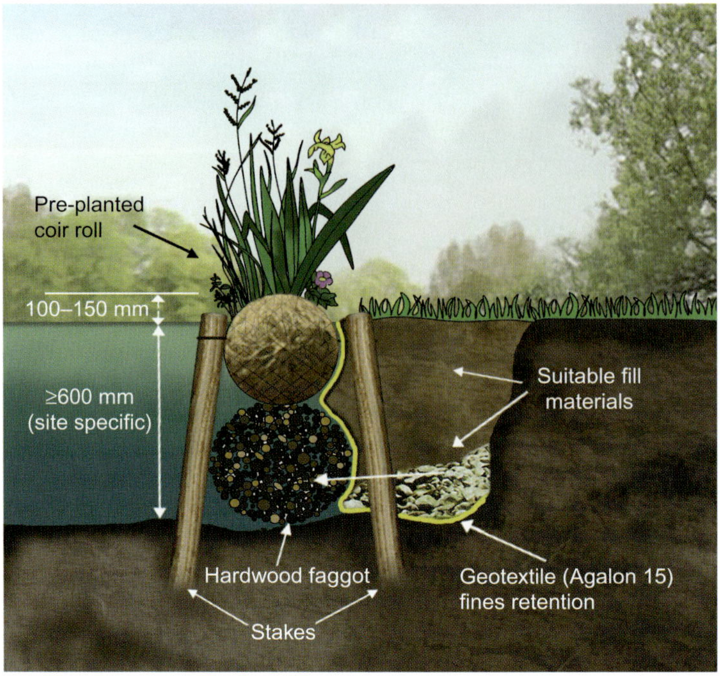

Figure 16.2 Revetment detail showing coir-roll placed on hardwood faggot.

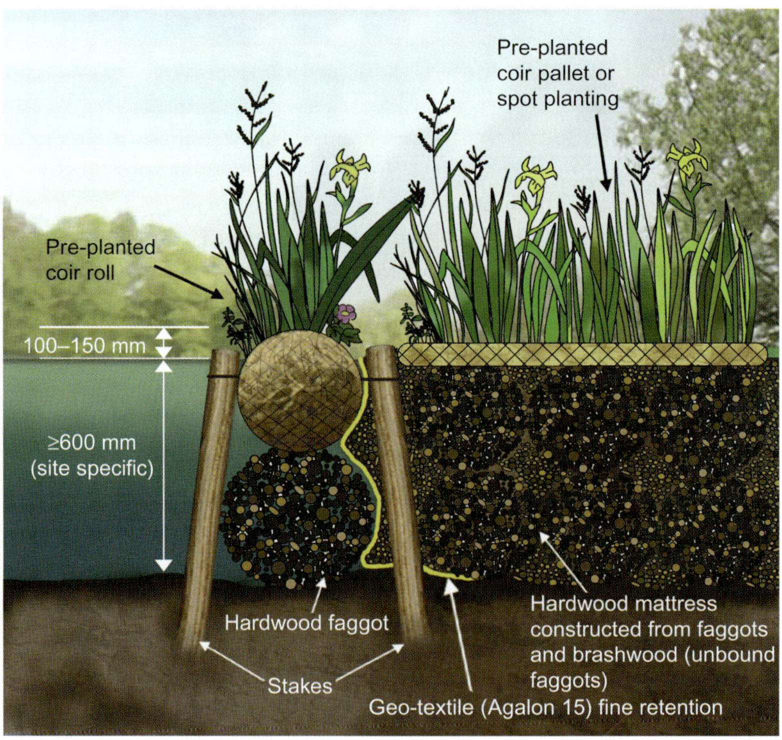

Figure 16.3 Spot planting behind a coir fibre roll revetment.

Figure 16.4 Live willow spiling.

The disadvantage to using the technique on still-waters is that when the growing season starts, the willow will soon grow to a height where it shades out other marginal vegetation that is beneficial to the ecology of the aquatic system. Shallow, warm margins with a variety of plants provide an excellent habitat for invertebrates of all kinds, and the use of willow spiling to some extent discourages this important habitat. In fluvial systems this is not so much of a problem, as shallow, warm areas are often naturally provided by riffles, or other areas of reduced velocity, and the use of willow spiling on fast-flowing, steep-sided sections of a river is an accepted method for providing environmentally friendly erosion control. In still-waters, it is likely to be used only on short sections of bank that are particularly susceptible to erosion by wave action.

Vertical erosion control (synthetic erosion curtain)

Synthetic erosion curtains (e.g. AGA Span) are a specialised product for vertical bank protection and provide a useful form of erosion control. The erosion curtain is a prefabricated, revetment fabric made from strong ultraviolet-stabilised yarns, which can accommodate strong currents and wave action. The curtain will also allow growth of grass through the fabric, giving a natural appearance to the channel.

The method of use is straightforward, whereby the curtain fabric fits over treated timber stakes that are driven into the ground, approximately half a metre apart so that they jut out from the existing bank (Fig. 16.5). To be more accurate, the fabric is stretched by

Figure 16.5 AGA Span nylon revetment curtain.

Table 16.1 AGA Span: approximate post lengths and spacing

Height of revetment (m)	Post length (m)	Post spacing (m)	Post diameter (mm)
0.5	1.67	0.75	75
1.0	2.44	1.0	85–100

installing posts as one moves down the bank. This ensures that the revetment is tight. As with the other methodologies, a suitable material is then used to backfill behind the new revetment, vegetation can be encouraged either by sowing grass seed in the soil on top of the bank or by placing turf immediately behind the span. Again, aquatic emergent species can be used with this method both to help consolidate the soil and improve the control of erosion.

The length and spacing of the posts (Table 16.1) will obviously depend on the prevailing soil conditions, and the depth to which the post must be driven for stability must be calculated. This is not always straightforward and Table 16.1 may be used as a guide.

It is important that the backfill is not subjected to turbulence or high flow velocities at the toe or ends of the revetment, and additional anchorage to the bank may be required. If the bed of the still-water undulates such that the curtain does not remain in contact with the bed, a woven geotextile should be placed vertically behind the curtain and horizontally on the eroded invert before backfilling.

Backfill material placed against the curtain should have a minimum average particle size of 400 μm: sandy gravel is suitable. When fine soils are present, it is recommended that a more sand-tight filter or geotextile is placed behind the curtain; however, this material should not be carried much above the normal water line as it may prohibit the growth of vegetation.

After installation and with time (a few months), the face of the curtain will start to green up, as it becomes colonised by natural algae and lichens, providing a more sympathetic aesthetic appearance. To encourage vegetation further, topsoil mixed with grass seed may be placed immediately behind the curtain.

Turf reinforcement

Turf reinforcement covers a broad range of methodologies and products, which are widely used and can be successfully applied to fisheries management.

Rolled erosion control products, consisting of temporary erosion control blankets (ECBs) and permanent turf reinforcement mats (TRMs) (Fig. 16.6), have been a popular choice for effective erosion control for several years in the construction sector.

The popularity of the fibre-based products has increased as many companies and government agencies are taking 'green' approaches, moving away from hard-armour techniques such as riprap and concrete.

Most ECBs and TRMs are manufactured from organic or synthetic fibres, such as straw or coir (coconut fibre), which are sandwiched between two or three layers of netting. In most cases, both types of product are applied over the top of a seeded or planted area and have a twofold purpose. One is to help the newly seeded areas maintain moisture, promoting germination; the second is to provide erosion control until the vegetation can provide adequate erosion protection on its own. Additionally, TRMs also provide effective stem and root reinforcement for the mature vegetation.

When it comes to choosing between an ECB and a TRM, the main factor is the type of erosion control that is required for a specific site. Temporary erosion control, which can be defined as anywhere from a few months to a few years, can be achieved with less expensive

Figure 16.6 Turf reinforcement mats over a seeded bank.

ECBs that break down after a specified amount of time. For permanent erosion control, TRMs offer a more robust choice and take longer to break down than the ECBs. To be classified as a TRM, the permanent portion of the matting must be at least 6 mm thick, which is adequate when providing reinforcement for the vegetation.

TRMs are a most effective solution for tough erosion control projects, even in the most environmentally sensitive of areas owing to their biodegradability (approximately 5–7 years). This allows for use in areas that would otherwise be inaccessible to erosion control projects, while not impacting the surrounding habitat. Indeed, they can be expected to be colonised by primary species within a matter of weeks, allowing for good natural regrowth of eroded banks.

The TRMs offer highly effective erosion control and protection for the root zone of the undersown vegetation. Owing to the material's flexibility, it maintains very close contact with the seeded surface, ensuring moisture retention. This creates a microclimate for quick seed germination thus ensuring better, speedy establishment for long-term erosion control by the vegetation.

16.4 Other considerations

The main advantage to the erosion control methods discussed in this chapter is that the technique of using plants (bioengineering) brings with it many other associated ecological benefits. The uptake of excess nutrients by plants is one of the main advantages in using bioengineering techniques in bank stabilisation projects, reducing the amount of nutrients entering the water. This nutrient removal is extremely beneficial as most still-waters in the UK can be considered eutrophic, which in turn may encourage algal blooms.

As has been previously mentioned, before control measures can be put into place the source of the erosion must be identified. It is only by understanding the process of sediment transport through the hydrological system that suitable options can be considered. In some instances this may be obvious. If a bank is being undercut by wave action then some form of revetment will be needed. In other cases it is less obvious and management may involve re-landscaping or planting of an area.

Overland flow inputs into still-water systems, although often overlooked, have the ability to transport large amounts of sediment and water and can easily undermine sections of bank. It is again worth reiterating that management of still-water fisheries should be undertaken in a holistic fashion, which incorporates the immediate terrestrial environment. The management of terrestrial vegetation around the site (buffer zones) can drastically reduce the risk of bank erosion and sediment transport. This is discussed in section 23.1.

The construction of a silt trap will obviously have a cost. However, analysis over an extended period of time will reveal that trapping and removing smaller amounts of silt from an accessible area will be far more cost efficient than removing it from the main basin of a still-water. The principle of settling or trapping silt is based on allowing the sediment that is being transported as part of a flow to slow and fall from suspension. There are many ways to construct such a structure and several examples are discussed in section 23.2.1 The common function is that they must trap the sediment in a suitable quantity and retain it until the trap can have its contents removed as part of the predefined management strategy.

16.5 Conclusions

The control of bank erosion and sediment transport is an essential element of fisheries management that, if considered at an early stage, can be relatively simple to achieve. The methodologies outlined in this chapter offer viable solutions to protecting against both sedimentation and bank erosion in still-waters. However, where lakes are 'on-line', the control of sediments into the system are more of a problem and every effort should be made to take the lake 'off-line' or, where this is not possible, to create areas designed to retain the silt (silt traps) which are easily accessible so the material can be periodically removed.

17 Control of Predators and Nuisance Species

17.1 Introduction

Many predators may take the types of fish usually found in a fishery. Most of these should cause a fishery no concern. The natural productivity of a lake can be between 200 and 600 kg of fish per hectare per year. Much of this productivity will be either consumed by predators or die from disease. If this were not the case the fishery would soon be overrun with stunted fish. In a natural situation, predators control populations of species lower down the food chain, which in turn control plant populations. In most systems these predator–prey relationships maintain a healthy prey population, as the predators tend to take the weaker fish. If the population of predators increases and they start taking excessive numbers of prey species, then they will of course eventually run out of food and die out themselves. Often therefore the presence of a predator is nothing to worry about and will enhance the fishery.

However, some predators can cause real problems. These mainly arise from some specific issues, which are as follows.

- When the number or amount of fish taken by a predator species rises above the replacement rate of the prey species. This is often associated with the introduction of a new predator to the system. Here the situation may become unsustainable because the predator will deplete the prey population to such an extent that it will quickly become unable to catch enough food to sustain itself. However, if the predator is highly mobile this may not be a problem as it can simply move on to another area with a better prey density.

- Where the predator is preferentially taking the higher value or larger fish, and is competing directly with the angler. However, it should be noted that most predators will generally not take the larger, healthy prey items.
- Where the stocking densities are very high, the fishery will actively attract predators to the easy pickings. Fisheries often enhance the biomass in the fishery artificially, sometimes to very high levels. This is obviously a very attractive situation for a predator. Also, the fishery owner has invested time and money in the fish and is much more aware of fish disappearing.

There is another possible consequence of fish predators visiting a fishery and that is the risk of disease transfer. As highlighted in Chapter 19, the most likely means of disease transfer onto a fishery is the movement of live fish; however, fish predators do represent a real threat in this regard, often far higher than the risk of movement on angler equipment for example.

17.2 Predatory species

Birds and mammals may impact on a fishery mainly through predation, which damages fish stocks and reduces catch rates. Several avian and mammalian predators are piscivorous (fish-eating) in nature and are fairly common within the UK. As an island, several seafaring, piscivorous avian species have migrated inland and now regularly predate on freshwater fish, especially in the colder months when freshwater fish are more docile, making them easier prey. However, potential impacts are not limited to predation. Geese for example can impact upon a fishery by overgrazing, fertilisation and puddling of banks and grass surrounds.

Mammalian predation is an increasing problem, especially as the otter becomes more widely dispersed after the reintroduction of captive bred specimens and the fact that the species has been afforded stringent protection. The non-native mink is now established throughout England and Wales. Mammalian predation has not been widely perceived as a serious threat as these predators remain largely elusive, and, as they are largely nocturnal, it is extremely unlikely that one can actually see mammals taking fish.

The conflict between fishery managers and predators such as piscivorous birds and mammals has increased over recent years, for several reasons. Primarily this is due to increases in populations, especially in the case of cormorants, whose numbers have risen rapidly over the past 20–30 years. For example, figures published by Defra for 2006 suggest that the cormorant population may have stabilised, with a UK wintering population of around 15 000–16 000 birds. Many birds migrate away to coastal breeding sites in the spring and summer. However, some cormorants now breed at inland sites so the species may be found at inland fisheries throughout the year. The number of birds breeding inland in England has increased and is now estimated at around 1500 pairs. There has also been a dramatic rise in otter populations.

The threat to fisheries posed both by avian and mammalian predation is of serious concern to both fishery managers and the industry as a whole; as such, Defra has published several guidance papers on how to protect against predation. The species covered in this chapter are considered a serious threat to fish stocks at the time of publication; however, these threats may well change over time.

17.2.1 Birds

Several avian predators actively feed on freshwater fish, the most notable being the cormorant. Like all wild birds, these piscivorous birds are protected under the Wildlife and Countryside Act 1981 and cannot be killed, or their eggs or nests taken or destroyed, except under licence.

Cormorant (Fig. 17.1)

The cormorant is perhaps the greatest predation threat to inland still-waters due to their increased numbers and the amount of fish they consume. It is estimated that there are about 200 000–250 000 pairs in Europe with approximately 7500 pairs nesting in the UK, of which 1500 pairs nest inland posing the greatest threat. Birds of the coastal race *carbo* are now increasingly nesting inland. It is not known exactly why these coastal birds are moving inland but possible reasons include the following: the creation of many reservoirs and gravel quarries since the 1960s and the stocking of fish in these waters; overfishing of prey species around the coast; legal protection; and a reduction in pollutant levels.

Cormorants are opportunistic predators that take a variety of species. In England and Wales, roach and perch are the most numerous prey, whereas rainbow and brown trout are consumed at put-and-take trout fisheries. Cormorants commonly take fish between 5 and 15 cm (2–6 inches) in length, but have been recorded eating fish of over 40 cm (16 inches) and eels of over 60 cm (24 inches) long.

Figure 17.1 Cormorant.

On average, an adult cormorant requires around 0.4–0.5 kg (about 1 pound) of food each day, although the weight of fish eaten on a particular day can vary considerably. Losses do need to be viewed in relation to the available fish stock, but the calculations are complex and need to take into account natural reproduction. Cormorants can also damage and scar fish, especially larger specimens which they fail to catch properly. This can increase the risk of disease, mortality, stress and induce behavioural changes (fish become more difficult to catch).

A range of measures can be used to reduce the impact of predation by cormorants, but their effect will vary from one site to another. Disturbance by people is consistently effective and visual or noise deterrents are most likely to work on still-waters. These methods are less likely to be useful or effective on rivers.

At any given site there may be potential to improve the fisheries habitat, offering fish greater chances of escape from attack, or to change fish stocking policies to make the food source less attractive to cormorants. Ultimately, good-quality habitat will assist fish recruitment and survival.

Shooting cormorants as an aid to scaring can work, but its effectiveness varies. At some sites, shooting, to kill or to scare, appears to make little difference to the number of cormorants present at the site. At others, birds leave for a short while, but return after several weeks.

The cormorants in the UK are part of the European population. Any large-scale cull of cormorants here may simply create a 'gap' for birds from elsewhere in Europe. To make any difference to the cormorant population, it is estimated that 30 000–60 000 cormorants would have to be killed every year throughout Europe. This is impractical; it simply would not be acceptable to the general public. A licence can be obtained from Defra to shoot a prescribed number of birds where economic suffering can be shown to exist.

Heron

Herons can be found throughout the UK, occurring almost everywhere except in the most mountainous regions. However, they are most abundant at lower altitudes, along major river systems and on the coast. At the present time, there are no studies that attempt to quantify heron predation at fisheries in the UK, other than at fish farms. However, the UK heron population is estimated to have stabilised over recent years at around 20 000 birds. Herons usually nest colonially, often at traditional sites (heronries) with extensive, suitable foraging habitat in the vicinity. They forage in a wide variety of habitats, from rivers, estuaries, ponds and lakes, to fields and pasture.

Herons take mostly small fish (i.e. less than 20 cm), so few of these are likely to be of commercial size, or of the size sought by anglers, except in the case of some cyprinids. Although herons occasionally come into conflict with particular fisheries, problems are not considered to be widespread and it seems unlikely that heron predation could cause 'serious damage'. The birds are more commonly associated with predation on fish farms and are not seen as posing a serious threat to still-water fisheries.

Sawbill ducks (goosanders and mergansers)

Goosanders and red-breasted mergansers, collectively known as 'sawbill' ducks, are not considered a major threat to still-water coarse fisheries, preferring riverine environments instead. The species breed on rivers and lakes in northern and western Britain, which is on the southwestern limit of their distribution in Europe.

Numbers have increased in the UK over recent years, which perhaps can be attributed to climatic changes, and the birds have extended their range into many parts of England. An estimate of the current UK wintering population is in the region of 9000 birds. Goosanders typically breed and inhabit freshwater upland streams, such as those found in the Scottish highlands, but do extend their range and can be found on larger still-waters. Adult male birds migrate away each spring to moult, but some birds are resident on rivers throughout the year. Highest densities of goosanders on the lower reaches of rivers tend to occur in midwinter, with other peaks in the summer in years with high breeding success.

The feeding technique and diet of both species is similar, with the ducks scanning the surface, head partially submerged, or diving from the surface. The backward-facing 'teeth' on the edges of each mandible help the birds to hold any fish that are caught, whereas goosanders also use their bill to probe for prey under stones.

Goosanders typically eat fish 8–15 cm (3–6 inches) in length, whereas red-breasted mergansers usually take fish around 5–12 cm (2–5 inches). Both species are opportunist predators, so their diet commonly reflects the range of prey available at a site. Because the birds mainly frequent more upland rivers, the main prey species tend to be brown trout, salmon parr, eels, minnows, bullheads and stone loach. Goosanders need 0.24–0.52 kg (10–20 ounces) of food daily, although the weight of fish eaten each day can vary considerably. Although it may seem simple to calculate the weight of fish taken by birds at a site or across the country, such figures are of limited value in isolation and predation must be considered in relation to available fish stocks.

Although goosanders and mergansers are not seen as being a serious threat to still-water fisheries, in the winter months the birds become more widely distributed across the UK and do roost on larger still-waters. The number of birds can also be increased by the arrival of birds from northwest Europe, particularly during harsh winters. Over the winter, goosanders tend to aggregate in flocks to roost, often on larger still-waters. These aggregations tend to be relatively short lived, with numbers at their peak in January and February. Over this period the birds are more likely to consume coarse fish species, representing a temporary localised problem for specific coarse fisheries. There is a concern from fishery managers and angling clubs that predation, especially by goosanders, can affect the viability of particular fisheries over winter months. At this time there has been no widespread effort to control numbers, but conservation groups are concerned that such an effort could threaten the birds' conservation status.

17.2.2 Mammals

There are two main mammalian predators that actively predate freshwater fisheries: the mink and the otter. Both can take fish of relatively large sizes, with predation rates increasing in colder months when the fish are docile and easier prey.

Mammalian predation often goes unnoticed at many fisheries for some time as the predators for the most part feed nocturnally; however, the problem is becoming increasingly well documented and there are measures that can be put in place to deter the predators.

Mink

Mink are a non-native species descended from animals that originally escaped from fur farms. The species is North American in origin; however, following successful breeding in the wild and with no natural predators, mink are now established throughout England and Wales. It is often necessary to control populations because of the damage they cause to wildlife, fisheries and game. Mink are nocturnal animals but are sometimes active during the day, particularly in colder months.

The ideal habitat for mink is close to water, preferring places with dense bankside vegetation, which provide sites for their dens. An individual will have several dens in its territory; these can be in holes in trees, in rabbit burrows or in gaps in rock or walls. Mink swim well but may move away from water to forage, and they can climb trees easily. An established adult male's territory is usually around 2.4 km (1½ miles) of waterway, whereas a female uses slightly less. Mink are opportunistic predators and take a wide range of prey including fish, birds, amphibians and mammals, but game and coarse fish do make up a large proportion of their diet.

The effect of mink on still-water fisheries can in some instances be significant. As mink are rarely seen, a decrease in catch rates or the local waterfowl population may be the first indication of their presence. If mink are suspected of predation, the corpse of a prey can often provide useful clues; fish are normally killed by biting the backbone, between the head and the dorsal fin, mammals and birds are killed by a bite to the neck, usually near the base of the skull.

Detecting the presence of mink is difficult, but finding droppings is a good indication of their presence. They can be firm or shapeless, varying from dark and tar-like to light brown, depending on what has been eaten. When firm, they are dark and sausage shaped, twisted along the length and pointed at the ends. They are usually 60–90 mm (2½–3½ inches) long and 9 mm (3/8 inches) in diameter, and they often contain fur, feathers and fish scales.

The dens of mink can also sometimes be located in warm weather by the smell from the faeces, decomposing animal remains and the mink's strong scent. Footprints show the marks made by five claws, but the complete print is often only visible when made in soft mud. The tracks are smaller than those of an otter and larger than those of a stoat. However, mink and polecat prints are almost indistinguishable.

Otter

In the late 1950s–1960's, the otter population underwent a dramatic national decline, mainly because of the toxic effect of chemicals associated with agriculture and industry, along with widespread habitat destruction. Otters disappeared from most rivers in central and southern England. However, since being afforded protection and with few predators in the wild, there is now evidence of their natural recovery. Figures published by the Environment Agency from 2000 to 2002 showed that 36% of the rivers that they surveyed showed signs of otter presence, an increase of 527% from 1977–1979.

Otters are elusive, generally nocturnal and seldom seen. They have dark brown dense fur, with a paler belly and a cream patch under the chin. Male otters average 1.2 m in length from nose to tail and can weigh up to 10 kg, whereas female otters tend to be slightly smaller. Otters favour all forms of wetland habitat: from rivers, lakes, streams, ditches and dykes to marshes, reed-beds, estuaries and coastland. Individuals have large home ranges, which may contain up to 30 lying-up places, or holts, in the form of natural hollows in tree roots or among dense vegetation.

Up to 40 km of river bank may be used as a home range (territory) for feeding, resting and breeding. However, where food is abundant, for example on the west coast of Scotland, ranges are much smaller. Fish, and particularly eels, form the major prey species. Small mammals, amphibians and crustaceans are also commonly eaten, with an adult otter requiring about 1 kg of food a day. Otters can breed at any time of the year, producing between one and four cubs, which remain with the mother for one year. The large home range and solitary nature of an otter mean that a healthy river can only support a few individuals, and will not affect local fisheries. Otters also may be beneficial to natural fisheries by taking weaker fish and the more easily caught eels.

Many fisheries and anglers groups work very closely with the wildlife trusts and otter groups, and welcome the return of the otter, an indicator of healthy fish stocks. There is, however, a growing concern among anglers and fishery managers that otters are taking fish, particularly large carp from still-water fisheries, where they are stocked at high densities. When some of these fish are worth several thousand pounds apiece, it is not surprising that managers may become hostile to the presence of otters.

Otters, being predators, will take advantage of easy prey, particularly when fish are torpid during the colder months, if they are left unprotected. Work has been done to look at ways of protecting fisheries, using fencing and decoy pools for instance. Fencing appears at the moment to be the most effective solution against otter predation, but this is not cheap and can run into thousands of pounds.

17.2.3 Prevention

For most recreational still-water fisheries there is likely to be a problem with one predator or another. These problems will vary, depending on the situation and the geographical location. There are, however, many possible methods of reducing or eliminating the problem. It is likely that for most fisheries a combination of tactics will be suitable. This may appear costly but consideration should be given to the loss of stock and the loss of revenue and the disease risk that can occur through predation; a cost–benefit analysis will quickly ascertain whether the methods are cost effective in any given situation.

We can split the prevention methods into four categories:

- preventing access;
- scaring and destroying;
- refuges;
- trapping.

Prevent access

If the predator can be prevented from getting onto the site or into the actual pool, then the problem is sorted. Although this is likely to be the most expensive of all the possible solutions, it has the advantage of being a long-term solution.

Fences

Fences can be designed and erected to prevent almost any land animal access to an area. If the design and construction is good they are also very effective. Unfortunately they are also very expensive. However, some consideration should be given to this option because fences can be useful to provide significant additional site security, deter poachers and act

Figure 17.2 Electrified otter fence.

as a general deterrent for a variety of other possible problems, including biosecurity (see Chapter 20).

There is very great public support for the recolonisation of the otter into the waterways and the protection of existing populations, so extreme care must be taken when dealing with otter predation. Ways of keeping them off the site without any harm are very useful (see, for example, Fig. 17.2).

Flight-path obstructions

Most birds, particularly cormorants, require some sort of flight access onto a pond. If this can be disturbed by tree planting and stringing ropes across the water then the birds will not land. In reality many birds can in fact land in very small areas but it is taking-off that causes them problems. For example, a cormorant can land relatively easily but needs quite a large area to take off, particularly when it has fed well. So the obstructions are often only an issue when the bird takes-off. Obstructions are particularly successful if such birds as do land on the fishery are forced to take-off again rapidly and in difficult circumstances. This is particularly successful when birds are scared by human disturbance, as the difficulty in take-off will cause the bird to panic and reinforce the difficulty.

Roost management

Predatory birds often like a roost close to the water, so removing these may help to deter the birds. This is not often successful as the look of the site may well deteriorate and trees often deter birds such as cormorants that like a nice easy flight path when they take-off.

Scare

Human disturbance

This has consistently proved to be the most effective predator deterrent available. However, frequent and extended periods of human presence may be required; particularly during the times the predator is most likely to visit the fishery, for most very early in the morning or late in the evening.

To be most effective a daily or near daily walk-round by at least one person, at first light and preferably accompanied by a trained dog for at least 30 minutes would be recommended. This is obviously not easy to achieve and may require some careful planning. One way of achieving the objective may be to encourage the local community to walk their dogs around the site. People often walk their dogs first thing in the morning and last thing at night and so could make ideal predator scarers. The increase in litter and intrusions may well be considered an acceptable price to pay.

Scarers

There are extensive examples of things that will potentially scare away predators. These can range from a fairly straightforward effigy such as a scarecrow through kites that are designed to look like hawks and have helium filled balloons to keep them aloft. Another example of this sort of scarer includes cars parked around the fishery. All these sorts of devices have the same problem in that the predators quickly come to realise that they represent no threat. This is called habituation and may take from several weeks to just a few days to occur.

Automatic scarers. There are a variety of automatic scarers on the market, which are easily available. Examples include pop-up effigies, helium balloons and gas cannons. All of these have been shown to be effective in the short term, but most predators will get used to their presence in time. Combinations of these automatic scarers can be more effective. The most effective appear to be pop-up effigies with gas cannons where the effigy is designed to appear just before the gun report from the gas cannon.

Shooting

All the above techniques for scaring predators are helped by reinforcing the inherent threat. For example, a car parked by a fishery may keep the birds off the fishery for a while but they soon get used to it. It becomes much more effective if every now and then someone gets out of the car and shoots at the birds.

Shooting has very obvious appeal as a way of dealing with predators that eat your stock fish. However, most of the predator species are afforded some sort of legislative protection. It should also be noted that shooting on its own is not very effective in dealing with predators. However, as a support technique, along with the others listed above, it can be helpful. A licence will almost certainly be required to shoot cormorants. In this licence application you should be very clear about the fact that it is a support to help reinforce other scare tactics.

Fish refuges

If you ever watch a fish when it is threatened, what does it do? Does it just scatter in a random direction or does it seek cover? Well, most will seek cover if it is available. Most prey species are not able to swim for long at high speed and if they cannot find cover they will soon tire and make easy pickings for a stronger predator. As a result, prey species are very good at knowing where the nearest cover is at all times and not moving too far from it. Unfortunately, many fisheries lack much cover so the fish are forced to forage far from cover, which makes them easy pickings for predators such as cormorants.

Marginal vegetation

Areas of marginal vegetation make excellent cover and provide habitat for food items for the fish. Planting between swims and leaving unfished areas to grow wild will provide excellent cover. Also, this will provide spawning habitat and natural food production as well as making the fishery a more attractive place.

Artificial fish refuges

Marginal vegetation still leaves significant areas of the open waters free of cover. Specifically designed fish refuges can help. Again, there are many products on the market that can be used to create artificial cover in the middle of a fishery. These tend to be wire-mesh frames, similar to gabion baskets, stuffed with things like willow sticks. These can then be sunk into the fishery wherever you choose. An excellent additional idea is to place them with a floating island attached to mark the submerged refuge (Fig. 17.3). A floating island not only marks the position of the refuge but provides additional cover. Figure 17.3 shows the extensive root systems that hang below the floating vegetation, providing excellent cover as well as spawning substrate.

Trapping

Trapping can be used to remove unwanted pests such as mink and rats. Traps can be land based and are usually baited. It is, perhaps, best to consult with a commercial pest control

Figure 17.3 Anti-predator refuge beneath a floating island.

firm should rats become a problem. Before live trapping is undertaken, the associated responsibility must also be considered; wherever live trapping is undertaken there is a responsibility to inspect the trap daily and to humanely dispatch any target species that has been caught.

17.3 Nuisance species

Many still-waters have some nuisance species in them. These are animals that affect the quality of the angling in some way. This may be through direct competition with the target species, predation on the target species or prevention of angling, for example through eating any bait before the target species has a chance. Control of these types of nuisance species can be fraught with problems. A further group of nuisance species is those that inflict an impact upon a still-water habitat; these include geese, ducks, cattle and dogs.

17.3.1 Native fish species

Pike

One of the most consistently problematic species in a still-water is the pike. It is viewed as a significant predator, and can make it difficult to catch other coarse fish because of the effect on the populations. When viewing pike as a problem, it is important to realise that they have a major part to play in the overall ecology of the lake. They will be the top predator in most circumstances, and therefore generally keep the populations of the other fish under some sort of control. In many cases this may actually help the job of the fisheries manager. The pike will generally eat smaller and less healthy fish, removing them from the population and leaving the stronger, larger fish with less direct competition for space and food. So, in many instances, control of the pike population may in fact make the fishery perform worse.

The best way to control pike numbers is annual electrofishing. Removal of all the fish under 4.5 kg (10 pounds) will leave the larger predators, which are notoriously cannibalistic. It is impossible to remove all the pike totally, so controlling the numbers is the only available option. When assessing the impact of pike on a fishery the fisheries manager should ensure this is done with due consideration and relative to other predators such as cormorants.

Undersized silver fish

For some fisheries, an over-abundance of silver fish can be a problem. This not only prevents the rapid growth of other fish due to the competition for food and space but also makes angling difficult. Although some anglers are happy to be catching small roach and rudd, many want to get their bait to the larger fish. This can be difficult when, no matter what hook bait is used, small roach and rudd immediately surround it.

Regular annual removal operations are the best way to deal with this issue. If there are no fish predators in the system then consideration could be given to the addition of a predator species to control the silver fish. However, as with the pike and the zander, these can quickly give rise to other problems.

17.3.2 *Ducks*

Ducks only rarely cause problems on still-waters, and in fact any impact they do cause is usually in the form of members of the public feeding the resident waterfowl. This can result in two major problems. The first is that the uneaten food thrown in will contaminate the water. This is made worse by the fact that often the ducks are fed with feed that does not match the required nutrient profile of the waterfowl themselves. Large quantities of rotting white bread can soon make a significant difference to the water quality. The second problem is that the ducks quickly associate human visitors with free food, so the arrival of an angler will often result in large numbers of ducks congregating in the swim. This will not only disturb the fishing but it is often difficult to get the bait into the water without it receiving attention from the ducks, with inevitable results.

On some fisheries with public access, it is impossible to ban this practice but it can be managed. Duck feeders and hence the ducks will congregate at areas of the bank if these are allocated to duck feeding and made safe and easily accessible. Signs explaining the consequences of overfeeding the ducks with bread can also help to reduce the quantities of feed.

17.3.3 *Rats and mice*

Rodents can cause problems, especially where there is a store of fish food on site. To avoid problems with rodents all food should be kept in secure and sealed stores or bins. Excess, spilled food should be removed and not left lying. Once rodents are established, infestation problems should be addressed by trapping or poisoning. If poisons are to be used, great care should be taken to keep these from entering the water or contaminating feeds or other products and equipment used in the fishery. The risk of Weil's disease (leptospirosis) should not be underestimated; rats are a carrier of this disease so the reduction of rats will reduce the level of risk. However, it is almost impossible to effect long-term control of rats, so the disease should always be considered as a potential threat to the health of those working and enjoying leisure near water.

Unfortunately, rats and mice live happily in rural locations as well as urban areas; their presence does not rely upon a human source of food. The fishery manager should be aware that a seine net stored in a 'lock-up' at the fishery will be irresistible to rats in particular: do not leave your nets in unsealed containers unless you are ready for the rodents to put even more holes in your net.

17.4 Invasive non-native species in the UK

Plant and animal species have been moved around the globe for centuries and, as people trade and travel more often, species get spread further afield. On a global scale, invasive non-native species are the second most significant cause of biodiversity loss, with only habitat destruction being more damaging. Invasive non-native species pose a risk to still-water fisheries in several ways, most notably from a biosecurity perspective. A non-native species is defined by the Defra (2003) *Review of non-native species policy* as: 'A species introduced (i.e. by human action) outside its natural past or present distribution'.

Invasive species can cause substantial problems for our native plants and animals and cost the British economy an estimated £2 billion per year. Once established the invaders can cost millions of pounds to eradicate, that is if they can be eradicated or even controlled. The manager of a still-water fishery should be aware of the variety of these organisms so that he or she can identify them if they appear in his lake or the surrounding countryside. This will enable the manager to take appropriate action for the early elimination of the pest from his own property and report its presence to the authorities.

17.4.1 Freshwater fish

The freshwater fish and crayfish species that fishery managers should be aware of, and may be considered potentially the most harmful, are the following.

Zander (Sander lucioperca) (Fig. 6.18)

Zander is perceived as a significant predator, which is likely to prey on weaker fish and generally predates on the smaller size groups.

Zander are good to eat so perhaps encouraging anglers to take a few home for the pot will reduce the problem.

Signal crayfish (Pacifastacus leniusculus) (Fig. 17.4)

The American signal crayfish is one of the worst nuisance species in many countries where it has been introduced. Given the right conditions, populations can reach incredible levels, as there is often an abundance of foods and habitat and few, if any, natural predators. The crayfish undermine the bank and bottom by digging substantial burrows. They will feed on any bait, giving false bites, and they will directly compete for food with many coarse fish species. They can grow large enough to predate directly upon many fish species as well. Although our native white-clawed crayfish rarely live in still-water fisheries, they are sus-

Figure 17.4 Signal crayfish.

Figure 17.5 Topmouth gudgeon.

ceptible to 'crayfish plague' that is carried by the signal crayfish. The assumption is that the more places that signal crayfish are found, the greater chance they will find a way into the few remaining strongholds of the native white-clawed crayfish.

Control can be quite difficult and every effort should be taken to prevent the further spread of this animal. Trapping is the only real option to control them once established, although a licence is required. Furthermore, crayfish are excellent food species and anglers should be encouraged to capture them for the pot. They can also be used as bait in trapping for pest predators such as mink.

Topmouth gudgeon (Pseudorasbora parva) (Fig. 17.5)

The topmouth gudgeon appeared in British waters some years ago and is spreading throughout the country. It is possibly transferred by anglers using the fish for bait. The species is known to carry a parasite that poses a direct threat to UK fish stocks. This species is a good example of how a species that finds a niche can increase its distribution despite regulation.

Grass carp (Ctenopharyngodon idella) (Fig. 6.8)

Grass carp have been introduced into many waters in England to control weed growth. The species does not breed under UK conditions but lives to a great age and may attain a considerable size. It is known to disturb ecosystems and change natural patterns of aquatic plant growth.

Wels catfish (Siluris glanis) (Fig. 6.21)

This European introduction grows to an extremely large size and is capable of taking wildfowl and small mammals. It does have considerable fisheries interest in its own right.

17.4.2 Mammals, reptiles and amphibians

Non-native mammals that are considered to pose a threat to natives and the environment include the following.

American mink (Mustela vison)

Mink populations can be considerable in some waterways and around some still-waters. They are a voracious predator, posing a threat to water vole and fish populations. The species has no natural predators in the UK. Control is primarily by trapping, although bait can be used.

North American bullfrog (Rana catesbeiana)

The American bullfrog is a voracious predator and a serious threat to British wildlife. Tadpoles of the American bullfrog were at one time widely sold in garden centres. Imports are now banned owing to the risk of this invasive species establishing in the UK. Vigilance is still required as there have been reports of accidental introductions of tadpoles included with consignments of aquatic plants.

Red-eared terrapin (Trachemys scripta)

The effects of this species are, so far, unknown but could include competition and introduction of exotic parasites.

Alpine newt (Trituras alpestris)

The Alpine newt is a known predator that competes with native *Triturus* species.

17.4.3 Vascular plants (terrestrial and aquatic)

Some of the most invasive of the non-native vascular plants, which have the ability to negatively impact on a fishery, are listed in Table 17.1.

Water fern/fairy moss (Azolla filiculoides) (Fig. 17.6)

Water fern, also known as fairy moss, is a floating plant that has the ability to double its biomass every 2–3 days. In still-waters particularly those that are static with little water movement, it can cover the entire water surface, effectively blocking the interface between the atmosphere and the water surface that is critical for gas exchange.

Japanese knotweed (Fallopia japonica) (Fig. 17.7)

This is a large, vigorous weed that appears to have no natural enemies in Britain. It can colonise most habitats and is regarded as a troublesome pest in many parts of the country because of its rapid invasion and domination of habitats, which results in the exclusion of other plants.

Australian swamp stonecrop (Crassula helmsii) (Fig. 17.8)

This species aggressively spreads in the wild and outcompetes natives in a range of habitats from damp ground to 3 m water depth. It grows for 12 months of the year in the UK climate, with no period of dormancy. This plant can cause oxygen depletion because it assimilates

Table 17.1 Non-native invasive aquatic plants

Species	Description
Riparian	
Giant hogweed, *Heracleum mantegazzianum*	Colonises banks and its sap can cause blistering of the skin
Japanese knotweed, *Fallopia japonica*	Highly invasive, colonises banks
Himalayan balsam, *Impatiens glandulifera*	Loves damp banks and shaded riparian habitat
Emergent	
Australian swamp stonecrop (New Zealand pygmyweed), *Crassula helmsii*	Can spread by fragmentation
Parrots feather, *Myriophyllum aquaticum*	Emergent and submerged shoots
Water primrose, *Ludwigia grandiflora*	Forms dense mats
Submerged	
Canadian pondweed, *Elodea canadensis*	Arrived in Britain in 1842; here to stay
Curly water weed, *Largarosiphon major*	Similar in appearance to *Elodea*
Floating	
Floating pennywort, *Hydrocotyle ranunculoides*	Floating stems are brittle; they break off, aiding the plants dispersal
Water fern, *Azolla filiculoides.*	Can form dense blankets at the expense of all other plants
Water hyacinth, *Eichhomia crassipes*	Free floating
Water lettuce, *Pistia statiotes*	Dense mats that shade out light

Figure 17.6 Water fern/fairy moss.

carbon dioxide for up to 20 hours a day when it is existing in a submerged form; this is because of crassulacean acid metabolism.

Giant Hogweed (Heracleum mantegazzianum) (Fig. 17.9)

Giant hogweed colonises many areas of wasteland and riverbanks. It forms dense colonies that suppress the growth of native plants and grasses, leaving the banks bare of vegetation in winter and increasing the risk of erosion and recolonisation from seeds washed

Figure 17.7 Japanese knotweed.

Figure 17.8 Australian swamp stonecrop.

downstream. Do not attempt to remove this plant manually as its sap is an extreme irritant and may cause severe blistering of the skin.

Floating pennywort (Hydrocotyle ranunculoides) (Fig. 17.10)

This species grows in shallow, slow-flowing, eutrophic water bodies and forms dense, interwoven mats of vegetation. These mats will quickly cover the water surface and can grow 20 cm per day, resulting in a growth of up to 15 m from the bank in a single season. The mat starves the water body of light, nutrients and oxygen, thus killing many of the species living in it. It also increases the risk of flooding, by blocking the waterway.

Figure 17.9 Giant hogweed.

Figure 17.10 Floating pennywort.

Himalayan balsam (Impatiens glandulifera)

Himalayan balsam is rapidly colonising riverbanks and areas of damp ground. It grows in dense stands that suppress the growth of native grasses and other flora. In the autumn, the plants die back, leaving the banks bare of vegetation and vulnerable to erosion.

Parrot's feather (Myriophyllum aquaticum) (Fig. 17.11)

This species is popular and widely grown in garden ponds, and is not usually deliberately introduced to the wild. However, fragments can be concealed in the soil of other pot plants sold at aquatic garden centres and it will quickly spread and colonise new sites.

17.4.4 Risk identification

The identification of risk should be incorporated as part of the biosecurity management plan of a fishery. This will identify all the biosecurity risks, including those posed by invasive non-native species. Identifying these risks and how they can potentially enter a fishery can be seen as the first phase of the management plan. The plan should include identification of the particular species that should be discouraged, depending on individual situations of the fishery. Different invasive species have the ability to have an impact on different fisheries, depending on their nature and scale. As such, it is worth carefully examining which species potentially pose a threat. On-line fisheries are particularly susceptible to invasive species owing to their nature and the difficulty in controlling what enters the fishery. Where practical, it is recommended that still-water fisheries be taken off-line to reduce the risk posed by invasive species. It is acknowledged, however, that this is not always an option, so control measures must be put in place to mitigate the risk.

Figure 17.11 Parrot's feather.

17.4.5 Control measures

Control measures that can be put in place to manage the spread of invasive non-native species focus initially around monitoring. By awareness of which species can potentially have adverse affects on a fishery, and a good knowledge of which species are present, a plan can be formulated about how best to control the unwanted species.

Risks should be identified during monitoring, including gaining awareness of any invasive organisms that are in the immediate vicinity of the fishery, or that are likely to be imported with new material. Actions can then be taken to control unwanted invaders. Numerous control measures are available to fisheries managers, depending on the species to be controlled. These can range from weed eradication programmes to controlling the introduction of certain unwanted fish species. Control of unwanted aquatic plants can be undertaken following the procedures discussed in Chapter 15.

Control of fauna, both aquatic and terrestrial, raises a new set of problems. Increased biosecurity measures should help to prevent the introduction of invasive species and associated pathogens and parasites. These parasites and pathogens may be introduced either directly by fish or in some instances by angling equipment.

There are several options if it is known that an invasive fish species is already present in the fishery, as for example a lake with a population of topmouth gudgeon. The first step should be to inform the Environment Agency and Defra, who may well be willing to assist in the removal of the fish. Fisheries consultants will also be able to offer advice about fish removal and other actions that could be taken to remove the unwanted species. This may well take the form of electrofishing the water body, seine netting or the use of toxicants.

17.5 Poachers

Poachers are included in this chapter because they bring many of the same problems as predators. They remove the more valuable of your fish and present a disease-transfer risk. Every effort should therefore be taken to stop these people from getting on your site. This issue is followed up in more detail in the following chapter.

18 Managing the Impact of Climate Change on Still-Water Fisheries

18.1 General impacts of climate change

It is now widely accepted that climate change is a reality, and that the 0.6°C rise in the Earth's temperature over the past 100 years is at least partly due to human activities. The impacts of climate change are likely to become increasingly evident in the coming decades.

During the twenty-first century this warming pattern is predicted to continue by between 1.4 and 5.8°C. It is also predicted that by the end of this century there will be significant increases in the average global sea level, of between 0.1 and 0.9 m. There will be changes in weather patterns, which will include increased frequency and severity of extreme events such as floods, droughts and storms. Regional changes are much more difficult to predict than global trends; however, as time passes, stronger trends of change will emerge from a more localised perspective.

Climate change has the potential to impact upon fisheries through a diverse set of pathways, which are both direct and indirect. Potential impacts upon inland still-water fisheries include the following:

- flood events;
- increasing temperatures;

- drought;
- changing water levels.

Legislation, including the Water Framework Directive, focuses national attention on these questions. However, there are few if any objective methods (tools) available to predict changes in fish communities, other than relatively simple techniques, and still less information on how to develop fisheries to make them more resilient to possible changes caused by climate change.

In still-waters, climate will certainly interact with overexploitation, habitat degradation and destruction, non-native species and pollution to increase the potential impact. Ecosystems will come under significant pressure to reconstruct naturally and to adapt to these changes. Climate change may bring about increased competition for many organisms and increased opportunity for non-native species.

Decisions made today by fisheries managers should consider the implications of climate change to enable our fisheries to evolve and adapt. Management decisions will continuously need to be evaluated in line with developing trends. The manager of a fishery needs to be aware of any potential impacts and use them to inform his management decisions. The two most important areas are the potential impact of non-native species, and the risk of flooding, not only from rivers, but direct local flooding from very high rainfall events. This requires the effective management of non-native species through good biosecurity and the use of emergent and marginal vegetation to control the banks, which will also help to maintain the still-waters during flood events.

18.2 Management of still-waters within a floodplain

The management of a still-water in a floodplain provides the fisheries manager with an additional set of considerations as well as those more generally. Most legislation and river basin management considers that where a still-water exists within a floodplain, it is an integral and important part of a connective network, therefore as part of a river basin. Certainly the Water Framework Directive views them in this way.

The fisheries manager in this situation is faced with many complex interactions which are generally more involved than the relatively simple decisions related to whether a still-water is on-line or off-line. The decision will often involve the physical and biological relationship between the river, its floodplain and the degree of connectivity with the still-water. There will be an expectation that the still-water provides a positive ecological benefit as connectivity within the system is increased. Connectivity does not, of course, mean constant water flows. For example, still-waters can provide suitable refuge habitat for many river species in time of spate or flood flows. There are several potentially negative impacts that are associated with still-waters situated in an area where the river, its floodplain and the still-water become temporarily connected. These may include the following:

- dispersal of poorly adapted cyprinid species from the still-water;
- influx of riverine species;
- influx of novel disease;

- influx of pollution (decreased water quality);
- increase in siltation;
- erosion of banks.

Climate change may well have far-reaching, localised effects on such fisheries. For example, a fishery that is only technically within a floodplain today may actually come to lie within the floodplain at some stage in the future.

19 Control of Fish Disease

Fish disease is one area where a level of control must be exerted or the consequences can be catastrophic. Although only a limited amount of control can be exerted once a pathogen has entered a fishery, we can attempt to control the arrival of fish disease into a fishery.

19.1 Investigating a disease problem

Fish are looking 'off colour' and some have died. What do we do?

It is best to follow a set routine in these circumstances so that all the different aspects of the investigation are covered. Initially, some simple questions need to be answered to obtain information on the history of the mortality.

Box 19.1 gives some of the questions that should be asked. They relate to the history of the fish, the history of the outbreak and the husbandry. It is very important to recognise that set questioning is essential as those close to the outbreak are sometimes may overlook obvious issues, such as recent husbandry activities.

Box 19.1 Questions on the history of the mortality

- How many fish have died?
- How long ago were the fish first seen to be having problems?
- What species of fish are affected?
- What species of fish are unaffected?
- Have there been any recent management activities that could have affected the balance of the fishery/system?

As a simple guide to initial diagnosis the following rules are helpful:

- If the mortality has been rapid and most fish have been affected, including a range of different species, then we can conclude that there may be a water quality problem.
- If only one species is affected and the mortality has been slowly building for several weeks then it is possibly a disease problem.

This is not infallible as some parasites, such as whitespot, do cause mortalities in multiple species. Most mortalities are caused by poor water quality, therefore water quality parameters should be assessed as a matter of course.

Observing the behaviour of the fish can reveal a lot about disease and other problems within fishery. Indicators can include poor feeding response, irritation or flashing, and crowding round inlets or aeration devices. If this is the case, then it is often wise to call in an expert, which will usually involve a site visit. If this is not possible then it is best to submit a sample of live fish (showing symptoms of the disease) direct to an appropriate laboratory for examination. The laboratory should always be contacted beforehand to discuss the submission of such a sample.

It is worth noting at this point that if mortality is occurring and a notifiable disease, such as spring viraemia of carp or koi herpesvirus, is suspected, the fishery manager is under a legal obligation to notify the appropriate government department. In England and Wales this would be Cefas.

Although it is desirable to seek professional help, it may not always be possible. As a last resort, the following techniques can be undertaken with limited training.

A detailed examination of the fishes external appearance should be carried out. Factors that should be considered include any abnormalities of the following:

- shape;
- skin;
- scales;
- mucus;
- fins and tail;
- eyes;
- gills.

A detailed check does require the removal of the fish from the water for some time. It is advisable when handling large fish to anaesthetise them.

Box 19.2 Dose rates for anaesthetics

MS222: one teaspoon per 5 gallons: 1:5000 to 1:20 000.
Phenoxyethanol: 5–10 ml per gallon or 1 ml per litre.

19.1.1 Anaesthetics

Although there are many types of anaesthetic suitable for use on fish, there are two common ones:

- MS222 (Sandoz) (ethyl *m*-aminobenzoate);
- phenoxyethanol.

These products are made into anaesthetic baths, into which the fish are placed for a set time until the fish becomes unresponsive but the opercular movements can still be seen.

The quantity of water, size of receptacle and dosage depend on the size of the fish, temperature and procedure being undertaken (see Box 19.2).

When anaesthetising fish it is essential to use water from the system and to aerate the holding tank. A receptacle with heavily aerated water taken from the system should also be prepared as a recovery tank.

19.1.2 Surface examination

Once the fish is anaesthetised, it is important to do the following:

- note down the length and weight of the fish;
- carefully examine the body surface and note any obvious abnormalities; it is an idea to photograph abnormalities as a permanent record;
- take a skin scrape;
- take a gill scrape;
- if any abnormality is detected take a scrape from that area;
- observe scrapes under a microscope at ×100 magnification.

Skin scrapes

These are taken with a blunt scraper (a wooden spatula) or a cover slip. The implement is held at 45° and drawn gently across the body surface from head to tail. The material collected must then be transferred onto a microscope slide or spread across the cover slip. A drop of water is placed on the slide and the cover slip gently lowered onto the slide. The cover slip should then be gently pushed down, squeezing out some water and any air bubbles.

19.1.3 Gill scrapes

A sample of mucus can be taken from the gill tissue by gently inserting a cotton bud under the operculum and rolling it over the gill filaments. This should be done with extreme care and only gentle pressure should be applied to the gill surface.

The mucus sample can be spread on a slide and dealt with in the same way as for a skin scrape.

These scrapes now need to be examined under a microscope at around ×100 magnification. If you do not have one yourself, try the local aquatics shops or a veterinary surgery. The scrapes should not be allowed to dry out and should be observed within 20–30 minutes.

These techniques will give you a great deal of information about the external parasites present on the fish. This may be all that is required to indicate a cause of the mortality. It is, however, advisable to continue a more detailed examination to make sure that there are no further problems.

A full post-mortem examination takes a great deal of experience to perform accurately. Again, however, it is possible to take some tissue samples for submission to a laboratory, if a site visit or live fish submission is not possible.

19.1.4 Internal examination

Killing the fish

Fish being used for post-mortem examination should be humanely killed by the following methods:

- anaesthetic overdose;
- anaesthesia followed by cutting through the spinal cord.

Dissection procedure

A full dissection procedure is complicated and is best not undertaken without training. However, the following samples can be taken relatively easily by following a normal dissection procedure. The incisions are shown in Figure 19.1.

Bacterial samples

Samples from internal organs, as well as any obvious lesions on the skin, can be examined for bacteria. A sterile swab, similar to a cotton bud, is pushed gently into the lesion, twisted and returned to its container. Good transport swabs contain a jelly-like substance that will help to keep the sample fresh during transport. You may be able to obtain swabs from your veterinary surgery.

Viral sampling

Internal organs can be examined for the presence of viruses. Take 1-cm chunks of all the internal organs, including the brain, and place them in a clean container with a small amount of distilled water. Label carefully and submit to a laboratory with full details of the mortality.

Histology

This is a technique that allows the examination of organ tissue at a cellular level. It may also allow very high magnification examinations (×1000+) as well as electron microscopy. Here small 1-cm chunks of the major organs are preserved in 10% neutral buffered formalin and again submitted to a specialist laboratory. Neutral buffered formalin may be obtainable from a chemist or again from your local veterinary surgery.

Figure 19.1 Incisions for post-mortem examination.

Diagnosis

After these procedures, the fishery manager is often left with a great deal of information about parasites and disease present in his stock but no conclusion. What is the diagnosis?

This is a difficult process and it is usually best to consult an expert at this point. There will often be an underlying factor or primary problem in the fishery that is the cause of the disease. It is therefore important to identify this rather than treating the pathogen, which may be secondary.

Basically the questions that need answering are as follows:

- What type of pathogen is causing the mortality?
- Is there a treatment that can be used to control the disease outbreak?
- What has caused my fish to become susceptible to this pathogen?

19.1.5 Factors influencing a disease outbreak

To reiterate some basic principles, a disease outbreak is rarely merely the result of a meeting between a pathogen and a host. Under most circumstances the two organisms live in a dynamic equilibrium, which is not significantly harmful to the fish. However, this can quickly change if the fish is stopped from reacting to the pathogen, or the pathogen is not recognised by the fish. In these circumstances the equilibrium is pushed in favour of the pathogen. Because of their huge reproductive rates, the pathogen will multiply quickly and a disease outbreak will ensue. Problems may contribute in this way either by damaging the fish's immune system, by helping the pathogen or by bringing together new hosts and pathogens.

Damage to the immune system

The immune system in fish is a relatively simple but effective system, including physical barriers such as skin and chemical barriers such as the mucus. These are to prevent entry or attachment of the pathogens. Internally the fish has the ability to kill pathogens with a variety of immune responses. Producing these various systems is difficult for the fish and a variety of things can prevent or slow down these responses.

Nutrition

Inappropriate or inadequate food can lead to a reduction in the effectiveness of the immune system. All fish have a requirement for a certain amounts of essential nutrients (see Chapter 22). Nutritional deficiencies occur if the essential nutrients are unavailable to the fish or the nutrient is available but not in high enough quantities.

Stress

Stress leads to a reduction in the effectiveness of the immune system. Fish are adapted to particular environmental conditions and any conditions outside this range can lead to stress. Also, any rapid change in the environment conditions can lead to stress, even if the change is to a preferable condition.

It can be seen that it would be easy to stress fish and compromise their immune system. Fisheries managers must make sure that they understand the conditions required for their fish and that, as far as possible, these conditions are met.

Overstocking

Overstocking can lead to an explosion in the numbers of parasites in two ways.

- It is far easier for the pathogen to locate a new host if there are more of them. This situation is usually made worse if most of the fish are a single species (e.g., carp or trout). Species-specific pathogens then have a very easy time, as not only is it easy for them to find a fish but it is usually the favoured species.
- When fish are held in overstocked conditions it produces a stress response. This has clearly been demonstrated for carp and trout.

19.2 Disease control

19.2.1 Location

There are two significantly different situations where a fishery manager may encounter a disease outbreak: on a fish farm (see Chapter 12) and in a fishery.

Fishery

On a fishery it is likely that once a disease outbreak is underway there is little or nothing that can be done. On any inland water that could be described as a fishery it is simply not possible to treat the fish *in situ* with any of the recognised chemotherapeutic drugs. Several obvious problems prevent this being a realistic option:

- size of water and therefore amount of chemical;
- considerable harm will be done to other inhabitants of the water body;
- inability to estimate accurately the amount of fish to be treated.

Therefore, only simple management of the outbreak can be achieved such as optimising conditions, maintaining good hygiene and collecting all the mortalities on a daily basis and disposing of them in an appropriate way. Angling and the use of keepnets should be suspended immediately. Under no circumstances should fish be moved from the still-water experiencing the mortality. Diagnosis should be rapidly undertaken. This may show where the disease has come from and provide information about what is the underlying cause. This will help to give the fishery the best chance of avoiding a repeat of the mortality.

So if there is usually no cure then what can be done? As the saying goes, 'prevention is better than cure', and where there is no cure then prevention is the only option.

There are two main areas to concentrate on:

- prevention of disease coming onto site;
- making sure conditions are optimal and the fish are happy and healthy so they can control any diseases that are present with their own immune system.

Prevention is simply a matter of trying to isolate all the times and places a pathogen could enter the site and controlling them as best you can. The most obvious time that pathogens may enter a site is on live fish. If you have a totally enclosed system then do not import any fish onto site.

Pathogens may also be introduced into the water by other means, including with visitors, particularly anglers' equipment. It is therefore a good idea to do the following:

- restrict the use of equipment such as keep nets.
- place disinfectant dips at the site, especially the car park;
- try to educate anglers in disinfecting their equipment between visits.

The introduction of disease can be minimised by insisting that anglers only use equipment such as keep nets and unhooking mats supplied by the site.

Fish farm

On a fish farm there is significantly more control over the fish and their environment. Owing to the generally intensive nature of most fish farms, they are particularly susceptible to disease outbreaks from direct life-cycle parasites. Again, the techniques of exclusion such as quarantine should be practised, but in the event of an outbreak some treatments are appropriate.

19.2.2 Chemical treatments

The treatment of fish using chemicals is a problematic area and one that we cannot cover in any detail. The reader is advised to seek expert advice before considering any application.

In most countries there are legal requirements for the use of chemicals in any farming activity, especially with the potential discharge of these chemicals into the environment.

Other factors that must be taken into consideration when choosing a treatment include the following:

- the species of fish;
- the system the fish are being held in;
- the route of administration;
- the disease;
- the age of fish;
- environmental parameters.

There are five major routes by which fish can be exposed to chemicals:

- water borne
    - prolonged immersion
    - bath
    - flush
- oral;
- injection;
- spray;
- swab.

Calculating dose rates

Oral

Oral dose rates are usually given in milligrams (mg) of active ingredient per kilogram (kg) of fish to be treated. Unfortunately, fish are very resistant to accepting tablets and so the product must be mixed in with their normal diet. The most usual way of achieving this is by coating a dry pellet with the treatment mixed with oil, known as top dressing.

Dose rates are usually given as *X* mg of active formula per kilogram of fish for *Z* days: for example, 10 mg of oxytetracycline per kilogram of fish for 10 days.

To mix this dose, the total weight of fish to be treated needs to be calculated. If there are 45 kg of trout in the tank, then 450 mg of oxytetracycline will be required per day for 10 days, a total of 4500 mg. This then needs to be mixed with the total diet for the 10 days.

Box 19.3 Calculating the volume of a tank

1. Calculating volumes in square or rectangular systems

Length × width × depth = volume.

If the measurements are in centimetres, divide the final figure by 1000 to give volume in litres.
If the measurements are in metres, multiply the final figure by 1000 to give volume in litres.

2. Calculating volumes in circular tanks

Radius of tank × radius of tank × 3.14 × depth = volume.

It is recommended that about half the routine ration is used, hopefully ensuring that the fish are hungry and will consume the diet rapidly. If this is not known then 2% body weight per day is a good guide figure, namely 0.9 kg per day or 9 kg for the 10 days.

Therefore 9 kg of pelleted diet should be mixed with 4500 mg of oxytetracycline and then 90 ml of sunflower oil should be added to bind the oxytetracycline to the pellet. The resultant mixture should be split into ten equal parts and fed to the fish once per day. The portions should be placed in a sealed bag in a fridge. It is important that the fish consume the pellets quickly so they should not be fed in any other way. Although the oil is effective as a binder, the coat will wash off the pellet fairly quickly.

Bath

A very effective way of getting treatment to a fish is by dissolving it in the water that the fish is living in. Many chemicals are aimed at ectoparasites so are easily used in this way. Some compounds are also taken up over the gills and can therefore get into the fish from the water. For these chemicals, dosage rates are given in milligrams or millilitres (ml) of active ingredient per litre of contained water. This requires the total volume of water in the container to be known (Box 19.3).

Again, it can be seen that unless the fish can be contained in a small volume of water, then this sort of treatment is inappropriate. The most significant problem in treating open waters, however, is the effect any treatment will have on the other inhabitants of the system. Besides this, it would be illegal to add most chemotherapeutics to an open system.

19.2.3 Antibiotics

The only suitable cure for a systemic bacterial infection is the use of antibiotics. These drugs come in a whole variety of forms, but only a few are fully licensed for use on fish. Antibiotics work by directly killing the bacteria, but require time to achieve this, and as such are usually administered over several days. If at the end of this course of treatment there are any surviving bacteria, they will reproduce rapidly to fill the space left by those killed by the antibiotic. These remaining bacteria will be those most resistant to the antibiotic. If this is repeated over many generations of bacteria then eventually completely resistant bacteria are selected. This is exactly what has happened in some areas of fish farming, where strains of bacteria, particularly *Aeromonas salmonicida*, are completely resistant to many types of antibiotic.

If you want to use antibiotics you must therefore assess which bacteria is causing the disease, which antibiotic it is sensitive to and you must administer it in such a way as to kill all the bacteria. Antibiotics are often completely useless in a fishery for many reasons, mainly associated with problems in administering the correct dosage to all fish. Antibiotics are usually 'prescription only medicines' and therefore require a veterinary prescription.

19.3 Fish disease legislation

For most countries fish disease legislation is based around importation controls. Pathogens and indeed particular fish species are prohibited from entering a country and consignments are randomly checked at the border to ensure that they are conforming to this legislation. In some countries there is also some control of fish movement into inland waters. Controls are placed on the introduction of any fish into the 'wild', recognising the irreversible nature of any problems and the potential ecological concern of introducing inappropriate species. The Legislative framework surrounding the control of fish disease is discussed in more detail in Part III.

There are also 'codes of best practice' that are important in this area. The ten-point code set out in Box 19.4 is an example of the sorts of codes of conduct that are useful in this area.

Box 19.4 Code of best practice for disease control

I. Do you need to stock?
Get professional advice in the first instance. Stocking is often not the way to improve your fishery. In fact, you may be doing more harm than good!

II. Make sure all relevant paperwork is in order
Personally make sure that all the consents have been obtained.

III. No supplier is officially recommended
Although there are many reputable suppliers of fish, no fish farms or dealers are recommended or approved by official agencies. Anyone claiming to be recommended is giving false information.

IV. Only buy from reputable farms or dealers
Follow the recommendations of fishery owners you know who have had a good service. Ask your supplier for customers' references and contact them. Beware of bogus customers. Shop around and always think 'quality first: cost second'.

V. Be a careful buyer
Buying fish is just like buying anything else. Protect yourself by paying by cheque, obtain a receipt and keep copies of all of the paperwork.

VI. If possible attend the removal
Try to be present when fish are collected from the source water. Make sure that the fish are delivered straight to your fishery and not held or mixed together with fish from different sources in transit. Make sure that health checked fish are not mixed with unchecked fish.

VII. Be there on stocking!
Always be present when the fish are stocked into your fishery. Insist on a delivery date and time which gives you time to prepare for the arrival of your fish.
Do not accept things you, the customer, are not happy with, no matter what the supplier says. Do not put your entire fishery at risk.
Never accept fish in a poor condition.
Never accept species you have not ordered.
Never accept fish at night – darkness can hide quality and quantity.
Do not be pressurized by the supplier: remember you are the customer.

VIII. Never accept fish unless you are satisfied that they are healthy
You cannot tell if a fish is healthy just by looking at it. Health checks are the best way of ensuring that fish are healthy.
Do the fish you are stocking have a valid health check?
Did a competent person (a recognized fish health expert) produce the health check?
If you are not sure or are confused, contact a professional fisheries officer before you introduce the fish.
Remember: a health check only applies to the fish species checked. If it is for carp, then it only covers carp!
Health checks do not guarantee the fish are healthy, but they do provide the best protection possible.

IX. Make sure you are the boss!
Ensure that the supplier is willing to comply with your requests.
If the supplier is not willing to comply, then ask why not. You are the customer. You can always take your business elsewhere.

X. Ask!
Ask local fisheries officials agency for help and advice.
Contact your local fisheries official, who will treat your enquiries in confidence. He or she will always be pleased to advise you.
The diseases you are trying to prevent represent a real and significant threat to your fishery. Once they are present they are nearly impossible to eradicate: prevention is the only method of protecting your fishery.

20 Biosecurity

Diseases represent a serious risk to the sustainable management of any fishery. Treatments are almost always completely useless in a fishery. Therefore fisheries managers must make every effort to avoid problems with disease, not only for the success of their fishery but also to prevent problems for other fisheries in the area.

Biosecurity is about being aware of the ways in which disease can spread and taking every reasonable measure to minimise the risks of this happening. It includes the practical things you can do on your fishery to help prevent the introduction and spread of disease to and from your fish. Good biosecurity should be practised at all times, not just during a disease outbreak. Taking the right measures in the early stages can help prevent or reduce the spread of a whole variety of potentially damaging diseases. These measures will also include the following:

- help to reduce the risk of other aquatic pests gaining entry to the fishery;
- protect your neighbours and the other fisheries around the country;
- keep new disease out;
- reduce the spread of disease;
- keep more fish healthy;
- cut the costs of disease prevention and treatment.

It is recommended that a detailed biosecurity plan be established for each site, which aims at identifying the disease risks and putting in place ways of minimising the risks. The plan should also incorporate good husbandry practice. Then, if the worst happens, and the pathogen does make it into the fishery, its spread can be reduced or stopped. The healthier the fish, the less likely they are to succumb to a pathogen and so a disease outbreak may be avoided. The plan should also help prevent the infestation of the fishery with invasive species of plants and animals which can have damaging consequences.

It is also recommended that this process is approached in a methodical manner to try to ensure that all the possible risks are identified. The best way to achieve this is to combine a site biosecurity plan with the general management plan.

20.1 Site biosecurity plans

The site biosecurity plan is designed to be an effective tool to limit the spread of disease-causing organisms. It should allow all the various health management issues to be integrated into one scheme and show how each aspect helps in achieving the final result. In most cases this plan will fall into two sections:

I. Preventing pathogens gaining access to the site.
II. Good husbandry to create healthy fish.

This chapter deals with the first section, preventing a pathogen from gaining access to the site.

20.2 Prevention

Obviously, the best way to prevent a transmissible fish disease from causing problems is to prevent its access to the system. To achieve this, all possible means of access should be monitored and controlled. Each site will be different and the amount of control that can be placed on potential points of entry will vary, but every attempt should be made. So the first question to identify how disease may access the site is as follows:

- introduction of new fish;
- movement of animals, people and machinery between and within fisheries;
- fishery visitors: customers and their fishing equipment and vehicles;
- contact with other fish from the wild;
- contamination by predators and wild birds;
- transfer of contaminated water, for example contaminated supply, such as rivers and streams.

Each of these possible routes needs to be identified and controlled or monitored.

The first aspect of the plan should be to review the site and secure the perimeter as much as possible. In most cases the perimeter can be secured, leaving three main points of association with the wider environment: the main fishery gate, the water inlet (abstraction point) and the water outlet (discharge point).

20.3 Restriction of access

Any site management plan should suggest that the fishery will need to block as many of the vehicle access points as possible without making life difficult for paying customers or members. This has significant advantages as far as the biosecurity plan is concerned, and the two systems should be run as one. All the work that is done benefits both areas of the work of a fishery. The restrictions of access should therefore be performed in the same way as for controlling fishing activity.

20.4 Controlling customers

The risk from customers is that they, or their equipment, may be contaminated. This is a real risk and needs to be reduced. Some equipment is obviously more susceptible to contamination than others; keep nets, landing nets, unhooking mats and waterproof net storage bags are of particular concern, as these are in close contact with fish for long periods of time, are generally wet at the end of a fishing session and are often put into net bags for transport and storage.

One way of reducing the risk, therefore, is to insist that customers disinfect these items before use on the fishery. This is acceptable practice and is common in many fisheries. A two-container system is usually best, with one container containing disinfectant and the other clean water to rinse the equipment. Wheelie bins are particularly useful for this task as they are easy to move and have an easily lifted lid.

The most commonly used disinfectants are iodine based (iodophores), which are easy to apply, have minimal toxicity to fish and are non-persistent. Another very useful product is Virkon S, which is also readily available and has been proven to be effective against many fish viruses. Other products available include peroxyacetic acid, para-acetic acid and sodium hydroxide. All these chemicals should be used, stored and disposed of according to the manufacturer's instructions. We would endorse the use of Virkon S as the best of the products available. The customer should be reminded to peg out their nets to dry while they set up in their swim.

The disinfection route works in several ways. First, it gets the customer to disinfect. Secondly, it shows that the fishery cares enough to undertake this process, which helps the customer to see its importance. Finally, some customers do not like dipping their keep nets, so if they know net dips are used they often will not bring their nets along, thereby completely eliminating the risk.

The disinfectant and water will need to be changed regularly to ensure the system is really effective and, unless you have the staff on site and available to do this work, you should consider other methods.

A second way of reducing the risk from equipment is to ban its use on the fishery. Where equipment is essential, then the fishery can supply it directly to the customer on loan, ensuring that the equipment never leaves the site.

It should also be noted that the owner, manager and workers on the site represent a significant risk. They are often anglers and may even work on other fisheries when not on yours. They must undergo the same restrictions as any normal customer.

20.5 Unauthorised visitors

Another source of disease introduction is through unauthorised visitors in the form of poachers and members of the public. This is one of the main reasons for restricting access to one gate, as the perimeter fencing will discourage poachers and random visitors such as dog-walkers.

20.6 Predator control

It would seem obvious that any fishery would attempt to control predators but in many circumstances it may not appear cost effective, owing to the relatively small quantities of fish consumed by the predator. However, many predators can either be hosts to parasites that infect fish, or act as transmitters of pathogens. Many predators such as herons and kingfishers have relatively small home ranges and so the risk is limited. Predators such as cormorants have significant ranges and can potentially bring pathogens from far and wide.

On most fisheries it would be overly detrimental to the business to use the sort of predator prevention methods seen on farms, such as netting. The best methods therefore are those that make it difficult for the predator to catch the fish. The creation of fish refuges, such as reed-beds and other natural shelters, allow the fish to escape more easily, making it more difficult for the predator to catch the fish, which deters them from visiting your fishery. Floating islands associated with fish refuge cages can be particularly effective in fisheries where natural refugees are limited (see section 17.3).

20.7 Water inlet and outlet

In addition to predators, other wild fish may enter the site from the surrounding environment through the inlet or the outlet from the site, which can be a source of extraneous and sometimes repeated infection. This may be prevented by installing and maintaining adequate screening on the intake and discharge points.

20.8 Fish introductions

Obviously, by far the highest risk of a disease introduction is through the stocking of infected fish. This risk should be assessed and reduced as far as possible by following the procedure detailed in Chapter 12.

21 Control of Fishing Activities

21.1 Introduction

21.1.1 The role of management

Key roles for a fishery manager include protecting the resource, and informing and enforcing any rules. The aim of this chapter is to describe the management of access and fishing activity on a fishery. This requires the activities of anglers to be monitored on a regular basis.

It is important to understand some basic principles about the monitoring and control of access and angling activities and how to deal with any incidents that arise. Of course, most of the time the underlying aim of the fishery manager is to encourage anglers to come and visit as customers, and to make access easy. This may seem self-evident, but it does require further consideration. The job of controlling access and fishing activities is as much about facilitating the customers as it is about enforcing rules. If the balance is wrong then enforcement activities may well discourage, or alienate, the very people that the fishery needs to encourage. This chapter mainly deals with the rule-breakers and how to monitor, prevent and deal with such activities. However, a good fishery worker will always consider his customer as he goes about his duties.

The basis of this is good assessment and planning, which will make the tasks described easier and less time-consuming. This will give the best protection possible while also allowing your customers to enjoy their fishing with the minimum of disruption.

21.1.2 The national context

National legislation controlling fishing activities will vary according to the country where the fishery is located. Some primary legislation, such as rod licences, is not a legal requirement in every country. For example, in England and Wales a rod licence must be obtained from the Environment Agency before fishing; in many other countries no national licence is required. The powers of individuals to act will also vary according to the country and the legal position of the individual employed to complete these tasks. For example, in some countries there are various categories of fishery workers, such as bailiffs, wardens or keeper/ghillies. Each has a specific role and specific powers that are granted to that role. It is important that an individual understands the limitations of what he is able to do within his role and at which point help should be sought from higher officials within the organisation (fisheries managers) or from warranted law enforcement officials (the police and the Environment Agency). This is a matter not only for the individual concerned but also the organisation. It is very important that every organisation trains staff to understand the role that they are to perform, the limitations of that role and where to seek help if the limitations are realised.

For any fishery manager it is important to build relationships with the local government officers. Early discussions with these bodies will help clarify the roles of private organisations and individuals and avoid crossover in enforcement or fisheries protection activities.

21.1.3 The role of bailiffs

The word 'bailiff' has a legal meaning that may differ from place to place; for example in Scotland a bailiff is a warranted law-enforcement staff member. However, the term 'bailiff' is commonly used to describe other fisheries staff working on fisheries who do not have a warranted law enforcement remit. For example on a private fishery the fisheries manager may employ someone to collect tickets and enforce the fisheries rules. This person may be called a bailiff, even though they have no warranted powers. The term 'water bailiff' or 'fisheries worker' used in this book does not mean a warranted bailiff.

21.2 Risk assessment

Preplanning can and will help to make effort and expenses in this area go farther. A useful tool in this is to complete a risk assessment. This will help the fishery manager understand the real dangers and target the effort in the most effective way.

21.2.1 Access

People will want to access a fishery for a variety of reasons. Predominantly this will be for perfectly legal fishing by the customers or members. However, a variety of people will want to access the site for other reasons, and these will often change the way access is managed and will vary depending on the type of fishery and the time of year. As long as this is kept in mind, a monitoring strategy can be adopted that will cover most eventualities.

Three main groups of people may pose a risk to, or cause problems for, a fishery:

- The first relate to illegal fishing and fish theft. They come under the heading of 'poaching or Illegal fishing'. This covers people who are on the site entirely illegally, and therefore monitoring and prevention can be dealt with as one.
- Secondly, there are people using unauthorised or illegal fishing methods. These are customers, as they will have bought tickets, but then break the rules of the fishery.
- Finally, there may be access by people you may not wish to be there, such as tourists or dog walkers. Although this type of access may be described as undesirable, it may be perfectly legal. So monitoring and controlling access will also require the careful handling of people who have a perfect right to access the land, but may still come into conflict with the paying customer and the owner of the land or fishing rights.

Problems associated with legal access

Many people will have entirely legal access to the fishery, not just anglers. In some cases this can lead to conflict. Owing to legislation, such as the Countryside Rights of Way (CROW) act (in England and Wales only) and the Land Reform Bill (in Scotland), more people than ever have access to the countryside, which may include some fisheries. Where it becomes difficult is where right of access is allowed, but the person then engages in an activity that may harm your fishery or fishing. Obvious examples of this conflict include water sports such as canoeing, but can be as strange as gold panning. If this is not incompatible with any law then another method must be considered to avoid damage or conflict. This will normally take the form of raising awareness of the problems associated with that fishery. The fishery must engage in reasonable dialogue as soon as possible with those who want to proceed with these activities. It is common for these situations to escalate quickly into arguments and even confrontation. However, with planning and equitable division of the resource, for example usage during certain times of the year and for certain days of the week, all users can reasonably be accommodated. On many water bodies the formation of 'stakeholder' or 'user' groups is now an accepted method of ensuring that all interested parties have a chance to input into management plans, and the ongoing activities that take place.

Controlling access

The fishery will need to block as many of the vehicle access points as possible, without making life difficult for paying customers. It is usually best to restrict routine access to one point of entry, associated with the car park if available. This establishes an obvious monitoring point for vehicle access, allowing vehicles to be checked, and it means that any illegal entry would have to be through the most active part of the site.

Restricting access in this fashion also helps in several other ways, as it will mean that all your customers must enter and leave through the one access point. This allows the fishery to place good quality, promotional or signature signs at the entrance. Providing information to the angler will help to establish a fishery of quality. The fishery can also place notices where they will easily be seen. The entrance or access point should be maintained to the highest possible standard. Not only does this help to establish the quality of the fishery to the paying customer, it also helps to establish the fishery as one that is likely to take the time and effort to apprehend rule breakers. Locking gateways with numeric or key padlocks and distributing number codes or keys to staff and regular anglers can also be a very effective method of ensuring only authorised vehicles can gain access to the fishery. This can even work in the case of day-ticket fisheries, where the gateway can be opened first thing and closed last thing at night, although this is obviously very expensive in terms of staff time.

It may also prove useful to have details of your customers' vehicles, or to issue window stickers to be used in the appointed parking/access areas.

If it is not possible to restrict access totally to one point, then it is recommended that a particular access point is recognised as the 'main gate' and all other access is restricted to a minimum. These other access points should not be used, or they should be maintained to the same standard as the main entrance.

Where access points have been discovered that are not required, these can be blocked in a variety of ways. Concrete blocks are a favoured method, but many natural materials such as logs or earth banks can achieve the same end. When blocking access, bear in mind, if it is remote, that a minute with a chainsaw can remove a wooden post and large four-wheel drive vehicles can easily pull a lot of things out of the way.

During the risk assessment, non-vehicular entry points will have been discovered. These will not be suitable for vehicles, but will allow people easy access. As many of these access points as possible should be removed. They can be blocked with normal fences if it is considered suitable. However, it may be better to consider less-intrusive methods such as the planting of gorse, bramble or other plants that can deter access. These make for a more pleasant look to the boundary, are a lot easier to maintain and do not encourage the 'I wonder what's over that fence' attitude.

We will cover the routine monitoring of a site later in the chapter, but there are plenty of sites and fisheries that do not have any routine monitoring. The simple act of notifying people that there is regular and routine monitoring occurring on site will put off some potential problems. Therefore notices should be placed around the site informing everyone that this is the case. These notices should be kept clean and maintained as this establishes that it is a current aspect of the management of the fishery. Routine, preferably daily, visits should have their timing varied as much as possible, so that it is not apparent that the fisheries employee visits at 2 p.m. every day! Even on fisheries where surveillance is often the main purpose for the visits, make sure that occasional high-visibility trips to the fishery are made, again establishing that it is a current aspect of the management of the fishery.

The signs (Figs 21.1 and 21.2) can also be used to inform people that there is a fisheries manager and where to contact him if they see anything suspicious.

Once the access points have been restricted it is possible to use some high-tech solutions to prevent and monitor access.

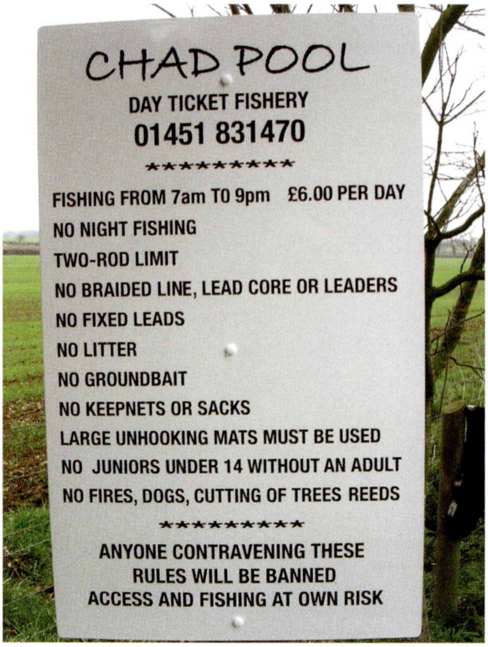

Figure 21.1 Contact details should be prominently displayed.

Figure 21.2 Localised fishery rules.

A cheap and easy solution is the use of passive infrared (PIR) sensors, which can detect and react to movement. These can be installed at likely entrance points. They can be rigged to switch on lights, turn on horns or even contact mobile phones. They work on the simple principle that if something breaks the beam the lights go on, or you get a telephone call. The major problem with these in fishery situations is that, unfortunately, the sensitivity of commercially available units at this time means that every fox, deer or badger can trigger them.

Closed-circuit television (CCTV) coverage can also deter unauthorised access. However, problems arise with the cameras not being sensitive enough to record sufficient detail for identification or to read details of cars. CCTV equipment can also be the actual target for theft.

For the above reason and the fact that both these methods require a power source, they are not generally suitable for remote areas.

Legal obligations of allowing access

The fishery needs to ensure 'safe' access for all those entering the fishery. There is a legal obligation to ensure that even those entering the fishery without specific permission do so in safety.

This obligation is important. It should be highlighted on permits, membership cards, and so on. that the individual is responsible for his or her own safety and that fisheries can be hazardous by nature, for example slippery banks, and so on. Specific hazards should also be posted on a notice board, as should a general warning. However, it should be recognised that this would only cover normal risks associated with that sort of area. Any structure that is established by the fishery, say an anglers peg, must be well built and maintained.

21.2.2 Procedures for drawing up a risk assessment

The first step is to draw up a risk assessment for the whole site. This will determine the risk to your fishery of any or all of the above groups of people. A good way to approach this is to start with a review of the whole site with a 'poacher's eye'.

Start from the beginning and check for the most likely access points. First, check access points that can be used for vehicles and equipment. How are people going to get close enough to be able to transport equipment to the fishery and then leave with the fish and get away? Reasonable access at the right time to catch fish is one of the most important factors for would-be poachers. This can be quite difficult, as there are only certain times and places where fish are easily taken, so determining peak risk times is important. The protection of a coarse fishery will involve the assessment of several different factors and may call for protection effort to be at its greatest during spawning times.

Where are poachers going to take fish? This will largely be dependent on the type of fishery. On a coarse fishery this may depend more on the method of taking. For example, where can a seine net be landed? Where can a fyke net be effectively laid?

This leads onto the next area of the review.

Poaching

Where and what form will fish removal take? If you needed to get a few fish out of your fishery, say for a health check, how would you do it? The best and easiest way is often going to be the way a poacher tries.

Poaching methods

There are numerous methods for poaching fish. The main methods currently used include the following: electro-fishing; seine nets; fyke nets; gill and trammel nets; rod and line – normal angling; fixed lines; foul hooking grapples; multi-hooks. Some of these methods are damaging to the fish, such as gill nets, trammel nets and foul hook grapples. Others, such as rod and line, are preferred for the capture of larger specimen fish. More details on the construction and operation of these gears are given in Chapter 11.

Poaching by anglers

A final method that is important to note is that of legal anglers poaching fish. This is one of the most prevalent methods of poaching coarse fish. A legal angler will be a person with a valid licence and permission to fish that water. However, this regular angler could be causing a poaching problem by occasionally taking 'one for the pot' without permission. There is also another type of angler who, when they catch a large, valuable fish, will have a number to phone. They keep the fish in their keep net or carp sac and then another person will arrive at the fishery and pay them cash for the fish. This is becoming a problem throughout the UK and has even led to cases of particular fish being targeted and stolen to order.

The actual poaching method chosen will depend on several factors, including the fish species to be taken and the lay of the land. Obviously, some of the techniques cause too much damage to the fish and are rarely used.

To prevent unauthorised fishing, you need to decide which is the most likely method, where the best place is to deploy that technique and at what time of the year. Identify and list all the methods that are considered likely to be used. Now list the likely venue for each of these activities, as each will work better in particular situations. For example, setting a seine net is relatively straightforward but landing it requires a clear, sloping bank and usually requires more than one person to set and handle the gear.

A poacher will have problems removing large quantities of live fish. As it is difficult and time consuming it requires very easy access. Although wheelbarrows can be used to transport fish to a waiting vehicle, it is much easier to get the vehicle to the bankside. Where the target is an individual large fish, this is much easier. However, moving them without arousing suspicion still presents problems.

Selling fish

Once removed from the fishery, how are the poachers going to get rid of these fish? It is true that without a ready market, nothing would ever be stolen. Selling the stolen fish is a problem and can be made even more difficult. Consideration should be given to marking large, valuable fish in such a way that they are verifiable as the property of the fishery. For large cyprinids, various marking devices have been developed such as PIT tags (see Box 21.1).

Other forms of unauthorised access

Many fisheries suffer from illegal access being gained by members of the public, with nothing more sinister on their minds than walking the dog. In some circumstances this can be a problem. There is the dog mess to deal with, and conflict with the legitimate angler to consider. However, there can be benefits, such as activity round the site making poaching more

Box 21.1 PIT tags

Passive integrated transponder tags, or 'PIT tags,' are small microchips (about the size of a grain of rice) that can be injected into a fish, often in the shoulder muscle, using a hand-held applicator gun. The fish can then be released back into the lake. The tag should have no impact on the fish.

If that fish turns up and the fishery needs to verify its identity, then a hand-held scanner can then be used to detect the PIT tags in the fish. The scanner reads the tag's electromagnetic code and displays the tag's number. This tag number is unique and allows the identification of that fish to be revealed beyond question as the fish previously injected with the tag.

PIT tags used are similar to the tags that are used to identify family dogs or cats permanently.

difficult and scaring predators. The fishery needs to develop a policy on what to do in these circumstances.

Prevention of poaching methods

The risk assessments will have highlighted likely methods, as well as times and places, where fish are likely to be taken. Depending on the likely fishing method, it may be possible to deter these efforts. For example, if seine nets are the likely choice then deliberate snags can be placed into the water. These can be stakes knocked in until they are just below the water line, or sunken obstructions. These will entangle the net and make it impossible to land. They will, however, represent a significant danger to other water users and all the health and safety implications should be considered before deploying these sorts of snags. Seine nets need a clear landing area, so snags or deterrents can be added into likely bank landing points. Take care that these do not interfere too much with anglers. Posts and sunken obstructions can make fishing difficult.

Part of the daily routine should include a look along the banks for fixed lines, fyke or trammel nets. These can then be removed. If the water is turbid then running a grapple through the likely areas may snag any nets.

Disposing of stolen fish

In reality, it is actually quite difficult to sell-on stolen fish. It is most likely that any stolen fish will be disposed of in the local area. It is therefore useful to raise awareness in the area local to the fishery, immediately after any fish theft. If the fish are marked, or photographs are available, this is even better.

Other considerations

Vandalism

Illegal access can lead to vandalism. Also, illegal access can lead to lots of litter, which in turn can lead to the fishing rights being revoked as anglers are often blamed for this litter.

Preventing unauthorised access and unauthorised fishing

Now the risk assessment has been completed, it should have identified weak spots for access, possible methods of fish removal, methods for getting the fish off site and so on. This can now be used as a template for developing strategies to prevent some of the highlighted problems.

Unauthorised fishing methods

In most countries there are some fishing methods that are specifically against the law. These will include fishing during a close season, or using illegal methods such as trawling from a powered vessel. In addition to this, most fisheries will also have their own set of rules. These should be included in the risk assessment and each one should be considered as part of this process. For example, if the fishery rules prohibit night-fishing, then the risk of this happening must be assessed and access for night-fishing should be considered. Furthermore, monitoring and prevention methods need to be established. Having, for example, a rule of 'no night-fishing' would certainly help with the management of resources, as any report of fishing at night could be assumed to be a poacher. A rule such as this needs enforcing and would require employees to visit and check that the rules are being obeyed.

It should be highlighted that any policy or rule that is established must be enforced. It is very counter-productive to establish rules that are either unenforceable or ignored by everyone.

Preventing the use of illegal methods by paying customers

When angling, there are certain methods of fishing that are illegal or that are prohibited by the fishery rules. It is likely that any people using illegal methods will fall into the category of paying customers. That is, they are members or have bought a ticket and have a licence, but they are still fishing illegally. Illegal methods will include methods enshrined in law, for example close-season fishing on rivers, trawling from powered vessels, use of certain baits, and so on (Box 21.2). These must be enforced as well as the fishery's own rules, so it is important that these are known.

Other rules can be set by the fishery and will generally include things like bait bans, gear bans, and so on. These restrictions will be relevant only to that particular fishery, and still need enforcing. The drawing up of the fishery rules and the way they are communicated to the members plays an important role in preventing rule breaking.

Considerable thought ought to be given to these fishery rules as they have the potential to make a significant difference to the financial viability of the fishery.

Box 21.2 Examples of national bylaws and illegal methods in England and Wales

The following are examples of issues covered by national legislation:

- Close seasons for coarse fish, non-migratory trout and migratory fish.
- The design, construction and use of keepnets, keepsacks and landing nets.
- The use of lead weights.
- The use of live baits and crayfish.
- National salmon byelaws in England and Wales.
- The number of rods that can be used.
- Salmon and sea trout catch returns.
- Unattended rods.
- Trawling from a powered vessel.
- The use of a gaff.

There are more aspects covered by national and local legislation, and these change from time to time.

Box 21.3 Example of a rulebook

Rules can be simple and straightforward, for example the following:
Rothley Marina

I. A total ban is in force on long-range feeders and floats.
II. Anglers must only fish to a maximum of 10–12 metres.
III. The use of catapults is prohibited.
IV. Anglers are asked to refrain from using foul or abusive language.
V. The use of bloodworm or joker is not permitted.
VI. All anglers are asked to be considerate when answering the calls of nature.

Put simply, there is nothing that puts off customers quicker than a perceived injustice. Anglers generally want to go fishing to enjoy the peace and quiet of a few hours spent on the bankside. The last thing anglers want is to read a long list of rules or to worry about whether they are breaking them, or for that matter to be harassed by an official who is also not sure what the rules are. It is important therefore that these rules are very clear, sensible and enforceable.

Setting fishery rules

Having said that fishery rules can be a problem, they are also important and will play an important part in the management of any given fishery.

Any problems can be eased if certain guidelines are followed:

- Keep the rules to a minimum.
- Consult stakeholders (e.g. the membership) about the rules.
- Do not include obvious rules such as litter. (Litter is an ongoing and serious problem but it is not best dealt with in this way as it can be assumed that littering is just generally socially unacceptable.)
- If the fishery has a membership list, make sure that all members receive a copy of the rules at least once a year. This can be achieved by sending them out with a newsletter or the membership reminders.
- Make sure that there is a full set of rules displayed at the entrance to the fishery.
- Enforce the rules evenly and fairly.
- Do not allow anyone to break the rules. It is surprising how often fisheries employees break the fishery rules. It is very irritating to the paying customer if the rules say no night fishing but employees or honorary bailiffs go night fishing as a perk.

Box 21.3 gives an example of a rulebook.

Know your customers

Preventing and monitoring unauthorised access and illegal fishing can be made easier if customers can be identified easily and quickly, preferably without approaching them. The fishery should therefore keep a log or diary of regular customers, or groups of anglers, and staff should be encouraged to get to know them. The log should include details of what cars they drive. It is obviously very useful to know who drives what car and therefore what cars are unusual. Furthermore, if there is an incident and the person involved gives false details, such as name and address, then they may still be pursued if there is a record of the car registration.

21.3 Conclusions

At this stage the fishery should now have a risk assessment, which should be written up in as much detail as possible. This risk assessment will have details of areas where work has been done to counter the risk. These should be noted as well as those areas where no work can be done at the present time.

Drawing a plan of the site, annotating the problem areas and completing a list of all the noted points, completes the risk assessment. This can be made available to all staff working at the site, so that everyone knows the details. An example of a risk assessment for a fishery is given in Appendix A.

22 Fish Nutrition

22.1 Introduction

Angling has an impact on the aquatic environment and on the fish that inhabit a recreational still-water fishery. By definition, the fish in a still-water fishery are captive and restricted to their environment, yet often very little thought is given to the vast amounts of bait that are deposited into the water in the form of hook baits, free offerings and groundbait. Baits are chosen with very little consideration as to their suitability as fish diet, or for any direct or indirect effects they may have on the aquatic environment. However, there is little doubt that, in many circumstances, anglers' bait provides most of the nutrition available to the fish stocked into still-water recreational fisheries, especially where the stocking is intensive and exceeds the natural carrying capacity. Fish nutrition is therefore very relevant to the fisheries manager because angling should be performed in such a way that it has, and can be seen to have, minimal detrimental impact on fish and their environment.

Fisheries managers should provide their fish with a balanced and healthy diet. To do this they need an understanding of the principals of fish nutrition.

Fish require nutrients for the following reasons:

I. Maintenance.
II. Growth and regeneration.
III. Reproduction (gonad development).

Nutrients can be broadly divided into two groups: major and minor, as shown in Figure 22.1. The main nutrient groups are summarised in Table 22.1.

All animals are heterotrophic, which means that they cannot synthesise their own food. They must therefore obtain organic carbon and nitrogen from other animals or plants (autotrophs). The process by which they utilise their food is described below.

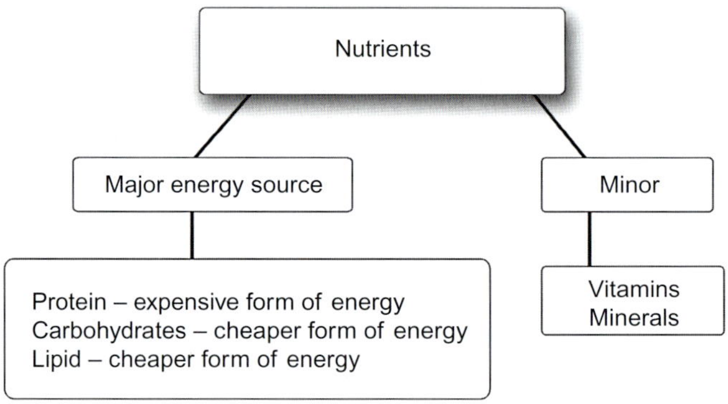

Figure 22.1 Major and minor nutrients.

Table 22.1 A short description of the major nutrient groups

Nutrient group	Description
Proteins	Complex chains of amino acids. Some amino acids are classed as essential: they must be provided in food as the fish cannot synthesise them. It follows that fish can synthesise the non-essential amino acids; they can do this only in the presence of nitrogen. A third group (semi-essential) can be synthesised but very slowly.
Lipids: fat	Lipids are complex chemicals, in their simplest form they are esters, glycerol and fatty acids. They occur in either liquid or solid form. Highly unsaturated fatty acids are regarded as essential (not all and not in all fish).
Carbohydrates	Carbohydrates are compounds ranging from the sugars to the starches. They are all composed of carbon, hydrogen and oxygen.
Vitamins	Vitamins are organic compounds, distinct from amino acids. They are only required in comparatively very small amounts. Nevertheless they are essential. They broadly fall into two groups: water soluble and fat soluble.
Minerals	Although many minerals are essential, most are not limiting except where the supporting or holding water is mineral deficient.

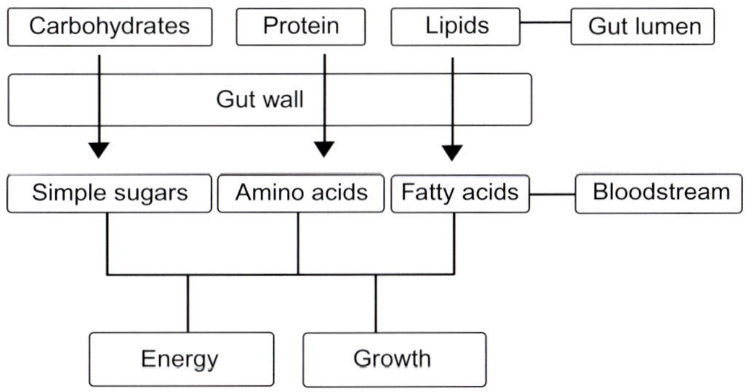

Figure 22.2 Metabolism of major nutrients in the fish.

Ingestion: taking in complex organic food (insoluble).
Digestion: breakdown of large, insoluble, complex organic molecules into smaller, simple soluble molecules. Examples include proteins to amino acids; carbohydrates to simple sugars; and lipids to fatty acids.
Absorption: uptake of soluble molecules into the bloodstream and body cells.
Assimilation: simple molecules used for energy or building into the constituents of organism.
Egestion: elimination of undigested food (waste materials and faeces).

These processes are summarised in Figure 22.2.

22.2 Components of the diet

22.2.1 Protein

The most important nutrient, on both a qualitative and quantitative basis, is protein. Protein is most important as it provides the basic building blocks for growth, although it can also be used for energy. Proteins are broken down to their constituent amino acids within the body and then reassembled into different useful proteins. Animals cannot manufacture protein without the requisite amino acids being present as part of their diet.

In aquaculture, the protein used for fish diet must be high quality and needs to be easily digested by the fish. For example, feather meal is almost 80% protein; it is, however, almost indigestible to fish as are many proteins from plants. Feed ingredients that provide protein are the most expensive.

Function of protein

Proteins have a structural function in the production of skin, scales, bone matrix and tendons. Keratin forms fibre protein, whereas collagen forms ligaments, although fish actually have only small amounts or low levels of collagen as they have no need to counteract gravity during movement.

Other important functions of proteins are the following:

- Movement – muscle.
- Hormones – insulin (chemical messenger).
- Immunity – antibodies (blood clotting protein/fibrin).
- Transport – haemoglobin (globular protein, membrane protein, receptor sites).

Fish require thousands of enzymes, which are proteins, to control chemical reactions within the body.

Protein structure

Proteins are built from chains of amino acids. There are 20 amino acids available to make up different proteins. All have a carboxyl group, COOH, of the alpha (α) carbon; they also have the amino group, NH_2. A third component that is common to all proteins is the R group, which is variable. It is the R group that makes the 20 amino acids different. It can be acidic, basic, neutral, hydrophobic or hydrophilic.

To form a dipeptide, a COOH combines with an NH_2 group. A condensation reaction can form peptides (Figure 22.3).

When a dipeptide bonds with another peptide, this is called a polypeptide, which can be a chain of perhaps 20 or more peptides. When chains over 50 are produced, these are known as proteins. Over 300 in a chain is not unusual.

Perhaps it helps to look upon amino acids as letters and proteins as words. The combination possible is a quite awesome: 20 to the power of 300. The sequence of amino acids in a chain is specific for each protein. This is determined by the genetic code in the nucleus. DNA codes for amino acids being formed into proteins. Protein molecules form very specific shapes that determine their function.

Gross protein

When looking at what ingredients are present in food, the level of protein is often quoted as 'gross protein'. The protein requirement for different species is sometimes difficult to ascertain, but it is influenced by the following factors:

- fish size;
- protein quality (digestibility and amino-acid balance);

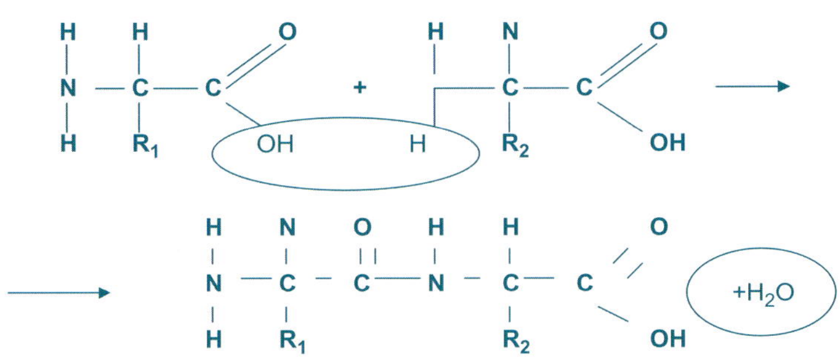

Figure 22.3 Synthesis of peptides.

- feed rate (fed less, require more protein);
- temperature.

Essential amino acids

Essential amino acids (EAAs) are amino acids that cannot be synthesised by animals. Growth will be limited if any of these essential amino acids are at lower levels than required. Non-essential amino acids can be synthesised from others. Essential amino-acid requirements are generally species specific.

In an aquaculture, complete diets are formulated to amino-acid requirements, rather than protein levels. When providing supplementary diets or bait products, on which fish stocked in a recreational fishery may become dependent, the fisheries manager might be best advised to adopt a similar strategy. This can be done by combining different protein sources, to balance the amino-acid profile, and by supplementing with synthetic amino acids. For example, soya bean protein supplemented with the EAAs methionine and lysine.

Simply feeding fish with excess protein is wasteful and often polluting.

22.2.2 Lipids, fats and oils

Lipids, fats and oils are a diverse group of organic compounds that are insoluble in water and polar compounds, but are soluble in organic, non-polar solvents.

Animals use lipids for energy and for storage of chemical energy. On demand, energy release from a lipid source is equal to more than twice that of a similar amount of protein or carbohydrate. Lipids are very useful for long-term storage, whereas the carbohydrate glycogen is very useful for short-term storage.

Lipids also performseveral essential functions in fish, including the following:

- cell membranes (phospholipids), bi-lipid layer, essential fatty acid (EFA) requirement;
- transport and storage of fat-soluble compounds (vitamins and pigments).

Lipid classification

The main types of lipid are shown in Figure 22.4. Triglycerides are the most abundant lipids. Their main function is as an energy store. They contain only carbon, hydrogen and oxygen.

Fatty acids

Fatty acids consist of hydrocarbon chains with a carboxyl group at one end. They contain an even number of carbon atoms. They may be:

- saturated single bonds of carbon;
- unsaturated double bonds of carbon;
- monounsaturated – one double bond;
- polyunsaturated – numerous double bonds.

Generally, the following applies:

- animal fats – more saturated;
- fish and vegetable fats – more unsaturated.

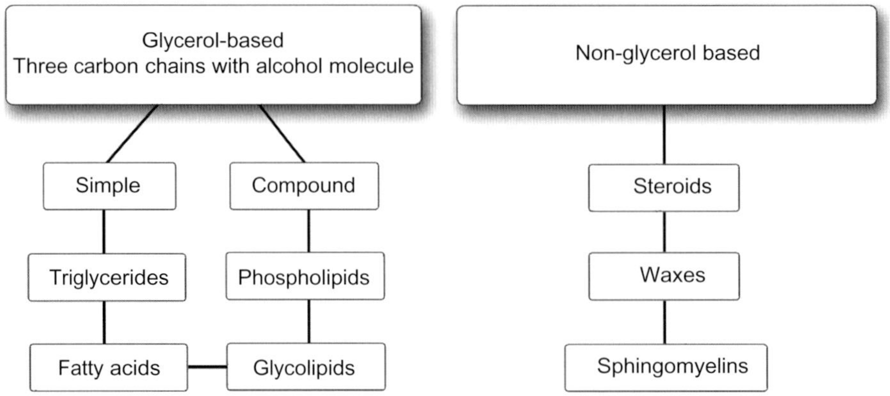

Figure 22.4 Lipid classification.

The more unsaturated, the more liquid the fat will be. Omega 3 (ω3) fatty acids are the most unsaturated fatty acids occurring in biological systems.

Phospholipids

Phospholipids are similar to triglycerides but are fatty acids that have the methyl group replaced with a phosphate and alcohol group.

The primary role of phospholipids is in the formation of biological membranes. These molecules are in a fluid dynamic state; cold-blooded animals need ω3 to keep lipids in a fluid dynamic state.

22.2.3 Highly unsaturated fatty acids (HUFAs) and Polyunsaturated fatty acids (PUFAs)

High-concentration unsaturated fatty acids, known as HUFAs and PUFAs, are essential to life in many fish species. Omega 3 fatty acids are more abundant than omega 6 (ω6) fatty acids. The quantity of ω6 fatty acids is higher in freshwater and warm-water fish.

Effect

The fatty-acid composition of the diet influences the fatty-acid composition of the fish. Feed that contains high levels of ω3 fatty acids will increase the level of this component in the muscle and viscera of the fish eating it. Therefore, in aquaculture, fatty acids can be fed to modify the profile of the fish to suit human requirements but all fish diets should have a suitable lipid profile.

Fish can synthesise saturated and monounsaturated fats, but they cannot synthesise PUFAs without a relevant precursor. The ability to desaturate and elongate fatty-acid chains varies among fish species.

Essential fatty-acid requirements

Fatty-acid deficiency leads to stunting, fin erosion and swollen fatty livers. Freshwater fish generally require both ω3 and ω6 fatty acids.

Lipids have a protein sparing effect on growth when present in fish diets. Different species have different optimum levels. For example, salmonids require between 15 and 30%, catfish around 10% and carp between 10 and 12% lipid.

Lipid digestibility

Generally, lipids are highly digestible and their digestibility increases with the degree of unsaturation; thus fish oils are most digestible, provided they are not oxidised.

Fat oxidation

Unsaturated fats are unstable in the presence of oxygen; they are readily oxidised and can become rancid. Fish oil and fishmeal must be protected with antioxidants to ensure that they are suitable as a high-quality component of fish diet. Vitamin E is a natural antioxidant.

22.2.4 Carbohydrates

Carbohydrates are not an essential nutrient in fish diet, but they do have a value as an energy source and are relatively cheap, being the least expensive source of dietary energy. Carbohydrates all contain carbon, hydrogen and oxygen, with the general formula $C_x(H_2O)_y$.

Principal types of carbohydrate

Monosaccharides

Monosaccharides are the simplest of the carbohydrates. They are the building units of more complex carbohydrates such as aldehydes, which have an oxygen molecule at the end of the chain, or ketones, which have an oxygen molecule within the chain.

Glucose is one of the most important of the monosaccharides. It provides an immediate burst of energy and is a major respiratory driver in vertebrates. Glucose concentrations must be kept at controlled levels in the bloodstream of any animal, or diseases such as diabetes may result. Glucose also combines to form long chains, which exist in two different forms called isomers. These are known as alpha (α: starch) and beta (β: cellulose).

Disaccharides

Disaccharides consist of two monosaccharide units joined by a glycosidic bond, such as sucrose. Both mono- and disaccharides are sweet, soluble and crystalline sugars.

Polysaccharides

Polysaccharides are formed from long chains of monosaccharide units. The chain may be hydrolysed to revert back to a monosaccharide. They are not crystalline in structure or sweet in taste.

Starch

Starch is derived from plants, in which it acts as a reserve of energy, being converted to sugars as required. Starch is stored as small granules within the plant cell wall. It is difficult to digest in its raw form and fish that can use starch as a primary portion of their diet are adapted to digest it. Cooking expands the granular blocks of starch and makes it more digestible; these partly broken down starch chains are known as dextrins.

Glycogen
Glycogen is a carbohydrate derived from animals, in which it also serves as a reserve source of energy. It is stored in the liver and muscle and is converted to glucose when necessary. This form is, however, limited in fish.

Cellulose
Cellulose is a structural plant polysaccharide and is very common. It is made up of long stable chains of glucose and cannot be digested directly by animals. Some animals, such as ruminants, have a gut flora adapted to digest cellulose. However, many fish species, such as common and grass carp, have evolved adaptations to make use of it.

Use of carbohydrates by fish
The various species of fish use carbohydrates differently, depending on their normal diet.

Carnivorous fish
Carnivorous species have a naturally very low proportion of carbohydrates in their diet, except for the chitin in insects. Carnivores use carbohydrates poorly because the digestive enzymes needed are relatively inactive. Many species have poor glucose tolerance due to low insulin levels, so high dietary content of carbohydrates does not increase lipogenic activity. This happens, for instance, in coho salmon, which cannot convert excess energy into lipids. In fact, high densities of carbohydrate increase liver size and liver glycogen. Because of this, the maximum dietary level should be approximately 25%, although cooked starch is more efficiently utilised than glucose and dextrin and it avoids glucose peaks.

Carbohydrates have a slight protein sparing action and therefore have a limited but significant nutritional role in commercial diets. They are also used to bind dry pellets together.

Omnivorous fish
Omnivores use carbohydrates more efficiently than carnivores. Polysaccharides are generally used more readily than monosaccharides. Carp can digest 85% of starch at a dietary level of 48%.

22.2.5 Vitamins

Vitamins are organic compounds that are required in minute amounts for normal growth, reproduction and maintenance. There are two classes of vitamins: water soluble and fat soluble.

Water-soluble minerals
Water-soluble minerals can be stored within the body and are required continuously. They include thiamine B_1, riboflavin B_2, niacin B_3, pantothenic acid B_5, pyridoxine B_6, biotin B_7, folic acid B_9, B12, inositol, choline and vitamin C.

Vitamin C (ascorbic acid) is vital as it is required for formation of collagen (connective tissue). Vitamin C deficiency leads to symptoms such as poor wound healing. it also plays an important role in the function of the immune system; damaged and stressed fish have an increased need for vitamin C. It is also an antioxidant.

In feed manufacture it is extremely labile (that is, easily broken down) so it is usually included as a protected form. For example, the normal form of ascorbic acid is best replaced by the protected forms: ascorbic sulphate and ascorbate phosphate.

All B-group vitamins function as co-enzymes (that is, biological catalysts). Vitamin deficiency causes metabolic pathway problems due to failure of enzymes that need non-protein organic molecules to function properly.

Fat-soluble vitamins

Vitamins A, D, E and K are fat-soluble vitamins and can be stored within fats in the body of a fish. They can even accumulate and become toxic (hyper-vitaminosis), although this is extremely rare.

Vitamin A is required for the formation of eye pigments, membrane transport and reproduction.

Vitamin D is central to the bone calcification process.

Vitamin E is a natural antioxidant that prevents the oxidation of HUFAs in membrane tissues. The requirement for vitamin E increases with an increase in dietary oil levels and if rancid oil is present in a diet.

Vitamin K is required in the blood clotting process.

Meeting vitamin requirements

Natural food will normally provide sufficient vitamins in fisheries.

In intensive culture systems, fish must be fed a complete diet and the vitamins are provided by the addition of a commercially produced vitamin supplement to the feed. Commercial feeds are formulated to provide an excess of each vitamin. Vitamins do not store well. Therefore the 'best by' date for the entire feed will often be based upon vitamin storage alone. The most significant vitamin losses are C, E, thiamine (B_1), riboflavin (B_2) and A.

Providing a supplementary feed based on a commercial feed can ensure that the fish in intensively stocked recreational fisheries receive a balanced diet with all the essential vitamins included.

22.2.6 Minerals

Minerals are simple inorganic elements, such as calcium (Ca), phosphorus (P), sulphur (S), sodium (Na), chloride (Cl), magnesium (Mg), iron (Fe), potassium (K), iodine (I) and selenium (Se).

The actual mineral requirements of fish are poorly understood. However, as minerals occur naturally in abundance and can be absorbed from the water, it is actually quite difficult to formulate mineral-free diets. Minerals can easily be stored in the body of the fish.

Minerals haveseveral functions, which includethe following:

- Formation of hard tissues such as bones and scales (Ca, P).
- Formation of soft tissues and molecules (protein – S; haemoglobin – Fe).
- Acting as co-factors for enzymes and hormones (for example thyroxine – I).
- Maintaining osmotic balance (soluble minerals – Ca, Na, K, Cl), pH and nerve impulses.

The requirement for minerals is met through diet and absorption from the surrounding water. Phosphorus is the most important, but must be supplied in the diet as 0.4–0.9% available phosphorus.

Plant phosphorus is limited in its availability and some accessible compounds, such as phytin, are toxic. Fishmeal used in diet formulation includes phosphorus in a highly available form, often as di-calcium phosphate.

Fishmeal and animal by-products are a rich source of minerals, but feeds that contain predominantly plant ingredients should contain supplementary minerals including selenium (Se), zinc (Zn), I, Fe and P.

22.2.7 Energy

Energy is the limiting factor on the capacity of an animal to do work. Work can be heat production, mechanical (movement), chemical (tissue formation) and osmotic maintenance. Simply surviving is work in its basic form.

Energy is measured in calories:

1 calorie = 4.2 joules.

1000 calories = 1 kilocalorie (kcal) or Cal = 4.2 kJ.

This is the amount of energy needed to raise the temperature of 1 kg of water by 1°C

Animal diets are measured in kilojoules per gram (kJ/g). Fish actually have a very low maintenance energy expenditure, some one-tenth that of terrestrial animals. The reason for this is that they are cold-blooded (poikilothermic), they excrete NH_3 (not urea/uric acid) and they do not have to contend with gravity.

Energy use and requirements depend on temperature, the degree of activity, the nature and quality of the diet, the growth rate, size and any stressors. A successful fish diet, whether manufactured or natural, will balance nutrients such as protein against energy requirement.

22.3 Digestibility

The digestibility of a feed or supplementary item in the diet is important because it is a measure of the ease with which the feed is incorporated into the fish.

Factors affecting digestibility include the following:

- Protein digestibility: this depends on the source and other ingredients. For example, the digestibility of fishmeal protein will depend on its ash level, overall protein level, fish freshness and processing temperature.
- Fat digestibility is usually around 80–90% efficient. However, unsaturated fats are more digestible than saturated and they are liquid at room temperature.
- Carbohydrate digestibility is very variable, with most mono/disaccharides being 95% digestible although the digestibility of starch will depend on its level and degree of cooking.

- Feed rate: continuous feeding lowers digestibility.
- Composition: very high carbohydrate content lowers digestibility of other nutrients.

22.4 Diet formulation

22.4.1 Feed formulation

Two main types of feed are available: nutritionally complete and supplementary.

Nutritionally complete feeds have been developed and sold especially for intensive farming. They are designed to meet all the nutrient requirements of the fish and are formulated to meet the requirements of different species. Often the requirements quoted are approximations. If the requirement is unknown, then the rule is to use the nearest related species.

Supplementary feeds are feeds developed and sold for semi-intensive farming, or for intensively stocked recreational fisheries.

More is known about the dietary requirements of aquaculture species than those found in most recreational fisheries.

22.4.2 Aims of formulation

The overall objective of formulation is to maximise production at lowest cost. This involves a compromise between what is ideal and what is actually practical in the formulation of any diet. Considerations might include things like the price and availability of ingredients; the acceptability and palatability of the formulation; whether the ingredients are dry or moist; the ease with which the formulation can be formed into pellets for feeding; the keeping qualities and ease of storage; and any anti-nutritional factors such as trypsin inhibitors that will reduce digestibility.

22.4.3 Feed ingredients

The cost of the ingredients is, of course, a major component of the cost to the end user. Ingredients may either come from by-products of food processing (cheaper) or be produced directly from raw materials (more expensive).

Composition and digestibility are most important for evaluating ingredients.

Protein sources

Fishmeal

Fishmeal is generally between 60 and 80% protein, although low-protein fishmeals with a higher ash level are also available. The highest-quality protein with an ideal essential amino-acid profile will have a digestibility of between 80 and 95%. The quality of fishmeal protein is very variable, depending on the type of source fish, how quickly it is processed, whether the oil is removed and the drying process.

Fishmeal has a good mineral profile with available phosphorus and calcium and is very palatable; however, it is very expensive and environmentally unfriendly.

Animal by-products

Meal and bone-meal typically have a protein profile of 50–55%, although the protein quality varies. It has a 10% fat content. High ash content at 28–30% can lead to high phosphorus content and therefore increased pollution, especially in less digestible diets. Most originates from dry slaughterhouse waste, but mammalian meal and bone-meal is illegal in farmed livestock feed.

Feather meal

Feather meal has a protein content of 80% but low digestibility (50–70%) and needs to be hydrolysed.

Shrimp waste meal

Shrimp waste contains 40% protein and is high in chitin, has low digestibility and is very palatable. It is a valued source of carotenoid pigments and as such is used to colour the flesh of some species such as salmon and trout.

Plant protein sources

Soybean meal

Soybean meal is 48% protein. It has the best essential amino-acid profile of any plant protein, although it is low in methionine. Commercial carp diets can be around 50–70% soybean. Soya is considerably less expensive than fishmeal, but if it is used as a replacement then consideration has to be given to other factors, such as the lower energy, essential amino-acid and mineral content, and the high carbohydrate, trypsin inhibitor and phytates (phosphorus) content.

Full-fat soybean meal

Full-fat soybean meal is highly processed, by heating to 175–200°C. It has a relatively high lipid content at 18%. Soybean protein concentrations can be 75% or more, so they are justifiably more expensive.

Oilseed meal

Oilseed meal is 45–50% protein and is high in carbohydrate but low in lysine and methionine.

Grains

Grains are a dependable carbohydrate source, being 60-70% starch. Carbohydrate derived from grains is digestible by most freshwater fish species, but is very much less digestible to marine or carnivorous fish. Carbohydrate can be important as a binding agent in the production of commercial pelleted diets.

22.5 Diet manufacture: industrial

The following steps are followed in the standard manufacture of batches of feed consisting of multiples of metric tonnes.

- Weighing;
- Grinding – improves mixing, pelleting and digestibility;
  - <2.0 mm screen for grower diet;
  - <0.5 mm screen for larval diet;

- Mixing – dry homogenous blend then liquid ingredients;
- Pelleting – steam-drying to less than 10% moisture;
- Extruded – drying to less than 10% moisture;
- Top dress with oil (2–20%). Coat heat-labile ingredients.

Dense pellets are formed by compression and steaming (16% moisture at 70–85°C) to produce partial gelatinisation. They require a binder, are cheap to manufacture and have a maximum oil content of 16% (8% in pellet and 8% maximum as a top dressing).

Extruded pellets are a very different product as they are manufactured in a pressurised conditioning chamber (20–24% moisture at 125–150°C). They can hold oil at above 30%, even up to 40% for some marine fish diets. The pellets expand during extrusion and complete gelatinisation takes place. The pellets are low density so they float initially and are slow-sinking later. They are expensive to manufacture.

22.5.1 Crumb feeds

Crumb and small pellets are made in a variety of sizes, simply by crushing larger pellets. There can be a loss of water-soluble nutrients during the process.

22.5.2 Moist feeds

Although most aquaculture installations in the developed world use commercial pelleted diets, supplementary feeding of recreational fisheries using moist feeds is still common. Generally, moisture content of the pellets is in the region of 18–20% and can be a mixture of wet and dry raw material. Moist pellets go off very quickly but are often very palatable to the fish.

22.5.3 Wet feeds

Wet feeds are not widely used and might now only be found in countries with developing aquaculture. Diets containing 45–70% moisture are made from wet ingredients.

22.5.4 Flake

Flake is a specialised diet generally used for juvenile fish and in specialised systems such as aquaria. It is made from dry ingredients that are then flaked or finely powdered. The mixture is often processed into slurry then thinly spread for drying. A further development of juvenile diets is micro-encapsulation

22.5.5 Micro-encapsulation

Micro-encapsulation is mainly used for weaning juvenile fish in hatcheries, between the phases of feeding 'live' feed and pelleted diet. It is very expensive because it is highly processed. Nutrients are encapsulated in a water insoluble coating that is digestible to the fish but protects the water-soluble ingredients inside.

22.6 Nutritional considerations in recreational fisheries

The fisheries manager needs to answer some question before deciding whether supplements should be fed to the resident fish community.

The first question is, "Can the fishery support the level of stocking?". In other words, does the fishery's carrying capacity have the ability to support the stock from a purely nutritional aspect? If the answer is yes, then there is no problem. However, if the answer is no, some further action may be required. This presents the fisheries manager with two options: firstly, to increase the carrying capacity of the water; secondly, to provide supplementary feed.

Before beginning supplementary feeding, two estimates need be made. Unfortunately, that is all they will be, estimates. The fisheries manager will need to estimate the contribution of natural food items to the diet and how much anglers' bait contributes. In addition to this, a manager wishing to reduce feeding-related stress in his stock may want to make supplementary feed available as anglers' bait.

When considering bait as a significant contribution to the diet of the stock, the fisheries manager should ask the following questions:

- Does it supply a balanced nutritional food?
- Are chemical attractants and additives a positive or negative addition to the bait?
- Can bait have a detrimental effect on the fish and the fishery?

Palatability is very important in anglers' bait. The effort put into bait presentation by anglers and bait companies alike may leave fish more predisposed to these baits than to more natural food items, such as plant matter and animal protein in the form of zooplankton and macro-invertebrates.

Although a great deal is known about the nutritional requirements of carp, *Cyprinus carpio*, there are many other native and introduced cyprinid species that currently inhabit still-water fisheries. There is little nutritional information available on many of these other species. One might assume that there is some difficulty in drawing generalised conclusions on the dietary needs of other cyprinid species; however, available data show that there is a remarkable similarity in the nutrition of teleost (bony) fish. The fisheries manager will not go far wrong in applying the rule of always using the nearest species when formulating fish diets to supplementary feeding and in assessing the suitability of bait items. It is therefore suggested that it is not unreasonable to use the common carp as the standard upon which to base the nutritional requirements for UK and European still-water fish generally.

The following examines the general nutritional requirements of fish, and those specific to carp.

In common with other animals, fish do not have a true protein requirement but need a well-balanced diet of essential and non-essential amino acids. The crude protein requirement of cyprinid fish grown in extensive culture is estimated to be about 35%. An extensive aquaculture system has many parallels with a recreational fishery in that the energy expended in fulfilling basic life functions, including foraging for food, are broadly similar in both systems.

Before the composition of anglers' bait can be critically compared with the nutritional requirements of cyprinid fish, it is necessary to compile a definitive description of the optimum cyprinid diet. Table 22.2 lists the requirement by the groups of nutrients already described in Table 22.1.

The quality of protein is determined by the proportion of the essential amino acids it contains. The ten essential amino acids for fish are listed in Table 22.3. The bioavailability of these amino acids is central to the quality of the protein.

The lipid requirement of common carp (*C. carpio*) include a proportion of either fish or vegetable oil containing the required essential fatty acids 18:3ω3 (linolenic acid) and 18:2ω6 (linoleic acid).

Once these essential lipids have been supplied, the remaining lipid content can be used by the fish as digestible energy. Dietary lipids are the only source of highly unsaturated fatty acids (HUFAs). Lipids are also important for the transport of fat-soluble minerals (see section 25.2.2). Phospholipids are the main lipid of biological membranes. The digestibility of the lipid decreases as the lipid content of a diet increases above 10%, so cyprinid diets do not fit the profile of many high lipid salmonid diets.

Carbohydrates act as an immediate energy source, or as a rapidly available energy source, which is normally stored as glycogen in the liver and muscle. Although not essential, carbohydrate has a protein sparing effect. So, for example, feeding grain as part of a supplementary diet might mean that a high protein pelleted proportion of the feed could all be used for growth. This might be particularly useful if the fisheries manager is feeding the fish

Table 22.2 Description of nutrient requirements of common carp (*C. carpio*)

Nutrient group	Source, requirement and description
Proteins	Fish meal, casein. 30–38% optimal protein level.
Lipids: fat	Fish oil, vegetable oil. 10–20%. Essential fatty acids. 1% 18:3ω6 and 0.5–1% of 18:3ω3. Phospholipids are not generally limiting in dietary lipids.
Carbohydrates	No specific requirements have been established.
Vitamins	Occur in dietary components. Some are synthesised from other essential nutrients. Required in trace amounts.
Minerals	Occur in dietary components and absorbed from the external aquatic environment. Many are essential. Phosphate and sulphates.

Table 22.3 Essential amino-acid requirements in common carp (*C. carpio*)

Amino acid	As a percentage of protein
Arginine	4.4
Histidine	1.5
Isoleucine	2.6
Leucine	4.8
Lysine	6.0
Methionine + Cys/2	2.7
Phenylalanine + Tyrosine	5.7
Threonine	3.8
Tryptophan	0.8
Valine	3.4

stock to achieve specimen proportions. Starch is the most efficient form of carbohydrate energy in carp compared with dextrin and glucose.

Eight of the water-soluble vitamins are required in relatively small amounts and have co-enzyme functions. These are known as the vitamin B complex, some of which, for example B_{12}, can be synthesised by carp. Three of the water-soluble vitamins are required in larger quantities. These are choline, inositol and vitamin C.

Table 22.4 lists known vitamin requirements of cyprinid fish.

Knowledge of mineral requirements is one of the least advanced areas of fish nutrition. Minerals differ from other nutrients in that they are neither used nor produced. About 20 inorganic elements are required to meet the structural and metabolic functions of common carp (*C. carpio*). Those that cannot be taken directly from the aquatic environment are shown in Table 22.5.

Other minerals such as sodium, potassium and chlorine are the most abundant electrolytes in the body and are readily absorbed from the aquatic environment. They may also occur in natural ingredients of baits.

Before investigating the nutritional value of bait, it may prove useful to consider factors that may affect how fish feed in wild or natural waters. Factors that may affect feeding and

Table 22.4 Vitamin requirements in common carp (*C. carpio*)

Vitamin	Requirement
Vitamin A	4000–20 000 IU
Vitamin E	100 mg
Thiamin	0.5 mg
Riboflavin	6.2 mg / 7 mg
Vitamin B_6	5–6 mg
Pantothenic acid	30–50 mg
Niacin	28 mg
Biotin	1 mg
Vitamin B_{12}	NR
Folate	NR
Choline	1500 mg
Myoinositol	440 mg
Vitamin C	No requirement known

Table 22.5 Minerals unavailable to common carp (*C. carpio*) from the aquatic environment

Mineral	Requirement
Phosphorus (P)	0.6–0.7%
Magnesium (Mg)	0.04–0.05%
Zinc (Zn)	15–30 parts per million
Manganese (Mn)	13 parts per million
Copper (Cu)	3 parts per million
Cobalt (Co)	0.1 parts per million
Iron (Fe)	150 parts per million
Calcium (Ca)	>0.028%

food preference may include species competition, temperature and environmental conditions, angling pressure and the palatability of a feed.

22.6.1 Species competition

Competition for food between species can induce stress among fish populations. Some species are often better adapted by evolution to feed in specific areas of an aquatic environment. These will have an advantage over other species when feeding in their own specialised area.

Where there is an abundance of food, some species will migrate into other spatially segregated areas to feed. Other species may well feed in the same area but naturally be less competitive. Anglers often refer to fish such as these as 'nuisance' fish; for example, specimen carp anglers often catch smaller fish such as bream and tench, which feed around areas dominated by carp. The bait chosen by the angler may therefore be selected to target the carp (*C. carpio*) while precluding other species.

22.6.2 Temperature and the environment

Because fish are cold blooded, they will be influenced by the temperature of their surroundings. In nutritional terms, this means that the speed at which they process items of feed will be reduced as water temperature decreases.

Other factors that have been observed to influence feeding behaviour include day length, incidence of parasitism, reproductive cycles and biomass. Although much research has studied feeding behaviour in aquaculture, very little has been done in natural, sport or recreational fisheries. Many anglers have spent a great deal of time studying this aspect of angling, but this information is rarely shared or published in a form that can have any confidence attached to the claims.

22.6.3 Angling pressure

It is known that fish can be stressed by being caught, or even being fished for. In aquaculture it is widely accepted that, where fish are stressed by handling and external disturbances, the energetic cost involved in dealing with these stressors will reduce growth rates. Stress also invokes a process of tissue breakdown in the animal. 'Fight or flight' are widely accepted as the only options available to fish as a response to stress. In the first instance it is likely, therefore, that fish will refrain from feeding in periods of high stress, both because of external disturbance or as a reaction to being caught. The struggle for freedom when hooked by the angler will stress the fish, which then requires a period of recuperation and regeneration, which will have a marked effect on feeding.

22.6.4 Palatability of bait

The palatability of bait is a very similar concept to the palatability of aquacultural feed. Both types of feed can consist of optimally balanced nutrients, but if the complete diet is not attractive to the fish it will not be eaten, regardless of what form it is in.

Both the angler and bait manufacturer must solve this problem to be successful in their own right. Good bait should be palatable to the target species. However, through research into feeding behaviour and chemoreception in fish, science has given anglers the possibility to enhance the palatability of their bait with attractants.

These attractants do not add to the nutritional value of the bait, but they can cause preoccupation with a bait, which may lead to nutritional deficiency. Chemical stimuli associated with food will initiate a search behaviour pattern in the fish. Successful chemical stimuli will deliver the message from the bait to the fish and the effectiveness of the attractant will increase with smoothness of the chemical gradient.

Having located a strong or smooth gradient, the fish will move along the chemical trail. In some species this can occur through chemically induced rheotaxis (that is, movement in response to a water current). Having found the area of origin, in this case the bait, the fish may assume a search visually or by taste, using external taste buds located on barbels or lips. Once the bait is found it will, under normal circumstances, be carefully mouthed, before sorting of the food and non-food particles is performed. At any stage during this process, if the fish associates the smell, taste or feel with a negative experience, it may well abandon the search, or eject the bait.

Chemical attractants can have a very positive effect on feeding behaviour, not only by attracting the fish to the feed, but also by increasing feeding. Chemical attractants such as dimethyl-β-propiothetin, for example, can lead to increased growth through feeding enhancement.

22.6.5 Nutritional value of anglers bait

Anglers will not need reminding of the vast array of items that are currently used as bait. A far from exhaustive list of some common feed items used or considered as bait is given in Table 22.6. The individual palatability is not considered with regard to these selected baits; a degree of palatability is assumed as they are acknowledged as effective bait items.

Although it is important for the angler to know that his chosen bait will catch fish, it is equally important for the fisheries manager to know that the bait is nutritionally beneficial to the fish and certainly does not have any anti-nutritional effects.

Table 22.6 Feed items used as anglers' bait for fish

Bait item	Comment
Boilies and pellets	Available as 'ready-mades' (shelf life), also ingredients to 'make your own'. Used as a species- or size-selective bait by carp (*C. carpio*) anglers.
Maggots	The most widely used general bait, not selective although readily available to smaller fish in the form of 'pinkies' or 'squat' (very small maggots).
Peanuts	A popular carp (*C. carpio*) bait until banned on many waters. Fish become preoccupied when feeding on peanuts.
Sweetcorn	Human quality maize (tinned or frozen). Very popular bait, particularly in the summer. Small fish will have difficulty feeding on it.

Table 22.7 Nutritional value of boilies shown as food conversion ratio for carp (*C. carpio*) in a controlled feed trial

Diet	Total weight gain (grams)	Total food fed (grams)	Overall food conversion ratio
Chicken food	155.4	568.74	3.66
Ready-mades	214.8	618.48	2.88
Four seasons	323.7	650.22	2.01
Big fish mix	454.2	762.78	1.68
Trout pellet	1307.6	1219.72	0.93

Nutritional value of boilies to fish

Boilies were introduced to carp fishing in the late 1970s and early 1980s with devastating effect. They were effective because they allowed anglers to target their quarry species, by excluding species that could not feed on large, hard-skinned bait. The nutritional value of a boilie is dependent on its ingredients, which are typically those shown in Table 22.7. However, the problem is that shop-bought or ready-made 'shelf-life' boilies do not carry a definitive description of their content.

In the early 1980s, all bait suppliers (manufactures) sold individual ingredients, but now supply ready-made or pre-mixed ingredients.

The boiling of the bait is done to put a skin on the outside so as to decrease the bait's palatability to smaller fish. This also has the effect of denaturing the protein, which makes it more digestible while not affecting its availability. Carbohydrates are also made more digestible by the cooking process.

Anglers who make their own boilies should consider the source of carbohydrate that they include. It has become very fashionable to include sweet flavourings and glucose in boilies, which can have very serious effects on the fish. Cyprinids do not have a high tolerance to oral glucose; in fact, they are essentially diabetic. The best source of carbohydrate is starch as it avoids glucose peaks.

Research into the suitability of 'shelf-life boilies' as anglers bait showed that they can have very serious effects. They may be anti-nutritional, reducing the appetite of the fish that eat them, if they become a significant and regular component of a carp's diet. Carbohydrate used as a preservative plays a large part in their anti-nutritional effects.

Fish oils will obviously supply the essential fatty acids required in the diet. Most are added to the bait as a means of adding additional attraction ability. The oil will float out of the bait carrying any lipid-soluble attractant with it. The danger of using rancid fish oils is as important in bait-making as it is in aquaculture.

Nutritional value of maggots to fish

Maggots are the most widely used of all anglers' baits, particularly in warmer weather. Fish readily take them, particularly those fish that feed in shoals such as roach and bream. Maggots that are not taken as they fall through the water column will be eaten from off the bottom by benthic-feeding fish, such as carp and tench.

Table 22.8 shows the nutritional value of maggots. From this it can be seen that fish that feed extensively on maggots are not taking in a nutritionally balanced diet. In fact, this may well be an explanation why fish feeding on maggots will inexplicably cease to feed; perhaps to feed periodically on food items that will supply essential constituent parts of the diet.

The nutrition supplied by the maggot, in common with most live foods, should be readily available.

Nutritional value of peanuts to fish

Peanuts can be very effective bait for specimen carp, although there is a stigma attached to their use. Like many other 'particle' baits, such as chick peas, and so on, they need to be very well cooked to avoid swelling in the gut of the fish. Peanuts are banned on many waters in the UK because of the fear of them being poorly cooked.

Comparison of the nutritional content of peanuts (Table 22.9) with the nutritional requirements of carp show that, if a fish eats nothing other than peanuts, it may lead to a nutritionally related health problem. This may take the form of a deficiency in essential fatty acids, which results in hepatic swelling due to severe lipid infiltration.

Nutritional value of sweetcorn to fish

Sweetcorn is another highly popular anglers' bait which, like the maggot, is used as hook bait. It can also be used as an attractor by throwing in amounts as 'loose-feed'. The most popular type is the small tinned or frozen niblets that are packed for human consumption. Tinned sweetcorn is pre-cooked and vacuum packed.

From the nutritional information given in Table 22.10, it can be seen why it is important that fish do not feed exclusively on sweetcorn. However, sweetcorn is effective as a bait, and although it should represent only a small part of the diet of cyprinid fish, it is an accept-

Table 22.8 Nutritional value of maggots fed, shown as proximate analysis

Composition	Percentage
Moisture	71.60
Crude protein	13.83
Crude lipid	2.30
Ash	2.55
Crude fibre	—
Nitrogen-free extract	—

Table 22.9 Nutritional value of peanuts fed to fish, shown as proximate analysis

Composition	Percentage
Crude protein	49.0
Crude fat	1.3
Crude fibre	9.9
Ash	5.9
Essential amino acids	All ten present and available
Minerals	Phosphorus: 0.61
Vitamins	B_{12} and K are absent
Essential fatty acid: 18:2ω6	32.0
Essential fatty acid: 18:3:ω3	—

Table 22.10 Nutritional value of sweetcorn fed to fish, shown as proximate analysis

Composition	Percentage
Crude protein	3.0
Carbohydrate	17.7
Crude fat	0.5
Fibre	3.4
Essential amino acids	All ten present and available
Minerals	
Vitamins	
Essential fatty acid: 18:2ω6	
Essential fatty acid: 18:3:ω3	

able form of vegetable protein. The proximate analysis (high carbohydrate to protein level) explains why sweetcorn remains largely indigestible.

22.7 Conclusions

Theoretically, the ideal bait, in terms of getting fish to eat as much as possible, would be of low nutrient value with very high digestibility and attractant qualities, combined with taste and texture that overrides the fish's natural instinct for nutritionally superior food items. Although this would not supply good nutrition, it would then allow the fisheries manager to supplement this with a high-quality, high-nutrient, high-digestibility commercial diet and encourage the growth of food organisms in the still-water to allow the fish to obtain good quality natural food.

It is probably a fair assessment that anglers' bait provides suitable nutrition in semi-natural (not heavily stocked) still-water fisheries where there are other sources of naturally occurring feed items.

The potential area of concern is in heavily or intensively stocked fisheries, where anglers' bait is a primary component of the fish's diet. Fishery managers should perhaps encourage the use of an array of bait items to provide as near a balanced nutritional profile as possible. This should be supplemented with a commercial coarse fish diet.

A further consideration in fisheries where anglers' bait is likely to be a significant component of the diet is that when there are no anglers on the water, perhaps because of a couple of days or more of wet weather during the summer, the fish will still require food. In such cases, a supplementary ration may be all that is required to look after their welfare.

23 Improving Conditions or Fishery Enhancement

The previous chapters have talked about control of various aspects of the fishery. Many of the sections have mentioned improving the habitat to benefit water quality, the spawning success of the fish or the general quality of the environment. We quickly realise that the habitat provided at the fishery is the most important aspect of fisheries management. Healthy and diverse habitat should be considered as one of the most beneficial components of a fishery and its creation and enhancement as critical to maintaining and improving a still-water fishery. Providing suitable conditions for the occupants of each of the trophic levels of the pyramid of production (Chapter 2) should be seen as the first step in creating a balanced ecology and optimal environment for the fish. So, although much of the habitat creation or improvement will be done to support and enhance the environment for fish directly, much of it should also be directed at the animals and plants that sit below fish on the pyramid of production.

Improvement to existing habitats is generally preferable to the use of remedial measures, such as the use of aeration systems. By creating aquatic systems that mirror, as closely as possible, natural systems, an ecological balance can be reached without the need for artificial solutions. For example, in resolving dissolved oxygen problems, habitat improvements by themselves may well be sufficient to increase dissolved oxygen levels, without the use of artificial aeration, through the natural process of photosynthesis by aquatic plants.

The carrying capacity of a water body may be increased by increasing the amount of aquatic vegetation. This is a delicate balance, however, because as aquatic plants die back in the autumn months they, along with leaf fall, input more organic material into the system. An effective management strategy is to reduce this organic loading, either by de-silting or by preventing or breaking down excessive organic inputs. Selective management of trees and shrubs surrounding a still-water can also help control the build-up of organic silt.

This chapter not only considers creating or enhancing habitat for the purpose of improved fishery management, but it also considers technology that improves the habitat as a legitimate way of enhancing the environment. Fisheries not only provide sport for anglers, but also are very important wildlife habitats. Most fisheries managers value the wider habitat as do the anglers, so promoting conservation interests is a valued part of fisheries management.

23.1 Plants: their role in habitat creation and enhancement

Chapter 4 discusses the importance of plants to the ecology of a still-water, including their role as a solution to bankside erosion. This chapter concentrates on the role of plants in the creation and enhancement of habitat.

Aquatic plants are normally broadly divided into four groups:

- riparian;
- marginal;
- submerged;
- floating.

At the outset, the suitability of an aquatic plant species needs to be considered. A decision needs to be made about the provenance of the plants: whether to establish only native plants or to accept introduced species. Given that there are over 200 taxa in 72 genera of aquatic plants in the UK and Ireland, it is difficult to understand why it would be necessary to use non-native plant species in the wild. Indeed, experience with non-native introductions has shown that these can pose enormous problems. If the decision is to use only native species it may be appropriate to decide upon the level of local provenance and, in conservation terms, there may be situations when it is appropriate to transplant or seed from local stock to retain genetic integrity. Individual species can then be selected. Considerations may include the following:

- availability;
- ease of establishment;
- speed of colonisation;
- biodiversity value;
- ability to self seed;
- aesthetic value;
- robustness;
- environmental benefits.

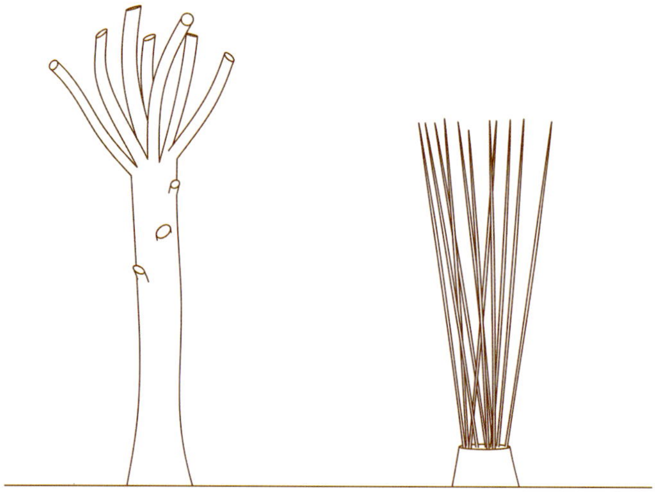

Figure 23.1 The difference between a pollarded and coppiced tree.

23.1.1 Trees

Trees form a valuable structural element around a still-water and form part of the riparian plant community. They provide shade and diversity but have to be carefully managed as they also have some disadvantages. If their numbers are excessive, they may create too much shade, and the leaves they drop can increase sediment and organic nutrient loading. Reducing the tree canopy has several effects, including increasing the light into the water margins, which in itself will power and energise the system. A warm marginal shelf is critical to obtaining a balanced ecology: it provides shallow areas that can provide a micro-habitat for the zooplankton community to develop and thrive.

From the viewpoint of the organic loading, less trees means less leaves.

It is often possible to replace trees that have been removed with others further back from the edge of the pond. Pollarding or coppicing are also considered to be effective measures to reduce the canopy, while prolonging the life of the tree in question. Figure 23.1 illustrates the principle of a pollarded and a coppiced tree. Another technique is to lift the crown of a tree, although this requires advice from an experienced tree surgeon to ensure that the tree remains in balance and that the operation remains positive for the tree. Finally, make sure you do not have any Tree Protection Orders (TPOs) on your trees.

Annual removal of floating leaves before they sink to the bottom is an extremely effective method of reducing silt accumulation. The simplest means of gathering the leaves at the water's edge is to use some form of floating boom: for example, a thick, floating rope, floating plastic booms or a flexible chain of interlinked, elongated buoys. This can be used to encircle the leaves when they have been blown to one end or one corner before drawing them to the bank for removal.

23.1.2 Marginal plants

Establishing plants can be difficult to achieve in practice as conditions in and around the still-water can be very harsh. The bankside of a still-water is often less than ideal for the establishment of marginal plants for the following reasons:

- suitability of substrate;
- suitability of the bank profile;
- suitability of plant species;
- fluctuating water level;
- grazing pressure;
- environmental conditions: waves, run-off.

Suitability of substrate

Fisheries managers are often faced with a difficult task when it comes to planting into an existing bank, because the substrate is unsuitable to provide security for the root structure that will anchor the plant in position. There are two extremes of this problem. At one extreme is the anaerobic, silt-type substrate, whose structure is insufficient to offer support to the roots. At the other extreme is the very heavy clay substrate that will be sterile by comparison and too stiff to accept the planting.

It follows that the most suitable substrate will have sufficient nutrients and will be well oxidised, with a stiff structure that will support the planting while allowing the infiltration of the root system. At some stage in the past, most banks have been cut into clay, by humans or nature, to form the bank profile.

Suitability of the bank profile

The suitability of the bank profile to accept planting is of equal importance to achieve good plant establishment. Of course, security from disturbance is essential once planting has taken place.

Sharply cut vertical banks that slope steeply into the water can make the establishment of marginal plants very difficult, if not impossible. When an unsuitable profile is coupled with a situation where no marginal shelf exists, a more involved system of establishment will be required, to avoid plants either being placed in water that is too deep or too shallow. Figure 23.2 shows the difference between suitable and unsuitable planting of 'bare root' marginal plants.

Figure 23.2 Comparing planting on a marginal shelf and a sloping bank.

Suitability of marginal plant species

To assist in the selection of marginal aquatic plants for creating and enhancing the habitat in a still-water, Table 23.1 lists ten of the most common aquatic plants useful in habitat creation. These are illustrated in Figures 23.3–23.12.

Fluctuating water level

It can be painful to establish marginal aquatic plants only to find that they are left high and dry during summer levels or that they are flooded and overwhelmed in winter by a high water level. So some consideration should be given to the level that plants are initially established at (Fig. 23.13).

Environmental factors

Just as fluctuating water level can stress and overwhelm newly established plants, so can other environmental conditions. Wave and wind action can cause very severe bank erosion and can wash out newly established or poorly rooted plants.

Table 23.1 Ten most common plants useful in habitat creation

Plant species	Description
Yellow iris (*Iris pseudacorus*)	A marginal plant that can grow to 1.5 m in height, Forms clumps rather than extensive stands so is not considered invasive.
Ponds sedge (*Carex riparia*; *Carex acutaformis*)	A marginal plant that adapts well to damp and boggy area. These two species grow to 1.0–1.4 m in height in clumps or in dense stands.
Reed canary grass (*Phalaris arundinacea*)	A marginal/emergent plant that can grow to 1.3 m in height. It does not easily adapt to water above 150–200 mm deep. It can form large stands.
Common reed (*Phragmites australis*)	A marginal/emergent plant that can grow to 3.0 m in height. Although it can grow in clumps, it prefers to form dense stands, It can grow in water up to 1.5 m deep.
Soft rush (*Juncus effuses*)	A marginal/emergent plant that can grow to 1.3 m in height. It forms clumps, although these can be extensive. Will tolerate damp riparian habitat beside water.
Purple flowering loosestrife (*Lythrum salicaria*)	A marginal plant that can grow to a height of up to 1.4 m. Generally it grows in small stands of one to five plants. It does not adapt to deeper margins.
Reed sweet grass (*Glyceria fluitans*)	A marginal/emergent plant that can grow to a height of 2.0 m. It forms dense stands and easily adapts to water that is 0.7 m deep, not usually invasive.
Water mint (*Mentha aquatica*)	A marginal plant that can grow up to 1.0 m in height. It readily colonises shallow margins and grows in small clumps among other marginal plants.
Reedmace (*Typha latifolia*)	A marginal/emergent plant that can grow to 2.6 m in height. It often grows in dense stands and can be invasive because it can adapt to establishing in water up to 1.0 m deep.
Bogbean (*Menyanthes trifoliate*)	A short trifoliate (three-leaved) marginal plant that can grow to a height of 30 cm. Often it will grow in extensive stands, although rarely invasive and simply removed.

Figure 23.3 Yellow flag.

Figure 23.4 Pond sedge.

Figure 23.5 Reed canary grass.

Figure 23.6 Common reed.

Figure 23.7 Soft rush.

Figure 23.8 Purple flowering loosestrife.

Figure 23.9 Sweet flag.

Figure 23.10 Water mint.

Figure 23.11 Reedmace.

Figure 23.12 Bogbean.

Grazing protection

Where there are animals on the bank or the water, grazing protection may be required to inhibit access to the plants. Where there is a significant number of waterfowl, there may be a considerable amount of grazing pressure. If the pressure is from geese or swans then protection will be required from both the bank and waterside. Figure 23.14 shows a professional protection system.

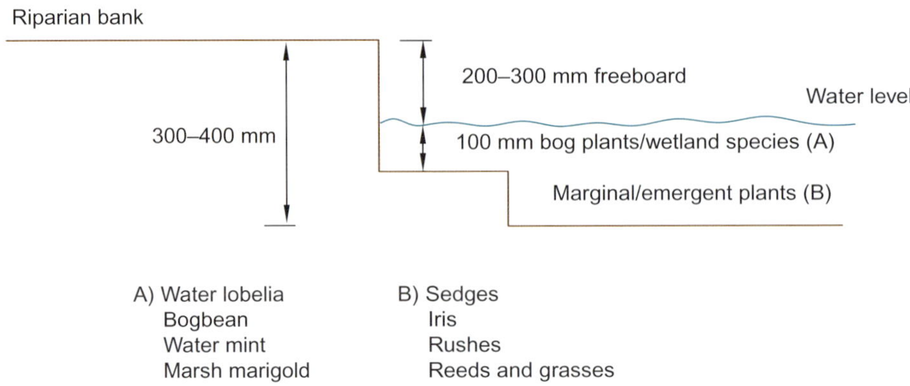

Figure 23.13 Planting depths for marginal vegetation.

Figure 23.14 Protection of marginal plants from grazing.

Establishing marginal/emergent aquatic plants

In a natural situation, marginal and emergent plants often have their lower parts submerged in water, as they grow through water from the substrate below. They are typically tall, narrow-leaved plants that offer little resistance to fluctuating water levels or high winds. Their height ensures that, in normal conditions, they are unlikely to be completely submerged during flooding, or submerged for extended periods. The stems have tough internal fibres and a hollow structure, which allows them to withstand severe winds without damage. Given the number of different physical, biological and environmental factors, no single method of planting will be able to resolve the requirements in all circumstances. The different planting systems that can be used are discussed below:

Plugs

Plugs (Fig. 23.15) are a small plant container, that are used as an interim, between seedling in a tray and whatever final size container will be used for providing a plant for use in the wild; this may be a root trainer or a potted plant of 1 litre or more. Plugs are not suitable for the establishment of marginal plants into 'wild' situations as they are small and comparatively fragile.

Root trainers

Root trainers (Fig. 23.16) are considerably larger in size than a plug. They train the plants root system down into the container. It is the smallest size plant that should be considered for establishment on an aquatic bank. Even at this size they are prone to damage from grazing wildfowl and wash out from wind and wave action.

Pots

Potted plants (Fig. 23.17) are available in several different sizes based on the soil volume of the pot. The most commonly sized containers are ½l, 1l and 2l. These are on-grown from smaller stock. The larger the pot size the greater the root system, providing that the plant has matured in the pot size that it is in.

Although a larger pot size can have benefits for increased robustness and ability to withstand the vagaries of the seasons, the root system will also require a much larger area of bank into which to be planted. Anchorage can therefore be a problem. However, in general,

Figure 23.15 Marginal plant grown in a plug.

Figure 23.16 Marginal plant grown in a root trainer.

Figure 23.17 Marginal plant grown in a 1-litre pot.

Figure 23.18 A bare root marginal plant.

the larger the plant is the greater is its ability to withstand environmental hardship and the greater its chances of become established.

Bare root

Bare root (Fig. 23.18) plants have the advantage that they do not have a large root ball and thus they require minimum disturbance of the substrate. They are easier to plant than translocated mature plants with large root balls. They are, however, very easily dislodged by waterfowl and wave action, and are more susceptible to climate, so can only be planted out in favourable weather conditions.

Translocation

Translocation involves digging out bare rooted, established plants by taking the root system in a clod of earth (Fig. 23.19). The plant may removed from one area of a still-water and translocated to another, or be translocated from an external source. Translocation within the same still-water satisfies the need to maintain local provenance and genetic integrity. By translocating within a fishery, rather than from outside, it is also possible to observe good biosecurity, reducing the risk of importing disease. Remember that you need the permission of the owner before you remove any plant material from the wild.

Figure 23.19 Clump of marginal vegetation for translocation.

Pre-established coir-fibre pallets

Pre-established coir-fibre pallets (Fig. 23.20) are an excellent way of ensuring the establishment of aquatic plants. Coir-fibre pallets are manufactured from coconut fibre, which is compressed into 1 m^2 mats. These then have aquatic plants established in them. The pallets are grown for at least 16 weeks before they are ready to be installed onto a bankside. Coir pallets are often installed in shallow margins to establish reed-beds (Fig. 23.21).

Pre-established coir-fibre rolls

Pre-established coir-fibre pallets (Fig. 23.22) are an excellent way of ensuring the establishment of aquatic plants. Coir-fibre rolls are similar but the coconut fibre is compressed into a 300 mm diameter × 3 m long roll. Aquatic plants are grown in the rolls and establish a good root system before the roll being installed. This means they can immediately start to grow into the substrate and bank. A typical installation is shown in Figure 23.23.

Coir-fibre planting systems are particularly useful in hostile environments, such as gravel pits. Figure 23.24 shows a detailed design for this.

23.1.3 Submerged aquatic plants

Totally submerged plants are the true water plants, or hydrophytes, in that they have adapted in many ways to life in water.

Figure 23.20 Pre-planted coir-fibre pallet.

Figure 23.21 Pre-planted coir-fibre pallet installed in a pond margin.

There is little or no mechanical strengthening tissue in their stems and leaf petioles. If these plants are removed from the water they hang limply, so it is important that they are always surrounded by water. Submerged plants lack the external protective tissues required by land plants to limit water loss.

Roots, which normally play a very important role in the absorption of nutrients and water from the substrate, are often reduced and their main function is anchorage.

Figure 23.22 Well established root system of a pre-planted coir-fibre pallet.

Figure 23.23 Pre-planted coir-fibre roll installed along a pond margin.

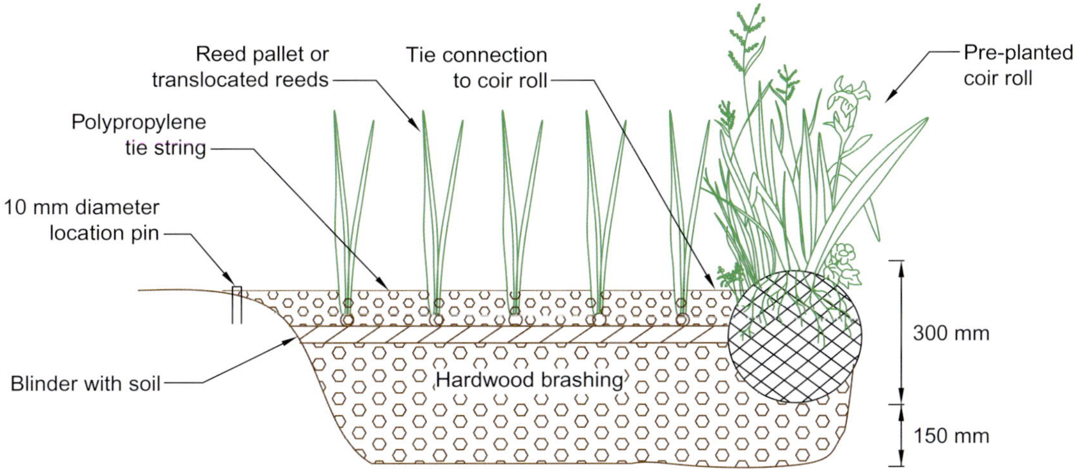

Figure 23.24 Pre-planted coir roll installed in a pond margin.

Many species have very specialised leaf shapes. The submerged leaves are often highly dissected or divided. This has the advantage of creating a very large surface area for absorption and photosynthesis.

As with marginal plants it is important to consider provenance before attending to other considerations.

To offer some assistance in selecting submerged aquatic plants for creating and enhancing the habitat in a still-water, Table 23.2 lists four of the most common submerged aquatic plants. These are illustrated in Figures 23.25–23.28. Several difficulties are often encountered by the still-water fisheries manager when considering establishing submerged plants these may include:

- water clarity;
- canopy shading;
- grazing by fish;
- grazing by waterfowl.

Table 23.2 Submerged plants useful for enhancing the aquatic environment

Species	Detail
Canadian pondweed (*Elodea canadensis*)	Excellent oxygenating properties
Hornwort (*Ceratophyllum demersum*)	Tends to be very weakly rooted
Water milfoil (*Myriophyllum spicatum*)	Can often be found with Canadian pondweed
Starwort (*Callitriche stagnalis*)	Provides exceptional habitat for aquatic invertebrates
Broad-leaved pondweed (*Potamogeton natans*)	Provides surface cover with its floating leaves

Figure 23.25 *Elodea.*

Figure 23.26 Hornwort.

Water clarity

Water clarity in a still-water is most often related to stocking density or to heavy sediment loads during high flows through the inlet. Where carp are the dominant species, their continued foraging into the hydro-soil may ensure that there is always a high level of suspended solids. So the answer may be to simply reduce stocking levels, the increased light being all that is required to kick-start the establishment of submerged plants. If a heavy sediment load is the issue then some form of trap or filter may be the solution (see section 23.2.1).

Figure 23.27 Water milfoil.

Figure 23.28 Starwort.

Canopy shading

A dense tree canopy over the riparian zone of a still-water will cut down light to the point that submerged plants are unable to grow. Thinning the tree canopy, or removal of some of the trees, may allow sufficient light to penetrate for the establishment of submerged plants.

Grazing by fish

Fish can have a considerable impact on the growth of submerged plants. This can simply be a consequence of foraging for food that muddies the water and cuts-down on the amount of light penetrating the water. Normally, however, when a water body is heavily stocked and/or heavily fished, the fish graze on the submerged plants. In such cases, heavy grazing may be the primary cause for a failure to establish submerged plants and it is suggested that any newly established planting be protected with an anti-fish grazing cage (Fig. 23.29). The submersed plants should be established using a planting container, or perhaps a hessian sack (Fig. 23.30).

Figure 23.29 Anti-fish grazing cage for submerged plants.

Figure 23.30 Establishing submersed plants in a hessian sack.

Grazing by waterfowl

When submerged plants are pressurised by waterfowl to such an extent that it becomes a management issue, there will usually also be an impact upon marginal plants and increased nutrient enrichment from the sheer number of birds. In such cases, experience would suggest that management actions should focus on reducing the number of both resident and visiting waterfowl (see Chapter 17).

23.1.4 Floating aquatic plants

There are two types of floating plant: those which are rooted, with floating leaves, for example water lilies; and those that are not rooted in the sediment but just float on the surface, for example duckweed. Unattached floating plants are generally not considered for habitat enhancement on most still-waters, although they may provide a spawning substrate in some species.

Floating leaved plants are generally tough, because they have to withstand the worst the environment can throw at them, in the form of the weather and water movement (wave action). In common with other rooted floating-leaved plants, water lilies are adapted to support floating leaves with the minimum of structure, allowing the water to support them. As with marginal and submerged plants, it is important to think of provenance before other considerations.

Lilies are considered the most important floating species for still-water fisheries management, as they provide refuge and cover for the fish. When they cover a significant amount of the surface area (greater than 40%), they can also inhibit the growth of nuisance submerged plants and algae. The two native lilies are depicted in Figure 23.31 and Figure 23.32.

Several difficulties are often encountered when establishing lilies these may include:

- water clarity;
- canopy shading;

Figure 23.31 *Nuphar lutea.*

Figure 23.32 *Nymphae alba.*

- grazing by fish;
- grazing by waterfowl;
- suitability of substrate.

Water clarity

Good, clear water is as important for the establishment of lily pads as it is for submerged plants. Lilies establish far more successfully as part of a managed activity, so there is always an option to ensure that they receive adequate light by placing them in shallow water where light can penetrate even when it is coloured.

Canopy shading

The effect of canopy shading can be as damaging to lilies as it is to submerged plants, although it is relatively easy to establish the plant away from shade and, once the rhizome is well established, it is not unusual to find that the pads then spread under the tree canopy.

Grazing by fish

Fish can be very partial to young lily shoots. Carp are particularly fond of lilies as a delicacy, and continued heavy grazing of the new shoots may eventually overwhelm and kill the plant. Newly established plants should be protected by a grazing cage (Fig. 23.33) which is simply placed above the lily on the hydro-soil. The young stems and leaves pass through a 75-mm mesh; thus the larger, more aggressive fish cannot get down to graze the small young shoots.

Grazing by waterfowl

Where waterfowl are present in significant numbers and are damaging the lilies, protection of the lily pad at the surface by a grazing cage (Fig. 23.34) is well advised. After perhaps three years establishment under the grazing cage it may be possible to remove the grazing cages. However, if you have only established a single lily pad on a still-water frequented by fifty waterfowl, then removing the cage would see the pad decimated in a very short period of time.

Figure 23.33 Lilies established in a crate with anti-grazing protection.

Figure 23.34 Protection for floating-leaved plants.

Suitability of substrate

The lily rhizome is at best neutrally buoyant but more generally is positively buoyant and needs to be secured to stay on the bottom. Establishing bare root directly into the substrate is difficult if it lacks sufficient stiffness. It may be nearly impossible to plant a bare-root lily rhizome in a silty hydro-soil.

There are several techniques to overcome this including the following:

- bare root and mature plant in turf;
- bare root in hessian sack;

- containerised 'potted' stock lily;
- containerised translocated lily.

Lilies benefit by being established in nutrient rich soil and, although it is not critical, it is recommended that some manure or 'slow' release fertiliser be added to the planting medium.

Bare root and mature lily plants established in turf

A very simple way to weight down and create a secure, nutrient-rich environment for lilies is to obtain some reject turf from the local turf centre. Either place one bare root rhizome at either end of the turf or a single mature rhizome sticking out of one end (Fig. 23.35).

Put about 1 kg of well rotted manure in the turf alongside the rhizome before rolling and tying it up. It is now ready to drop into position as it will find its own level and the leaves will be guided skywards by the light.

Bare root in hessian sack

An alternative method of bare root planting of lilies is to fill a hessian sack (sand bag) with soil and manure (70/30 ratio), seal the end and then cut three slits into one side of the sack into which a bare root rhizome is placed (Fig. 23.36), closing the slit by stitching or stapling

Figure 23.35 Establishing a lily using turf.

Figure 23.36 Establishing a lily in a hessian sack with grazing protection.

Figure 23.37 Lowering an established lily sack into the water.

to ensure that the rhizome is secure. Placing the planted sack into position requires more care and attention than does the turf as it is critical to ensure it is the right way up. It often works well if the sack is lowered onto the hydro-soil using a rope at either end of the sack (Fig. 23.37).

Containerised 'potted' stock lily

This method uses lily plants that have been grown-on in pots so the plant is significantly more advanced than is the bare root stock. Several 1-litre lily pots are combined into a larger planting container. The planted lilies should be secured by covering the surface of the planting container with gravel (Fig. 23.38). The containers need to be carefully placed in position to avoid any spillage.

Figure 23.38 Weighting the soil down with gravel.

Containerised translocated stock
This method is identical in principal to that described above, other than the size of the plant. A translocated lily will ideally have a mature stem system that is comparatively more robust and developed with longer and more numerous stems, including young shoots.

23.1.5 Floating rafts (eco-Islands) and reed-beds

A novel development and an incredibly valuable asset in the habitat enhancement toolbox is the floating raft (known as an eco-island), consisting of marginal plants supported on a floating framework (Fig. 23.39). The plants on the rafts grow hydroponically, taking up nutrients and improving water quality, however, their function in providing and enhancing habitat is to provide cover and refuge, the root structure provides:

- spawning habitat;
- micro-habitat for zooplankton;
- food larder for fry;
- fry refuge.

In addition, it is possible to install an anti predation cage under the island as illustrated in Figure 23.40. These cages allow the fish that are naturally attracted to the cover and habitat provided by the floating eco-islands to seek refuge within the cage, but inhibit entry by avian predators such as cormorants and saw bill ducks.

23.2 Maintenance of the hydro-soil

The hydro-soil often declines as an effective area for the treatment of organic material and the demand for oxygen by the decomposers found there then rises. Measures such as dredg-

Figure 23.39 A floating island as additional habitat.

Figure 23.40 Floating island with a fish protection cage.

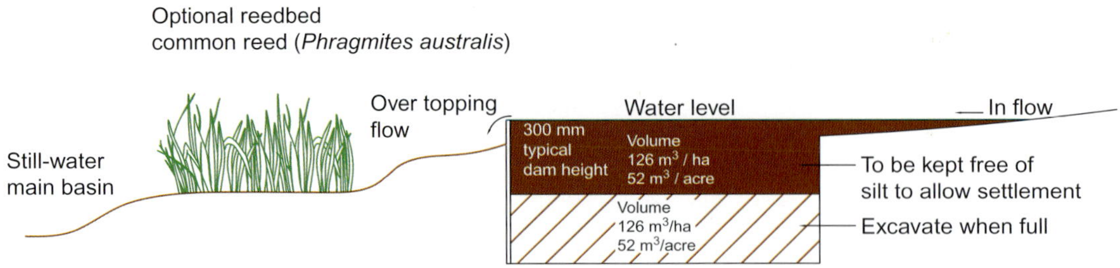

Figure 23.41 Diagram of a silt trap.

ing the accumulated sediments, or draining the water to expose the sediments to the air for a period of time, may have to be considered even though they are expensive. A more affordable method of conditioning the hydro-soil is to use a calcium compound in the form of calcium sulphate or calcium carbonate (see section on liming).

Once you have a functioning hydro-soil, it will be beneficial to investigate ways to reduce the organic loading on the system, this might include the following:

- installing silt traps;
- installing reed-beds on inlets;
- reducing the tree canopy: decreasing the amount of leaf litter;
- encouraging responsible land management in the still-water catchment.

23.2.1 Silt traps

The installation of a silt trap at the inlet, or slightly upstream of the inlet, can serve to reduce the amount of sediment that enters the fishery. This sediment adds to the silt load. It also carries fertilisers and organic compounds, which have bound chemically to soil particles as water travels across and through the soil to the watercourse that feeds the still-water. The principal of a silt trap is shown in Figure 23.41. A silt trap can be simple in construction but it will only be as effective as is the regime to keep it cleaned out, once it is full it will fail to function efficiently.

23.2.2 Liming

The term liming comes from the field of aquaculture, where liming is used as part of the annual cycle, to prepare ponds for the growing season. Where ponds are used for semi-intensive or intensive aquaculture, it is highly recommended for use as a precursor to fertilisation. The technique has since been transferred to still-water fisheries management, serving similar functions with the same associated benefits.

The authors promote the use of lime only as part of a wider fishery management plan where all factors impacting on the fishery are considered holistically. It should be acknowledged that trying to change the chemical composition of a water body is at best problematic, and the consequences, should it go wrong, can be severe. The authors therefore only advise the use of liming after consultation with a professional fisheries consultant.

Table 23.3 Calcium compounds used for liming ponds and lakes

Basic chemical	Common name	Toxicity to fish	Disinfectant properties	Relative price	Preferred
Calcium carbonate: $CaCO_3$	Limestone (90–95% $CaCO_3$ Dolomite (double carbonate of calcium/magnesium) Marl (20–80% $CaCO_3$ others: basic slag, coral, shells, etc.	Low	None	Low	water pH: >4.5 Fish are present
Calcium sulphate dihydrate: $CaSO_4.2H_2O$	Gypsum A.G.A. Aqua Bio	Low	None	Low	water pH: >4.5 Fish are present
Calcium hydroxide: $Ca(OH)_2$	Hydrated lime, caustic lime, slaked lime (approximately 70% CaO)	Medium	Strong	Medium	water pH: <4.5 No fish present Pest control
Calcium oxide: CaO	Quicklime, unslaked lime, or burned lime	High	Strong	High	Drained pond Pest control

NV, neutralising value of the pure salts, as a percentage, with reference to $CaCO_3$ (NV = 100%).

The term 'lime' is often casually used to describe various different types of calcium compounds, used for very different purposes (Table 23.3). Not all waters will require liming or will significantly benefit from it. Indeed, in certain circumstances it can be harmful. It should also be noted that calcium in some forms can be toxic to fish.

Liming can be done with four basic compounds, depending on the management objective of the application. They are as follows:

- calcium carbonate, $CaCo_3$: naturally occurring, quarried as limestone or chalk; used as fine particle powdered chalk;
- calcium hydroxide, $Ca(OH)_2$; commercially manufactured as hydrated or slaked lime; also known as 'builders' lime;
- calcium oxide, CaO: known also as quick or burnt lime;
- calcium sulphate dihydrate, $CaSO_4.2H_2O$, or gypsum.

Potential benefits

Liming may benefit pond health and fish populations in recreational still-water fisheries in many ways.

Improvement of the hydro-soil

The addition of calcium, a vital nutrient, and the release of phosphorus as phosphates from the hydro-soil, promoted by the application of lime, are both beneficial to the aquatic environment.

Calcium is an essential nutrient for many invertebrate organisms that provide the base of the food chain for fish.

Phosphates stimulate the growth of plankton and water plants, providing increased food for the fish.

However, if too much phosphate is released it can lead to eutrophication; a common problem in still-waters in the UK. In such circumstances algae and water plants may grow wildly, choke the water body, and use up large amounts of oxygen. This can be seen as an instance where the application of lime may not be advisable.

Using calcium compounds can act to sterilise the pond substrate and have the effect of increasing mineralisation.

The key benefits to the hydro-soil are the following:

- it improves the soil structure;
- it accelerates the decomposition of organic matter, particularly cellulose;
- it increases pH.

Improvement of the water

One of the most important effects is that which calcium has on total alkalinity. Water with a high alkalinity will have a good buffering capacity; it will be chemically quite stable, with a small diurnal variance in pH.

Increasing the alkalinity of a water body may bring about improvements in fish health. Fish in acidic waters are commonly stressed, have lower resistance to disease, and grow more slowly, to a smaller maximum size, than fish in alkaline waters. High acidity, and associated toxic metal levels, also kill fish eggs and larvae and reduce spawning success. However, raising the pH in itself will have little effect without the chemical stability, or buffering capacity, which reduces the diurnal and seasonal fluctuations.

Liming can help neutralise acidic waters. It is worth noting that different fish species have different tolerance levels and, although some species perform well in neutral pH waters, most notably carp (*Cyprinus carpio*) which prefer pH 7.0–7.5, others prefer a more alkaline environment.

The application of lime can help buffer against diurnal and longer-term fluctuations of pH. Using lime to increase the alkalinity to a level where daily fluctuations will not fall into a suboptimal zone ($pH < 6.5$) will chemically stabilise the pond and bring about benefits for the health of the fish.

Production within the entire food chain (plankton–insects–fish) is stimulated by liming. Liming can also enhance the growth of rooted aquatic plants that serve as nursery areas for young fish.

The key benefits to the water are the following:

- it increases and stabilises ph;
- it increases total alkalinity;
- the calcium content will increase, to be used by plants;
- it helps neutralise and precipitate certain toxic substances, such as iron compounds, as ph increases;
- it helps precipitate excess organic matter, decreasing the demand for dissolved oxygen in the water (this is discussed below).

Chemical de-silting of lakes

The use of lime to de-silt lakes chemically has several benefits, not least it is a lot cheaper than mechanical removal. Where there are deep layers of organic rich hydro-soil, oxygen cannot penetrate effectively and this stops aerobic bacteria from efficiently decomposing the material. Potential nutrients are thereby ‘locked up’.

An application of calcium carbonate or calcium sulphate will revitalise, or kick start, the beneficial fauna. A greater abundance and diversity of bacteria is encouraged, which can then break down the organic matter. This is critical to the efficient operation of the natural mineralisation cycles that drive the aquatic ecosystem. As the microorganisms break down the material, so the silt content is reduced and the still-water may be deepened to some extent. A decrease in the activity of anaerobic bacteria will reduce the build up of methane and hydrogen sulphide in the sediments.

However, in the authors' experience, the reduction of great depths of silt by using calcium compounds is not realistic. Medium- to long-term treatment of the hydro-soil with calcium compounds will have many advantageous benefits, but deep silt reduction will not be one of them.

Calcium compounds can perhaps be most useful in stopping the build-up of silt in the future, while having the benefit of revitalising the hydro-soil if used annually as a conditioning treatment.

Powdered calcium carbonate and powdered calcium sulphate have a very small particle size (a teaspoon will have in the region of 500 million particles). They are therefore very efficient in their mode of action, which is to adhere to suspended particles of sediment, forming a clump, or 'colloidal mass', which sinks. The powdered chalk will gradually sink through the water body binding on to any suspended fine particulate matter, carrying it to the bed. This will often cause the water to go temporarily clear. The clarity is desirable as sunlight is an important part of the mineralisation process. However, the clarity may also increase the growth of submerged macrophytes as a response to increased light. Calcium should also encourage those macrophytes that are calciferous by nature, for example hornwort, chara, and Canadian pondweed.

Sterilisation before stocking

If the objective is to use lime to sterilise a water body before stocking, then a different compound is required. In this instance the 'lime' used is either calcium hydroxide or calcium oxide, which are significantly more potent than calcium carbonate and calcium sulphate.

Calcium hydroxide and calcium oxide are more expensive than the other calcium compounds but act as very effective sterilisers, raising the pH quickly and dramatically above tolerable levels for most aquatic organisms. They should be used very carefully, avoiding contact with the operative, and never be used in water bodies containing fish.

History and effectiveness

Liming still-waters is a tried and tested method for treating acidified surface waters, with notable studies in Sweden in the 1980s.

In all waters where liming was tested it was found to result in a sufficient and durable pH increase, the fish started to reproduce again, even if the populations were composed of very few and old individuals at the time of treatment. Lime treatment is not a cure all for all acidified waters, but may protect fish populations in lakes with long turnover times, in areas with relatively low acid deposition. Areas of high run-off, and lakes with short turnover times, are a lot more difficult to treat with lime and the application is not recommended.

In areas with a high acid load and acidified watersheds, liming of lakes and running waters will not prevent acidic groundwater, with elevated concentrations of toxic metals, from entering surface waters.

Identifying liming needs can be accomplished by taking either water or soil samples from the water body over a period of time (ideally over a few months). Measuring the total alkalinity of water is the most effective way to determine if liming is necessary.

When should lime be used?

Lime is most commonly used when fisheries experience periods of depleted nutrients. This is especially common in older, established lakes where nutrients become locked within the sediments, causing a decline in plant growth and productivity. The sediment may have become increasingly acidic, slowing the breakdown of organic matter and releasing fewer nutrients into the water for use by plants.

Caution must be taken, however, as there may be a different reason for the symptoms attributed to acidification; so make sure acidification is responsible before applying lime. The pH should ideally be tested at various times of the day over several weeks. In addition, seek advice from a professional fisheries consultant, or the Environment Agency, before adding lime in any of its applicable forms.

New lakes created in rich soil are generally considered productive, but lakes in gravel pits, clay pits and sandpits contain few plant nutrients and are therefore not productive to begin with. Nutrients may also be lost from lakes if a stream or groundwater flows through them.

Fertilisers can be added to enhance the productivity of still-waters, but care must be taken to avoid pollution, damage to protected species or rapid removal or outflow from taking place.

Adding lime and fertiliser are conventional fisheries management practices for enhancing fish production in still-waters. Lime and fertiliser should not be applied simultaneously, however, because limestone may precipitate phosphorus and reduce its availability.

There are conditions where the application of lime is inadvisable. Generally, still-waters should not be limed if the following apply:

- fertilisers will not be used subsequently, unless the water is very acid;
- the water pH reaches more than 8.5 by the end of the day.

Making chemical improvements to a water body is the easiest fishery management strategy to get wrong! If in doubt, always seek professional advice.

Application methods

The liming product of choice is ideally applied directly to the bottom of the pond before filling with water and should be spread evenly over the entire base. Of course, in fishery management, the focus is most usually on improving existing water bodies and it is impractical to apply the lime in this fashion.

Applying the chosen liming product to still-waters that are full of water is more difficult, but can be achieved without fear of harming the fish (Fig. 23.42). The application of lime should be a staged process, with no more than a quarter of the area being treated at one time; the exact process, and rate of application, are dependent on the size of the water body and the type of product. This allows the fish to adjust to the changes in environmental conditions and reduces the chance of stress.

The efficiency of liming materials increases as their individual particle size decreases. The powder to be used should ideally pass through a 0.25-mm sieve. Granules and larger par-

Figure 23.42 Treating a still-water with a calcium compound.

Table 23.4 Dosage rates for pond liming

Form of lime	Dosage (kg/hectare)
Calcium carbonate: $CaCO_3$	750–1000
Aquatic chalk	4000 (over two installations per annum)
Calcium sulphate dihydrate	750–1000
$CaSO_4.2H_2O$ as Aqua Bio	3000 (over two installations per annum)
Calcium oxide: CaO	500–750 (thick mud)
	200–300 (thin mud)
	1000–1500 (disinfectant)
Calcium hydroxide: $Ca(OH)_2$	750–1000 (thick mud)
	300–450 (thin mud)
	1500–2250 (disinfectant)

ticulates can be used for the disinfectant of drained pond bottoms (calcium oxide, calcium hydroxide) if this is the objective of the application.

The only entirely suitable (safe) forms of calcium for the liming of ponds with fish present are calcium carbonate or calcium sulphate.

The application of the more concentrated and toxic forms require more gradual application and careful monitoring of water chemistry during and after application. Dosage rates are shown in Table 23.4.

Further considerations

When undertaking the liming of a still-water the whole area should be treated, not just those areas perceived to be 'silty'.

It is suggested that, after any form of liming, there may also be significant benefits to be gained in pond production by carrying out a fertilisation programme.

Fertilisation should only be done if there is a clearly identified management objective. It should be noted that inappropriate fertilisation can increase population densities of nuisance algae, such as blue-green species, and lead to eutrophic conditions.

The pH, organic content and aerobic status of the hydro-soil are critical factors in deciding upon whether to lime a pond or lake. Without this knowledge it is difficult to estimate the appropriate dosage and almost impossible to predict the outcome. To ensure a beneficial, controlled and measured outcome, consult a professional about the suitability of any pond or lake for treatment with a calcium-based product.

Health and safety

The application of all forms of liming material should be subject to health and safety considerations. Although calcium carbonate and calcium sulphate are not caustic, the granular form can be a mild irritant to the eyes, so the use of goggles is recommended. An eye bath should be part of the first-aid equipment on site for all liming operations.

Hydrated lime and slaked lime can cause serious burns to the skin and eyes. A one-piece coverall, face mask and goggles should be worn when applying this product.

Conclusions

The use of lime in fishery management is now commonplace and can be extremely beneficial to both water quality and fish health. By applying lime in the correct quantities it is possible to encourage diversity of aquatic life.

Nonetheless, there is a fine balancing act between nutrient deficiencies and nutrient excesses that can lead to eutrophication.

That said, all liming methodologies have the following advantages:

- they decrease organic matter;
- they increase oxygenation and stimulate activity of aerobic microorganisms;
- they improve water clarity by settling suspended solids;
- they reduce methane production in silt bodies;
- they counteract acidity in water and hydro-soil;
- they provide essential calcium, increase alkalinity and conductivity;
- they increase water transparency;
- they reduce orthophosphates, and iron ammonia salts;
- they reduce disease in aquatic organisms.

As this chapter has highlighted, liming can be used to fulfil several fishery management strategies. However, although applications can be beneficial in the right circumstances, caution must be taken.

23.2.3 Reed-beds

A reed-bed is an area of submerged land, within the perimeter of the still-water basin that is intensively populated with emergent aquatic plants. The installation of a reed-bed at the inlet to a still-water can benefit by reducing the amount of sediment entering, and also by having an effect on water quality. Reed-beds can remove hydrocarbons, phenols, pesticides and heavy metals, among many other substances. They have been shown to reduce biological oxygen demand (BOD) by 50%. A reed-bed can both improve the water quality and protect against a one off accident.

The most common planting is with Norfolk or common reed (*Phragmites australis*) as it is considered to be the most efficient, although several other plants are also suitable.

However, many of the nutrients and pollutants that enter a fishery do so at times of high flow after rain; there will be reduced retention time of water in the reed-bed, so the filtering efficiency will be reduced at the most important time.

Where a significant water quality issue exists, consideration could be given to constructing a specialised reed-bed. These systems are known as constructed wetlands.

Constructed wetland

A reed-bed is much more than a facility that has the ability to purify water; it is a community, comprising aquatic macrophytes (reeds), fungi, algae, bryophytes and aquatic invertebrates. The reeds are an integral part of the community, providing oxygen and surfaces on which other organisms can grow and proliferate. It is this complex community that, in combination, provides a range of removal mechanisms. For example, certain species of fungi are capable of degrading a range of synthetic chemicals, including pesticides and chlorinated hydrocarbons. Fully mature reed-beds have been used to remove phenols, methanol, acetone and amines from a range of industrial effluents, as well as to reduce BOD and chemical oxygen demand (COD).

Design of a constructed wetland

Where the substrate is permeable, reed-beds are normally constructed over an impermeable liner, manufactured from HDPE or butyl rubber. A gravel bed is installed within the liner. This should be of graded gravel, into which the reeds are planted and established into the top 100 mm of the substrate (Fig. 23.43). The type of media used can be specific to the role

Figure 23.43 Common reed established in a gravel bed.

the system is intended for; for example, a mixture of soil and seashells has been used to remove phosphorus, as have iron-rich sands. Subsurface flow systems (i.e. those systems where most of the flow is below the reed-bed surface), are the most commonly used in the UK and are recommended for use in fisheries where suspended solids (silt) can be minimised before entry into the wetland (see section 23.2.1).

A graded gravel media is used because it allows the reed-bed to produce several habitats, which facilitate the numerous biological processes. There are two generic types of reed-bed systems: a horizontal flow reed-bed treatment system and a vertical flow reed-bed system. Only the horizontal flow system is appropriate for the treatment of a fishery inflow. However, details of a vertical system are given in case there is an upstream discharge of sewage into the feeder stream, from a septic tank, etc. In these circumstances, a vertical system in combination may be appropriate.

Horizontal flow reed-beds

Horizontal flow systems normally have a gradient of 1–2% and a depth of 0.6–0.7 m. The direction of flow through the system is, as the name suggests, horizontal, with designed inlet and outlet areas. The flow through the bed is continuous. Horizontal flow systems tend to become oxygen limiting and may not achieve significant, let alone complete, nitrification.

Vertical flow reed-bed

Vertical flow reed-bed systems are normally constructed of similar materials to those used for horizontal flow systems. The difference is that the effluent is applied in batches, effectively flooding the bed, and is allowed to drain, unimpeded, out through the system. Air is allowed to refill the voids within the gravel and soil, replenishing the oxygen supply. Perforated pipework may be installed, which allows air to penetrate the bed, providing oxygen both for biological and chemical reactions. Vertical flow systems are capable of producing a fully nitrified effluent and may be more favourable when treating high-strength industrial effluents and meeting stringent discharge consents for ammonia. However, to obtain the discharge standards required, it is often necessary to construct a combination system relying on both horizontal and vertical reed-beds.

Mode of action of a combined reed-bed system

In a combined system water passes through the vertical bed first. Batch loaded liquid trickles through the gravel where it is exposed to aerobic bacterial communities.

By the time it reaches the bed base, most of the suspended solids have been removed, the BOD content has been significantly reduced (by more than 50%) and nitrification (breaking down ammonia to nitrates) has also occurred. The water is then collected by drainage pipes and fed to the horizontal bed.

The horizontal bed is anaerobic. Water enters one end of the bed and moves horizontally through it, effectively filling the bed and displacing all air, until it is allowed to escape at the far end. The microbial fauna converts much of the nitrates to nitrites and nitrogen gas (denitrification) and consumes any pathogens in the water.

Floating reed-beds

A floating reed-bed is a further adaptation of these systems. Although floating reed-beds are less efficient at removing pollutants, because the water flows under rather than through

Figure 23.44 A modular floating reed-bed.

them, they are very good at removing nutrients as the plants grow hydroponically. The major benefit of a floating reed-bed is it is easy to locate and install and is available in a modular design. A modular floating reed-bed is shown in Figure 23.44.

23.2.4 Land management

Management of the land surrounding the fishery can be difficult if it is not under the direct control of the fisheries manager. Nevertheless, every effort should be made to encourage your neighbours to practice an environmentally sensitive approach that considers appropriate use of fertilisers and a reduction of run-off. The indiscriminate use of fertilisers by the farming community is not widespread. They are very expensive and their inappropriate use is rare for that reason if no other. The larger issue is that of land management and run-off from fields. Here deployment of buffer strips and appropriate working of the land (methods of ploughing) can have a significant effect on both the quantity and quality of the run-off. So although the actual construction of buffer strips, which are a unique habitat in their own right, undoubtedly falls into the responsibility of the land manager, the fisheries manager must, at all opportunities, raise these issues both locally with neighbouring farmers and land owners and more widely with agencies, in forums and with conservation bodies.

24 General Administration

General administration includes all the routine activities of a fishery, the day job as it were. All lot of this is to do with enforcing the control of fishing activities that was covered in Chapter 21 but also the other miscellaneous duties that it is all too easy to overlook.

24.1 Monitoring

Having completed the risk assessment described in Chapter 21 and as many of the preventative measures as possible, the next stage is to plan a method of monitoring the fishery to prevent or apprehend anyone committing an offence. This area of work is included in the chapter on general administration because it is important to recognise the connection to the day to day routine work on any fishery.

24.1.1 Monitoring the site for signs of unauthorised access

Vehicle access points should have been restricted by this stage, so illegal entry should only be possible at a few points. These points should be monitored for signs of activity. The whole boundary should also be monitored in case new entry points have been made. These checks should include obvious signs of entry such as vehicle tyre tracks, as well as trampled vegetation. Visits should often be made to the fishery at unexpected times. Access areas should be checked as well as the surrounding area. These checks are to look for anything unusual; records should be kept of any observations made. For example, strange or unusual vehicles parked close to the fishery.

Closer to the fishery, inspections should be made of the banks. There may be signs of illegal taking of fish in the form of bankside disturbance. This will generally include things like waterweed on the bankside, trampling, drag marks from nets, fish scales, discarded tackle, and so on.

From all that has been said and done so far, the routine monitoring of access should be very obvious for each particular case. Some golden rules should be followed, however:

- Keep it going. It is easy to allow other duties to become priorities, particularly if no problems are encountered. This is the time that the fishery is at risk.
- Record everything. Everyone hates paperwork so keep it simple, but it is easy to forget things so do make a note. Even if it says 'nothing seen'.
- Regularly review and change the monitoring programme. Too steady a routine can mean people become complacent.

Fishwatch

Communication with other local fisheries can help enormously with routine monitoring. A system can be devised where fisheries regularly notify each other of any problems or suspicions. This can be anything from the sighting of a strange vehicle to suspicious behaviour. It should also include people offering cheap fish for sale with no questions asked. Communication should occur at least once a week even if this is to say there is nothing to report. Obviously e-mail systems work well here, but text-messaging services can be just as effective.

This system can then be run in a similar way to the Neighbourhood Watch scheme, not only just by regular contact with other fisheries, but by the use of posters and by setting up awareness campaigns aimed at members of the public, encouraging them to report suspicious activity. These campaigns can be jointly funded by several fisheries and can be very effective. The police should be involved in the establishment of any scheme like this and this has the added advantage of raising the profile of fish thefts with the local police in the area. Also, ensure that the local government officers are involved in the scheme. They can help in all sorts of ways.

These schemes have been successful in other countries and in marine fisheries, although there are no good examples in freshwater fisheries.

Monitoring illegal fishing

It should be recognised that illegal fishing takes other forms, not just poaching. Unauthorised fishing is still a problem even if the angler has no intention of removing fish, and he is just there for the sport. A completely legal angler will have all the required permits, including a rod licence (England and Wales) and the written permission of the owner of the fishing rights, normally in the form of a day or annual ticket or club membership. Anyone fishing without these is fishing illegally and needs to be dealt with. These types of illegal angler tend not to take the same access routes as poachers. The routine monitoring duties should therefore include routine monitoring of tickets. For day-ticket fisheries this is simple, but it is easy in membership fisheries to take the view that tickets do not need to be checked regularly. Some procedures must be established in all circumstances where routine monitoring of tickets may be expected to pick up anyone fishing without a permit. This can be a person just asking everyone for their tickets, or could include permits to be displayed in car windows or on the back of jackets.

Monitoring illegal methods

As already stated, the problem in this area is people using illegal methods, who in all other respects fall into the category of paying customers. Monitoring all the rules will usually fall into the routine duties of the bailiff. Generally a bailiff will visit the fishery at least once a day. During this visit most of the anglers on the site will be visited to obtain day ticket money or to check their membership information. It is sensible for the water bailiff to monitor them for adherence to the rules at the same time. This may just be a case of having a sneaky look at the angler's tackle or bait box while chatting and looking at the ticket, but they may be required to wind in their rods so that the terminal tackle can be checked.

Some bait bans for coarse fisheries are put into force at the whim of the fishery owner, and the thinking behind the rule can be difficult to understand. When introducing new rules and restrictions it is advisable to consult your anglers and to explain why new, or revised, restrictions are required. If the angler understands this, they are far more likely to respond positively.

Monitoring duties

Having completed the risk assessment and reviewed the monitoring options it is now possible to introduce a scheme of routine work that will enable all the highlighted points to be covered. A sample visiting schedule is given in Box 24.1.

Box 24.1 Sample schedule for inspection visits

Day	Times	Work	Notes
Monday	8.00	Visit and inspect all points of entry.	
	9.00	Contact FishWatch and record any new information.	
		Check anglers.	
Tuesday		Check anglers.	
		Check perimeter and whole fishery.	
		Check water for illegal fishing; drag for netting and lines.	
Wednesday		Check anglers.	
Thursday		Check anglers.	
		Check perimeter and whole fishery.	
Friday		Check perimeter and whole fishery.	
		Check anglers.	
		Check water for illegal fishing; drag for netting and lines.	
Saturday		Clean signs at access points.	
		Check anglers.	
		Check water illegal fishing; drag for netting and lines.	

24.1.2 *Legal obligations of allowing access*

Finally in this section, it is important to note than when the fishery allows and even encourages access to its land it also has corresponding legal responsibilities to ensure that anyone coming onto the land is safe and will not be injured owing to the negligence of the fishery. That is to say, the fishery needs to ensure 'safe' access for all those entering the fishery. In fact, there is a legal obligation to ensure even those entering the fishery without s pecific permissions can do so in safety. This is covered in terms of the legislation in Chapter 25.

To discharge this responsibility, the fishery must point out on permits, notices and membership cards, etc. that the individual is responsible for their own safety, and that fisheries can be 'hazardous by nature', for example slippery banks, deep water and so on. Specific hazards should be posted on a notice board, as should a general warning. Specific hazards such as deep water can be signposted.

The fishery committee, manager or organisation management must also be aware that they have a duty to inspect the fishery at reasonable intervals to assess health and safety and to make any adjustment to the risk assessment. If this inspection reveals a change, a new risk, or a change in the status of the risk, they must notify the membership and any members of the public that may have both authorised and unauthorised access. This can be achieved by posting of a notice, or by letter, as is felt most appropriate.

All structures in place for the use of the paying customer must be maintained to the highest standard. For example, if a peg constructed by the fishery collapses or causes an injury the fishery will be liable to prosecution.

24.2 Dealing with incidents

The fishery now has a risk assessment, highlighting problem areas. Preventative work has been done to restrict some of the problems. A scheme has been developed to monitor all of these problems and catch anyone who is breaking the rules. The next and final step is to consider how rules are checked and what happens when members of staff come across someone breaking the rules. This area needs to be considered from the very basics.

When dealing with an incident there are several factors to be taken into consideration. In most cases the same person has several jobs on a fishery. They can be bailiff, keeper, ghillie, litter picker, and so on, depending on the day. More importantly, they are the face of the fishery; they are the one the customers see. Their behaviour will often decide whether the customer returns.

24.2.1 *Training*

First, there is the appearance and training of staff. Personnel should be smart and clean. Any identity cards should be clearly displayed, or easily accessible. Personnel should identify themselves to anyone they approach as soon as possible. A responsible fishery will ensure that staff are trained to deal with the situations that they are confronted with. This training should include ensuring that all staff are aware of the fishery's rules and regulations. All staff should be aware of the risk assessments outlined above. Training in this respect should be an ongoing process and should be the responsibility of the personnel as well as the fishery.

24.2.2 Planning

A successful fishery will have considered many different situations, so that a variety of plans, to fit most circumstances, can be drawn up. This will form a set of rules that all personnel can follow in any given circumstance. It should include rules for first contact. Usually there is a need to pause when a situation is first discovered, to take in details of what is actually happening. Watching and recording details at this stage can be very useful later. If the situation is a single fisherman, who is not recognised and who is acting strangely, then it is likely that after a brief period of observation the person can be safely approached without a problem. If there is a group of men, at night, handling a net, then it may be very unwise to approach them unless there are two or more personnel available.

For obvious reasons, poachers and other persons of ill-intent should not be able to anticipate the movements of a water bailiff; a set routine is unavoidable for certain duties such as feeding stock fish or taking water quality readings, but provided that other areas are kept flexible then the risk should be minimal.

An important part of any water bailiff's work is patrolling his or her area to ensure that no-one is fishing illegally or poaching. It is here that the danger of his movements being anticipated lies. To lessen this possibility, patrols should be planned on a random basis, to vary routes and times of starting and finishing each day.

The whole of the area needs to be covered in a given period, but the main effort should be concentrated on those areas where, and at times when, trouble is expected. Experience will probably show that early morning and late evening are the periods of greatest activity.

This does not imply that patrolling at other times should be reduced or ignored. Weather and water conditions can also dictate the pattern of a day's work; a sudden flood will alter one's plans at short notice. These factors, and a correct interpretation of their effect upon angling or poaching activities, play an important role in deciding which duties to perform.

24.2.3 Fisheries responsibilities to employees

Before appointing a fishery worker, the owner should interview them to find out if they meet the standard desired. The fishery has several responsibilities to the person appointed. These include:

- Issuing them with a written authority. This should identify them, preferably with their photograph, and empower them to act on the fishery owner's behalf.
- Ensuring that full risk assessments are in place for all the employees' tasks.
- Providing any clothing or equipment necessary for him to carry out their duties.
- Issuing clear instructions and guidelines (a code of practice) as to their duties, responsibilities and rights. These should set out any conditions under which he would be allowed to fish on club waters, or receive other benefits.
- Fully supporting them in any legitimate action they take to protect the organisation's interests.
- Insuring them against any injury or claim arising from their work for the club.

In addition, the organisation should arrange training, where necessary, to back up a water bailiff/keeper's knowledge, or to help overcome any particular weakness that is apparent.

24.2.4 Standardising procedure

It is important that all employees in any one organisation act within a defined set of guidelines or 'code of practice'. This requires that employees have a set of defined rules to enforce and uphold and a clear outline of their duties and responsibilities in a 'job description'. The instructions given to an employee on his appointment should be clear and unambiguous and should establish his role within the organisation. As discussed in the previous section, fishery rules and procedures will have been laid out for routine monitoring of the site and anglers. These should be included in any job description. Although this may seem proscriptive to both employer and employee, some guidance and structure must be laid down. This can and should be discussed and reviewed routinely.

A fishery employee tends to have other responsibilities that go far beyond enforcing fishery rules. These may include the following:

- To help implement the club's policies and to protect its interests.
- To prevent unlawful fishing using all legal means at his disposal. In the context of his work, the term 'unlawful' is taken to mean fishing without permission or flouting the club's rules.
- If appropriate, to account for all permit monies paid to him by anglers.
- To maintain and improve the fishery and advise the club when necessary.
- To keep the club fully informed of all matters that may affect the fishery.
- To promote the good name of the club or fishery.

The code of 'best' practice does need to be based upon the ability of the fishery worker to act within a sound knowledge of the club rules, statute and common law. This does not mean that the fishery worker should be able to recite the rules chapter and verse. In fact it is far better that they have researched in the rulebook either before approaching the angler, or indeed while they are with him.

24.2.5 Penalties

For offences covered by legislation this is straightforward: the suspect must be informed of his or her alleged offence and information sent to the relevant authorities.

For transgression of the fishery's rules, there should be a formal procedure set in place. This will include notification of the rule broken, evidence-gathering to report the offence, who the offence is then reported to, the process for deciding the penalty and any appeal procedure. For membership waters, the full procedure must be laid out in the members' handbook so that all the members know what to expect if they are caught breaking the rules.

Most angling clubs or associations will support even the most draconian of punishments (see, for example, Box 24.2).

The fishery worker should point out to the angler where the rule appears in writing (rule book, fishery sign, etc.) and the area of the rules that they believe may have been contravened. The fishery worker must have a working knowledge of the rules, in fact the same knowledge that the angler should be expected to have if they are to keep the rules. It is important therefore that the fishery worker is supported by interim or temporary rules

Box 24.2 The rules of the Wartnaby Angling Association

The rules of the Wartnaby Angling Association are laid out below. If a member is caught contravening any of these rules the following procedure will occur.

- The bailiff will approach the member and inform him of the rule he has broken.
- Evidence will be collated and submitted to the fishery manager.
- The fishery manager will decide on the appropriate punishment in line with the following procedures:

First incident: the member goes on report for 1 year.
Second incident: the member is banned from site for 2 years.
Third incident: the member is given a lifetime ban.

- In certain circumstances where the rule breaking is particularly serious, or the member has not co-operated with the bailiff, then a 2 year or lifetime ban can be enforced straight away.

being posted at the fishery, on a very easily observed notice board. Wherever possible, written notification should be sent to the members (newsletter, etc.). The fishery worker should remember that the close interpretation of the rules should be left to any disciplinary hearing.

As part of the 'code of best practice' there might be some guidance in the way that members or potential non-members are approached. It is important to make a good first impression, so the fishery worker should identify themselves with a bailiff card or other authority that bears their photograph and is signed by an official of the organisation. The approach should be made in a positive way; the fishery worker should smile and try to sound cheerful. They should introduce themselves: 'Hi, I am ... , I am the club bailiff for this water'. The angler should be presented with the bailiff's authority; let him have time to read it and see the photograph. Do not give over possession of the authority or you might find it floating in the lake! The angler should be asked their name and if it would be possible to see their permit. The fishery worker must always be polite.

It is very important to be polite and courteous. The presumption should always be that the illegal fisherman is a visiting tourist and so unfamiliar with the legislation. If gently put right, they may then go and purchase their permits, etc. Being polite and courteous will not aggravate an already delicate situation and may therefore prevent any danger of conflict.

If the fishery worker considers the situation is likely to escalate to a potential conflict or there is evidence that what is going on is serious (for instance, there is a gang of poachers present on the fishery), then there needs to be clear guidance on how to act. This should always be considered before approaching. A full contact list of all help should be readily available. This list should include the police, the local authority bailiffs and the senior company officer on duty. It is very important that anyone working in these types of situation is very clear on exactly what they can and cannot do in any given situation. This advice should also include instruction on what information to relay to the person contacted. Telephoning the police and giving vague directions and an unclear message about what is going on may not elicit the required response.

24.2.6 *Reporting*

There should be a set procedure and preferably a specific form for making a report for rule breaking (see Box 24.3). The employee should try to reach agreement with the angler that a rule has been broken. The water bailiff should then fill out a notification of the pending report to the committee; this should be done in front of the offender.

Examples of the information that is required for the club or organisation offence report are listed in Box 24.4:

It is not unusual for members and friends to change their story after the event, but before any hearing. A written witness statement taken at the time of the offence can be very helpful.

There should be a set procedure embedded within the code of practice giving guidelines for water bailiffs who have to ask people to leave the fishery. If the water bailiff asks a person to leave, they should be sure on what grounds they are doing so.

24.2.7 *Asking an angler to leave the fishery*

A member should be accorded the courtesy of being assumed innocent until found guilty by a disciplinary hearing. Asking a member who did not have a landing net to stop fishing would be appropriate, whereas requesting he leave the fishery might not be, although he might well choose to leave.

Box 24.3 Offence report form

Name: Joe E. Bloggs
Address:
17 The Street
Townsville
Devon
EX17 1PT.
Offence:
Pike fishing using barbed hooks.
Not using a wire trace.
Date of Offence: Saturday 20 Oct 2004
Cautioned Yes/No
Delete as applicable

Mem N0: 173
Location:
Riverfield Fishery
Long Lake
Church Bank
Exeter
Rule Nos: 17, 17a
Time of Offence: 15.20

Witnesses:
A.N. Other
5 The Close
Exeter
Devon

Statement
I fully acknowledge that I was fishing for pike using barbed hooks and without a wire trace contrary to the rules as stated above.
Joe Bloggs
Additional information:
Car reg no: K768 VTR

Issued by:
-R. Smith-
Received by:
Joe Bloggs
Date: 20/10/2004.

Box 24.4 Example of the information required for an offence report

- Full name and address: verified if possible, easy if club cards have photographs and signatures. Always check against car number plates if possible.
- Details of the offence: wherever possible this should be linked to a rule from a rulebook or a specific legal offence.
- Exact location, including the bank and even as precise as the exact swim is useful.
- The date and time of the offence.
- Take the name and address of all who witness any act of rule breaking. A brief statement would also be useful.
- A copy of any statement signed by the defendant.
- Information if the accused was cautioned under PACE.
- Any other actions should also be recorded.

The most likely situation for asking an angler to leave is if they are a poacher or indeed a member who has no way of proving they are a 'paid up' member because they do not have their permit on their person. It might be considered advisable that this in itself would be against club rules. In the case of a member without a permit, they should be asked to leave the fishery. Of course, if it is a club rule that they carry a permit, they have in addition broken a rule by fishing without one.

Part of any 'code of best practice' should endeavour to link good practice in the field with equally good systems for dealing with the offender in a quick, impartial and even-handed ways.

24.2.8 Disciplinary procedure

Disciplinary hearings and procedures should be in writing and be available to all members. It helps to ensure that members believe that the system is fair and impartial. It will be worth supplying all fishery personnel with an advisory sheet that can be given to any offender.

The sheet might outline the procedure following the report to the committee, what will happen and the timescales involved. An example of a disciplinary procedure is shown in Appendix B at the end of the book and examples of recommended penalties are given in Box 24.5. In this document club bailiff is a term used for the fishery worker.

Disciplinary procedures also need to take account of some other factors. Juvenile members must be dealt with in a suitable manner. This will usually mean that a parent or guardian, or other appropriate adult, must accompany any members under 17. However, they may only interfere or get involved in the proceedings in mitigation and only at a time when invited to do so by the chairman.

If a committee member has a conflict of interest they must declare this, at the very latest when and if asked to sit on the disciplinary subcommittee. An example of a conflict of interest might be a family relation or business partner. The chairman's decision about whether a conflict of interest exists or not is final.

Any defendant, if found guilty, should have some sort of appeals procedure available to them. For example, a member found guilty of an offence may be given the right of appeal

Box 24.5 Examples of Sentence Recommendations

Causing a disturbance
Examples: running, shouting, bad language, radio, lamps and misuse of lights.
3–6 months ban, may be suspended.

Abuse of privileges
Examples: fishing without pre-booked ticket, illegal night fishing (no night permit, fishing on restricted water), occupying a swim without fishing.
3–6 months ban, may be suspended.

Litter
6–12 months ban, may be suspended.

Abuse of environment
3–12 months ban, may be suspended.

Cruelty to fish stocks
6–12 months ban, may be suspended.

Serious offence
Examples: poaching, theft of another angler's equipment; wilful damage to another angler's equipment.
12 months to life ban, no suspension.

to the full executive committee. The member wishing to appeal should inform the fishery in writing within 14 days of the committee's notification of the subcommittee's disciplinary findings and sentence.

Any punishments should take immediate effect. Any close season is included in all suspension or bans, i.e. it counts toward the period of the suspension or ban. As stated earlier, members of angling clubs tend to be happy with quite draconian punishments. However, these should be seen to be consistent and they should be widely advertised. Suspended sentences may be used as a form of warning. These sentences will come into effect if another offence, however minor, is committed.

Appendix A gives a detailed example of a disciplinary procedure.

24.3 Routine (miscellaneous) duties

No chapter on the management of still-water fisheries would be complete without a section on routine duties. As anyone involved in a small business will know there are always a multitude of 'other things' to be done, which appear to have little or nothing to do with fish. This section will attempt to include some of these.

24.3.1 Fishery maintenance

The most obvious, and probably most important of these miscellaneous routine duties should be general fishery maintenance. This will include jobs that are done as often as possible and should certainly be part of a weekly routine. There is no doubt that a clean, tidy and well-maintained fishery will be more successful than one that is not. It is a simple fact that customers expect certain standards, and they will not visit fisheries that do not meet these standards.

Of the general maintenance tasks the worst is litter picking. This is the scourge of all fisheries managers and it consistently comes out as the biggest complaint about the job. In the wider context it is probably the most common complaint that is made about anglers, and has certainly led to leased waters being removed from anglers and fishing being banned. As fisheries managers, we have to accept that anglers will leave litter, but it should be made as easy as possible for them to dispose of it properly. So bins must be provided and just as importantly emptied regularly. Furthermore, regular litter picking must be done, as there is no doubt that anglers are less likely to leave litter on a clean fishery; it also becomes much easier to identify persistent offenders.

Easy access from the car park to the pegs is vital, so banks and paths must be maintained. For footpaths during the summer, this means that they must be mowed regularly (once a week) to keep them clear. The actual pegs must also be kept clear of weeds and maintained to a good standard.

When completing fisheries maintenance activities, it is worth considering previous surveys completed by anglers and angling enjoyment. When questioned, there are two things that anglers routinely rate as the best thing about their chosen fishery. The first is that their car is secure and safe, preferably close to the peg to be fished from. The second is that the fishery is clean and tidy and provides toilets that are also clean and tidy. When developing a fishery these factors will have been considered, but it is also very important to consider them when drawing up a list of routine duties. The car park must be safe and well maintained. The paths to the pegs must also be well maintained with easy access to a safe and secure fishing platform.

24.3.2 Health and safety

There are quite a few responsibilities on a fishery manager in health and safety legislation. The provision of a clean, tidy and well-maintained fishery will go a long way to helping prevent any harm coming to employees and customers on a fishery. However, a weekly log of health and safety issues and an up-to-date accident book are essential parts of a schedule of routine duties. Clear records will also help in the, hopefully rare, event that there is a legal claim against the fishery.

24.3.3 Paperwork

Much of the routine work of a fishery will be the completion and organisation of the dreaded paperwork. However, the difficulties associated with this sort of paperwork can be alleviated by ensuring a good system is in place beforehand.

Whatever system is in place, it must record receipts, invoices and payments. A good accountant is expensive but can save easily as much as they are paid, both by making the paperwork easier, thereby allowing the manager to get on with other work, and by understanding and explaining the tax system.

24.3.4 Advertising

It pays to advertise, particularly when a fishery first opens. Word of mouth is a great way of getting customers through the gates, but first some customers must arrive and have a good experience. When a fishery first starts, it is useful to make a bit of a splash by producing leaflets and placing targeted adverts in local papers or the angling press. However, nothing seems to work as well as the essentially free advertising that you can obtain with a few, regular stories in the angling press or local papers. So part of the routine duties should be taking pictures of large or unusual catches and sending them into the press.

24.3.5 Water quality

Any fishery, but particularly one that has been stocked to improve the fishing, must keep some record of the water quality in its lakes. It is normal for this record to include dissolved oxygen and pH levels as well as a record of the ammonia, nitrite and nitrate concentrations (see Chapter 2).

24.3.6 Diaries

A few of the previous chapters have discussed the use of a log or diary to record specific information like the names of regular customers and the results of routine water quality monitoring. The idea of the book should be that if the normal fishery manager is unavailable for any reason, then anyone with some experience should be able to find all the information they require to pick up and run the fishery effectively. The other major use is that the diary will provide concrete evidence of what has happened on the fishery in recent times. For example, the water quality and mortality record will provide invaluable reference material if the worst should happen and the fishery suffers a large-scale mortality event.

With these tasks in mind, the diary should include a section of useful information at the start. This will include telephone numbers of useful contacts. There should also be a list of regular customers and their car model and registration number. Done well, this should mean that you can always welcome your customer by name, which is great for making the customer feel welcome and respected.

On a daily or weekly basis the diary should also include the following:

- Visitor numbers and a rough catch record.
- Details of any mortalities, even if it is only one fish.
- Water quality results.
- Predators observed, for example the number of cormorants seen.

The diary should include any unusual events. These may be weather events or could just be the registration number of a suspicious vehicle that drove onto the site.

Part III
Legal and Social Framework

Angling, and the running of fisheries for angling, is an important part of the social framework of many countries. Whether the pursuit of angling is simply for sport, or whether the fish is taken home to eat, makes little difference to the core values of the angler or the management of the fishery. In fact, recreational fishing is the most important, and in many cases the sole use, of the freshwater stocks in most of Europe and North America. It is important therefore not only to recognise the value of recreational fisheries but also to recognise the responsibilities that go with management of recreational fisheries. It is the duty of every fisheries manager to know and understand the laws that affect the work of a fisheries manager and to live up to the expectations of codes of best practice to the best of their ability. Furthermore, failure to comply with the law because of ignorance of its provisions is generally no excuse and does not exempt the practitioner for prosecution. Thus familiarity with the law and compliance with its requirements is very much in the interest of the manager of recreational fisheries.

It should also be borne in mind that laws and regulations change in response to differing requirements and views of the natural environment. In Europe particularly, there is a need to respond to the centralised directives of the European Union as well as to national and local regulations.

Part III the book concentrates on how some of these responsibilities are enshrined in legislation, and how others are covered in codes of best practice. It concentrates on the laws that apply in England and Wales; however, readers in other countries can be sure that similar laws will apply in their own countries and should inform themselves of them.

25 General Legislation

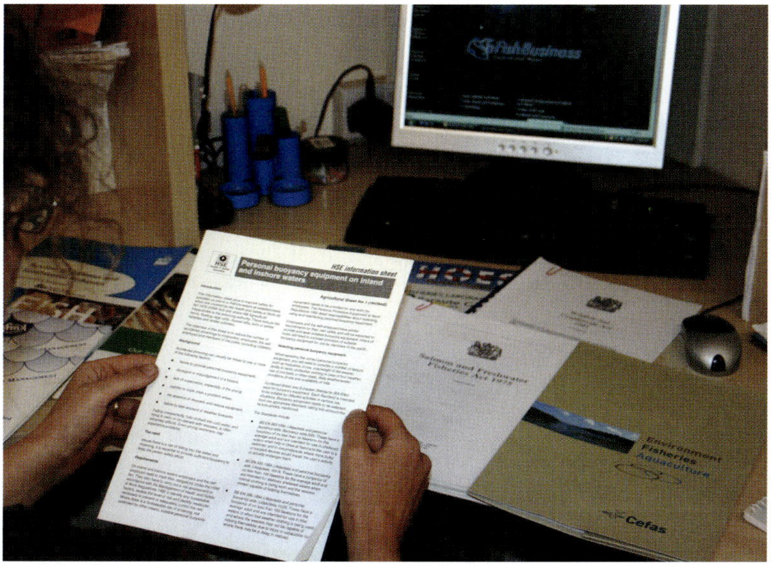

25.1 Health and safety legislation and risk assessment

Running a fishery presents the employer, employees and visitors with a multitude of potential risks. For example, monitoring access and fishing is an important part of the job of any fishery worker, but it is also a potentially dangerous one. There is a potential for conflict and probably a requirement for working alone or at night. It is important that all the risks associated with fisheries work are acknowledged, assessed and mitigated for. It is also important to understand that this work to reduce risk is everyone's responsibility, from the owner or senior managers to the worker on the bank.

Risk assessment is really a careful examination of what, in your work, could cause harm to people, so that you can weigh up whether you have taken enough precautions, or should do more to prevent harm. The aim is to make sure that no one gets hurt or becomes ill. Most risk assessments rely on a system of assessing the hazard and then the risk that the hazard may be realised. 'Hazard' means anything that can cause harm (e.g. chemicals, electricity, working from ladders etc.). 'Risk' is the chance, great or small, that someone will be harmed by the hazard.

Although all the known major risks may have been addressed by a policy, it is the responsibility of all employees continually to assess the tasks they do and bring to the attention of their supervisor, or safety representative, any new hazards and the level of risk. In this way the policy can be continually updated, to provide a greater level of protection at work.

In fisheries there will also be a requirement for a system to deal with incidents and emergencies. An example of a particular risk in this area is the need to work alone in isolated locations; as a minimum, a check ought to be made by the person in charge that all bailiffs have returned safely at the end of each work period. Where a bailiff is a sole employee, they should make their plans and whereabouts known to someone that they trust, such as a family member, an employer, a neighbour or the police. In the event of an incident, an accident or illness, it is vital that a bailiff's whereabouts are known and that they have the means to communicate, either by radio or mobile telephone. When it is not safe to perform particular tasks alone, arrangements must be made to ensure that an assistant is available.

In the UK, the Health and Safety Executive leaflet 'A guide to risk assessment requirements' should be consulted if there is any doubt about requirements under health and safety law. Another useful Health and Safety Executive publication is 'Health & Safety Guide for Gamekeepers'.

25.2 Personal protective equipment

Employers have basic duties in the provision and use of personal protective equipment (PPE) at work, and employers need to meet the requirements of the Personal Protective Equipment at Work Regulations 1992.

PPE is defined in the Regulations as 'all equipment (including clothing affording protection against the weather) which is intended to be worn or held by a person at work and which protects him against one or more risks to his health or safety', for example safety helmets, gloves, eye protection, high-visibility clothing, safety footwear and life-jackets.

Hearing protection and respiratory protective equipment provided for most work situations are not covered by these Regulations because other Regulations apply to them. However, these items need to be compatible with any other PPE provided.

The main requirement of the Personal Protective Equipment at Work Regulations 1992 is that personal protective equipment is to be supplied and used at work wherever there are risks to health and safety that cannot be adequately controlled in other ways.

The Regulations also require that PPE is properly assessed before use to ensure it is suitable. It must be maintained and stored properly. Any equipment that is provided must come with instructions on how to use it safely and it must be used correctly by employees.

25.3 Control of substances hazardous to health

Using chemicals or other hazardous substances at work can put people's health at risk, so the law requires employers to control exposure to hazardous substances to prevent ill health. Employers are required to protect both employees and others who may be exposed by complying with the Control of Substances Hazardous to Health Regulations 2002 (COSHH) (as amended). This control will range from appropriate storage to personal protective equipment to be used when handling the chemical.

For most commercial chemicals, the presence (or not) of a warning label will indicate whether COSHH is relevant. For example, there is no warning label on ordinary household washing-up liquid, so if it is used at work you do not have to worry about COSHH; however, there is a warning label on bleach, so COSHH does apply to its use in the workplace.

25.4 Occupiers Liability Act

Where a business actively invites members of the public onto their site, it is obvious that the health and safety regime should address the risks to customers, as well as employees. The Occupiers Liability Act lays out the liability of occupiers and others, for injury or damage resulting to persons, or goods, lawfully on any land, or other property, from dangers due to the state of the property, or to things done, or omitted to be done there.

Put simply, if people are on your land lawfully and they are injured, or their property is damaged, owing to the condition of the land, then you may be liable to prosecution under this Act. An obvious example would be the condition of the pegs and angling platforms. If these are provided they must be in reasonable condition and safe to use. If they are not, and a customer has an accident and injuries themselves, then the owner/manager has not lived up to their responsibilities under the Occupiers Liability Act and may face prosecution. As a fishery, the business is to provide a safe place for people to fish; therefore maintaining the fishery in good condition should be a high priority.

25.5 Disability Discrimination Act 1995

The Disability Discrimination Act 1995 (DDA) aims to end the discrimination that many disabled people face. This Act gives disabled people rights in the following areas:

- employment;
- access to goods, facilities and services;
- buying or renting land or property.

Under this Act the employment rights, and first rights of access, came into force on 2 December 1996. Further rights of access came into force on 1 October 1999 and the final rights of access came into force in October 2004.

The provision of angling facilities comes under the second area defined in the list above. Those that are involved in operating fisheries are, under the terms of the DDA, service providers. The duties of service providers under the DDA are shown below:

- since December 1996 it has been unlawful to treat disabled people less favourably than other people for a reason related to their disability;
- since October 1999 reasonable adjustments have had to be made for disabled people, such as providing extra help or making changes to the way services are provided;
- since October 2004 reasonable adjustments have had to be made to the physical features of premises to overcome physical barriers to access.

The most important date is perhaps that of the final phase of introduction, which calls for service providers to 'make reasonable adjustments to the physical features of their premises to overcome physical barriers to access'.

What is reasonable depends on several factors, in particular the size and resources of the organisation. For example, if you own a corner shop the changes you are expected to make are different to those expected from a supermarket chain. Equally, a village hall will have different requirements to the town hall or the banqueting suite in a large hotel. Installing a lift or new toilets may be inappropriate for a village hall or corner shop but an absolute necessity for the hotel or town hall. It may be reasonably considered impractical to put a disabled access path and disabled angling platforms along the entire length of a 40-acre gravel pit, but perfectly reasonable to make access available beside a still-water in the centre of a major British city for example.

It is important that service providers, who have not already done so, take reasonable steps to make their services accessible. Failure to do so could lead to loss of reputation or even litigation. Where it is reasonable, fishery owners may well find that funds for this type of improvement are readily available from external funding sources. Central, regional and local governments have an obligation under the Act to provide funding.

25.6 Planning legislation

The planning process in the UK is a large and complex system and varies somewhat in different regions. For fisheries it is safest to assume that planning will apply to most of the building work that the fishery may wish to do. For example 'change of use' planning will apply for the digging of a lake. Planning permission may apply for almost all structures that the fishery may wish to build on site, such as toilet blocks. Storage can be provided using metal transport containers, but almost everything else will require planning. On the positive side, planning is often granted for fisheries development as they are viewed as a positive development for farm land and a positive diversification for farmers to obtain new income streams. Furthermore, angling is seen by many district and county councils as a positive and healthy pastime.

General advice would be that if a development is being considered for the fishery, then it is best to assume that planning permission is required and run the plans past the local council officers as soon as possible, to get their opinion.

25.7 Theft Act 1968

Schedule 1, paragraph 2, of the Theft Act 1968 deals specifically with taking or destroying fish from private property.

> '(1) Subject to paragraph (2) below, a person who unlawfully takes or destroys, or attempts to take or destroy, any fish in water which is private property or in which there is any private right of fishery is liable on summary conviction to a fine; for an offence committed after a previous conviction for an offence under this paragraph, to imprisonment for a term not exceeding three months or to a fine or to both.

(2) Paragraph (1) above does not apply to taking or destroying fish by angling in the daytime (that is, in the period beginning one hour before sunrise and ending one hour after sunset), but a person who by angling in the daytime unlawfully takes or destroys, or attempts to take or destroy, any fish in water which is private property or in which there is any private right of fishery is liable on summary conviction to a fine.

(3) Any person may arrest without warrant anyone who is, or whom he, with reasonable cause, suspects to be, committing an offence under paragraph (1) above, and may seize from any person who is, or whom he, with reasonable cause, suspects to be, committing any offence under this provision anything which on that person's conviction of the offence would be liable to be forfeited.'

The power of arrest without warrant does not apply to angling during the daytime. The Theft Act 1968 stopped the 'apparent right' of owners, club bailiffs and gamekeepers from confiscating fishing tackle instead of going to court.

25.8 Trespass

A person who enters upon another's land without his consent or acquiescence, or without lawful authority, is trespassing. Trespass to land is interference with the possession of land. Where land and water are leased to an angling organisation, it is the leasing anglers who have possession, not the owner of the land. It is not necessary for the plaintiff to show actual damage in order to commence proceedings in a civil court. It is possible to obtain an injunction without proof of damage.

In the case of trespass it is suggested that if a trespasser is found, they should first be made aware of their trespass, politely asked to depart and be given time in which to leave the land. This law really would only apply to very persistent offenders.

25.9 Public Order Act 1986

This Act defines criminal offences ranging from violent disorder, affray or provocation of violence, to harassment or causing alarm and distress. For example, a magistrate's court may find a person guilty of using threatening, abusive or insulting words or behaviour, with intent to cause the threatened person to believe immediate unlawful violence will be used. Unlike trespass, which may be dealt with in civil courts, offences under the Public Order Act are criminal, involving the police.

The Act has been used successfully to deal with abusive and threatening people trespassing on private fisheries.

26 Environmental Legislation

26.1 Invasive and non-native species

Some of the most devastating ecological disasters of the past century have been caused by invasive species. It is thought that the cost of controlling invasive species runs into trillions of dollars worldwide. It is not surprising therefore that there is a range of legislation attempting to control invasive species. There are two areas where a fishery needs to acknowledge this legislation. First, the legislation controlling non-native fish species is the most obvious and will be covered in some detail later in this chapter. Secondly, it is also important to understand where else this may affect the running of the fishery business. For example, much of this legislation is designed to control invasive plant species.

The existing legislation dealing with non-native species in the UK is widely dispersed over many Acts that have historically developed in isolation, to deal with specific problems. The most important legislation governing non-native species is The Wildlife and Countryside Act 1981.

26.1.1 The Wildlife and Countryside Act 1981

This provides the primary control on the release of non-native plant species into the wild.

It is an offence under Section 14 (2) of the act to 'plant or otherwise grow in the wild' any plants listed in Schedule 9 part II.

26.1.2 *The Countryside and Rights of Way Act 2000*

This introduced stricter enforcement provisions for wildlife offences committed under the Wildlife and Countryside Act 1981.

Other legislation relating to invasive non-native plant species includes the following:

- The Environmental Protection Act 1990.
- The Town and Country Planning Act 1990.
- The Water Resources Act 1991.
- The Landfill (England and Wales) Regulations 2002.

Practical guidance is available from several sources and some of these publications are listed below:

- The invasive non-native species framework strategy for Great Britain – protecting our natural heritage from invasive species (2007), Defra.
- Guidance for the control of invasive weeds in or near freshwater (2005), Environment Agency.
- Strategy for non-native species research (2005), UK Biodiversity Research Advisory Group.
- Review of policy on non-native species (2006), Defra.
- Risk identification and assessment of non-native freshwater fishes: Concepts and perspectives on protocols in the UK (2005), Cefas.
- Audit of non-native species in Britain (2005), English Nature.

As with much of the legislation, new acts are coming into force, particularly from the European Union, that will impact on this area. For example, Defra is proposing to ban the sale of certain non-native species. If this proposal is agreed, it will result in an amendment to Section 9 of the Wildlife and Countryside Act 1981.

26.1.3 *The Environment Act 1995*

The Environment Act 1995 was responsible for the establishment of the Environment Agency, which combined the National Rivers Authority, the Waste Regulation duties formerly performed by County Councils and Her Majesty's Inspectorate of Pollution. It provided a more comprehensive approach to the protection and management of the environment, by combining the regulation of land, air and water.

26.1.4 *Control of Pollution Act 1974*

The control and prevention of pollution of rivers and other protected waters is the responsibility of the Environment Agency. Through this legislation it is an offence for a person to cause, or knowingly permit, any poisonous, noxious or polluting matter to enter a stream. It is an offence, except with the consent of the Environment Agency, to cause a deposit accumulated from a dam, weir or sluice to be carried away in the water of a stream when cleansing the channel, or wilfully and without consent allow a substantial amount of cut or uprooted vegetation to remain in the stream. By-laws can be made by the Agency to deal with certain types of pollution.

Each Environment Agency region keeps a register containing details of applications to discharge, consents to discharge and water quality information obtained from samples of rivers and effluents. These registers are available for inspection by the public.

26.1.5 Waste regulations

The UK produces around 330 million tonnes of waste annually, a quarter of which is from households and business. The rest comes from construction and demolition, sewage sludge, farm waste, and spoils from mines and dredging of rivers and lakes. For fisheries, therefore, any dredgings from a water body are considered waste. In some circumstances this waste may even be designated as hazardous waste if there has been industrial leakage into the water body in the past.

Waste regulations also control the burning of certain types of waste on site. This has been common practice in the past, but is not allowed under waste regulations legislation. This will relate particularly to the burning of plastic waste and animal products.

26.1.6 Water Resources Act 1963

The Act is to promote measures for conserving, redistributing, augmenting and securing the proper use of water resources, or for transferring those resources to another area. Water is a valuable and scarce resource, and this Act attempts to ensure its proper use. A person who abstracts water from a watercourse, or underground strata (groundwater), requires a licence to abstract. The construction or alteration of impounding works in a watercourse also requires a licence.

This legislation may affect a fishery in two ways. First, in some circumstances a fishery owner may wish to abstract water, either directly from the local watercourse, or from the ground water through a borehole. If the fishery wants to take more than 20 cubic metres a day then a licence will be required.

Secondly, the owner of a fishery can, under certain circumstances, apply for an existing licence to be revoked where they can prove damage or loss due to an existing abstraction. This is particularly relevant in water bodies naturally fed by groundwater, where the groundwater is being over-abstracted and resulting in the deterioration of the fishery.

26.1.7 Drought Act 1976

Any person who takes or uses water in contravention of a prohibition or limitation order, or discharges water contrary to any condition or restriction, commits an offence. An order may be made to measure the flow of water and keep records.

26.1.8 Land Drainage Act 1976

The purposes of land drainage can be summarised as the protection of life, land, urban areas and property from flooding; the improvement of agricultural land; and the conservation of water for riparian use. Nothing in the Land Drainage Act 1976 can affect or prejudice the provisions of the Salmon and Freshwater Fisheries Act 1975, and due regard must be taken of fishery interests. The Environment Agency must exercise a general supervision over all matters relating to land drainage in its area. Internal drainage boards were not affected by the 1976 Act. The Environment Agency has powers to make by-laws for such things as planting of trees alongside watercourses, blocking of watercourses and the like.

26.2 Reservoirs Act 1975

A large reservoir, any part of which is above the level of the adjacent ground and which has a capacity for holding in excess of 5.5 million gallons (25 megalitres) will require periodic inspection by a qualified (Panel 1) civil engineer. A qualified civil engineer must also design and supervise the construction or enlargement of a reservoir, which can only then be filled in accordance with his certificate. A certificate of safety is required by the competent authority. At the time of writing, the Government in England and Wales is considering significantly reducing the capacity of a reservoir before it comes under the Act.

26.3 National Parks and Access to the Countryside Act 1949, Countryside Act 1968 and Wildlife and Countryside Act 1981

These give power to the local Planning Authority to enforce access to a public path, and to stop notices deterring the public from using footpaths. It also gives the Agency powers of entry upon land for the purpose of surveying the land.

26.4 Further conservation legislation

There is, of course, much more legislation surrounding the conservation of natural habitats. Much of this does have to do with the aquatic environment, as water and wetland habitats are considered very important. Furthermore, much of this is not directly relevant to many fisheries managers as it is possible to run successful fisheries that are also important wetland habitats and therefore worthy of conservation.

Designations include Sites of Special Scientific Interest, Biodiversity Action Plan habitats, Special Protection Areas and Ramsar sites.

27 Direct Fisheries-Related Legislation

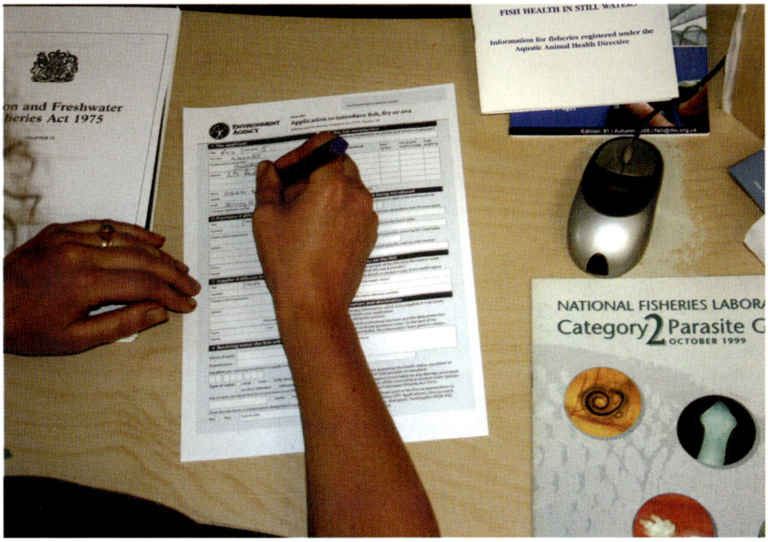

27.1 Salmon and Freshwater Fisheries Act 1975

This Act contains the major fisheries legislation and is largely enforced by the Environment Agency. It lists the fisheries duties of the Environment Agency, controls for the methods of fishing, the close seasons during which fishing may not take place, and in certain circumstances the size of fish that may be lawfully caught. It covers the licensing of users, the administration and reinforcement of the Act, and movement and introduction of fish into inland waters. It also gives the Environment Agency powers to make by-laws. The Environment Act 1995, with the Water Resources Act 1991, has modified and, in some cases, extended the 1975 Act powers. The following provisions of the act are of interest to still-water fisheries.

27.1.1 Section 28 Salmon and Freshwater Fisheries Act 1975

Before using fishing instruments (other than rod and line) and removing fish from inland waters, the written consent of the Environment Agency must be obtained. The written consent is usually a one-off 'permission' issued for a specified date and site.

The Environment Agency regulates the use of certain fishing instruments, such as seine nets and electrofishing equipment. This legislation is designed to protect fisheries from activities such as poaching.

The Environment Agency regulates the removal of fish from inland waters in England and Wales, to protect fisheries and the environment from aspects such as the impacts of inappropriate fish transfers and disease transmission.

There is a national application form that enables an application for the removal of fish to be made at the same time as the application for using the fishing instruments to catch the fish.

27.1.2 Section 30 Salmon and Freshwater Fisheries Act 1975

Under Section 30 of the Salmon and Freshwater Fisheries Act 1975 'a person shall be guilty of an offence if he introduces any fish or spawn of fish into an inland water, or has in his possession any fish or spawn of fish intending to introduce it into an inland water, unless he first obtains the written consent of the Environment Agency or the inland water is one which consists exclusively of, or of part of, a fish farm and which, if it discharges into another inland water, does so only through a conduit constructed or adapted for the purpose.'

Under the current guidelines for enforcement of this legislation, the Environment Agency considers the effects that the fish stocking will have on the receiving and adjacent water, before deciding whether to grant the consent. This may include things like the health of the fish, fish ecology and the ecology of other animals and plants. It is the potential impacts of these environmental concerns that drive the enforcement of Section 30.

One of the biggest risks when stocking fish is the possibility that the fish will carry new pathogens into the receiving water. Therefore the fish intended for introduction may need to have a health check. This depends on the location and type of site to which you plan to introduce them. Mandatory health checks will be required where fish are to be introduced into rivers, streams, drains or canals, or where there is a likely risk of fish movement into or from the receiving water. Health checks will not normally be required in waters where the risk of fish escape is minimal (e.g. enclosed waters). Such health checks are normally non-mandatory (voluntary), but there may be occasions where the Agency will still insist on a health examination. Obviously, to protect your own fishery, it is important to get a health check no matter whether the Environment Agency requires one or not. At the time of writing, changes to this legislation are ongoing. However, written consents for introductions and removals will still be required.

27.2 Non-native fish species

If the fish species to be introduced is non-native to the UK, then further legislation comes into play.

27.2.1 The Wildlife and Countryside Act 1981

Section 14 of this Act makes it an offence to release, or allow to escape, into the wild any animal that is not ordinarily resident in, or a regular visitor to, the UK, or that is established in the wild and listed in Schedule 9 of the Act, without a licence.

27.2.2 Import of Live Fish (England and Wales) Act 1980 – Amendment 1994

This Act gives the Minister power to make an Order to restrict the import, keeping or release of live fish, eggs and gametes of non-native species.

27.2.3 The Prohibition of Keeping or Release of Live Fish (Specified Species) Order 1998) made under the Import of Live Fish (England and Wales) Act (1980) (ILFA).

ILFA, as it is often known, was introduced to provide further controls on the keeping and release of particular species. The list of species covered by the measures was subsequently extended, with regard to England, by the Prohibition of Keeping or Release of Live Fish (specified species) (amendment) (England) Order 2003. An equivalent Order that covers Wales was issued in 2003.

27.3 Fish health regulations

Apart from Section 30, there is a raft of other legislation that is concerned with protecting fish populations from the spread of serious fish diseases. These diseases are known as notifiable diseases, and are the ones identified as potentially having the most serious economic consequences.

27.3.1 Diseases of Fish Acts 1937 and 1983

These Acts enable the Minister of Agriculture, Fisheries and Food to take measures to control the spread of disease by making certain diseases 'notifiable'. The Acts also control the importation of live fish and the eggs of fish.

27.3.2 The Fish Health Regulations 1997 (SI 1997 No 1881)

These Regulations implement Council Directive 91/67/EEC and control the movement into Great Britain, from elsewhere in the European Union, of all live fish, their eggs and gametes and certain dead fish. They also implement European Union rules on marketing, transport and identification of fish and fish products through the Animals and Animal Products (Import and Export) Regulations 1998 (SI 1998 No 190) and The Products of Animal Origin (Import and Export) Regulations 1996 (SI 1996 No 3124).

These Regulations implement EC directives about veterinary inspections of consignments imported from other Member States.

27.3.3 The Registration of Fish Farming and Shellfish Farming Businesses Order 1985 (SI 1985 No 1391)

This Order makes it a requirement for anyone carrying on a business of fish farming to register the business with the Ministry, and to keep movement records and a record of mortalities.

The system works by preventing the free movement across national boundaries of any fish possibly infected with notifiable diseases. However, because this objective goes against the objectives of free trade, before a country or region can declare itself free of a notifiable disease, it must put in place a monitoring programme to prove that it is free of these diseases.

The registration of fish farms enables these premises to be recorded and routinely checked for symptoms of notifiable disease. The act of making the diseases notifiable makes it a legal obligation for people to inform the government if they suspect an outbreak and further allows the government powers to take samples to check for the presence of notifiable diseases. If a notifiable disease is identified, then movement controls can be placed on the site to prevent further transfer of the disease.

Once these controls are in place then the country or region can legally restrict imports of live fish into that area from places that are infected, or cannot prove that they are not infected.

In the UK this regime has enabled the UK government to declare freedom from most of the notifiable diseases and therefore restrict imports of most types of live fish, particularly salmonids. This has helped to maintain the generally high fish health status of the UK, and allowed our fish farmers to trade their products abroad.

27.3.4 The Aquatic Animal Health (England and Wales) Regulations 2008

Recently the Aquatic Animal Health (England and Wales) Regulations 2008 came into force across Europe. Under this legislation several significant changes have taken place.

Under the new Aquatic Animal Health Directive (EC Directive 2006/88/EC), all stocked fisheries must be registered with the competent authority, the Fish Health Inspectorate (FHI) based at the Centre for Environment, Fisheries and Aquaculture Science (Cefas). The registration will require a named person, normally the owner, lease holder or manager, to take responsibility for the waters. That person must provide all reasonable help and access in the event of a fish mortality problem and must report any suspicion of a notifiable disease to the government authority (in England and Wales this is the Fish Health Inspectorate). It is a legal requirement for fisheries to be registered.

Under the new Aquatic Animal Health Directive (EC Directive 2006/88/EC), all aquaculture production businesses must be authorised by the competent authority; in England and Wales this will be the FHI based at Cefas. The definition of an aquaculture production business includes all fin-fish farms, traders in live aquatic animals, shellfish and crustacean farms, shellfish purification and dispatch centres, some processors and small-scale producers for the local market.

Authorisation conditions require that aquaculture production businesses must keep a record of the following:

- all movements of aquaculture animals and products into and out of the site or business;
- mortality in each epidemiological unit as relevant for the type of production unit;
- mortality during transport.

To ensure that record keeping requirements are fit for purpose, the minimum acceptable format for movement and mortality records is prescribed under the Regulation.

In addition, businesses must:

- participate in a risk-based surveillance scheme;
- operate in accordance with a documented biosecurity measures plan;
- inform the FHI of any material changes to business or fish farming practices.

The FHI will apply controls on authorised and registered waters in line with the seriousness of the non-compliance. For minor non-compliance, written advice will be provided. For more serious non-compliance, enforcement notices will be issued, with the potential to progress to prohibition notices and prosecution. Ultimately, authorisation to operate an aquaculture production business could be revoked.

There will be a right of appeal against any of the enforcement measures.

27.4 Transport of fish

The European Union Welfare of Animals During Transport Council Regulation (EC) 1/2005 sets out requirements on the transport of live animals. It states that any transporter of animals in connection with economic activity needs to hold a transporter's authorisation issued by the State Veterinary Service. Fish are included in this legislation, so operatives will be regulated when transporting fish from a source to a destination.

These authorisations have been free in the past, and last for 5 years. There are two types of authorisation: for short journeys under 65 km (approximately 40 miles) or up to and including 8 hours; and for long journeys, which covers all journey times and distances above this.

A journey is defined as the time at which the first animal is put into the means of transport at the premises of departure until the last animal is unloaded at the final destination.

To be eligible for authorisation, you must meet the following criteria:

- The individual or business must be established in the United Kingdom. Those operatives based in other European Union Member States are expected to be authorised in their respective Member State.
- Only one authorisation may be held.
- The individual or business must demonstrate there is the appropriate equipment and operational procedures in place to transport animals in compliance with the regulation.
- That the operatives performing the transportation must be trained and competent to transport animals.
- Anyone likely to be involved in transporting animals under the authorisation must not have a record of serious animal welfare offences in the 3 years preceding the date of the application.

28 Agencies and Organisations

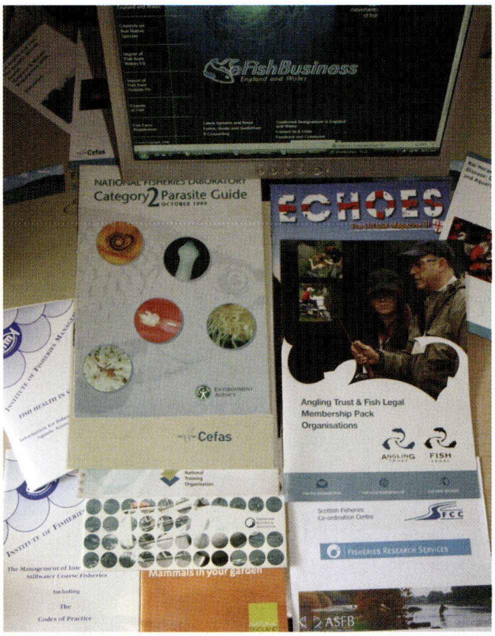

28.1 Fisheries enforcement agencies

Fisheries enforcement agencies such as the Scottish District Salmon Fisheries Boards, the Environment Agency (England and Wales) and the Centre for Environment, Fisheries and Aquaculture Science (Cefas) can help private fisheries in all sorts of ways. It is recommended these organisations be contacted at the earliest opportunity. A formal discussion with these bodies can help clarify the roles and responsibilities of organisations and individuals. They can provide much needed support and advice for some of the more difficult situations that arise in the running of a fishery, such as monitoring access and fishing activities.

28.2 Accreditation schemes

In the UK a Fisheries Accreditation Scheme was established by the Institute of Fisheries Management (IFM) in 2008. The aim of the scheme is to drive up the skills of those managing fisheries. Furthermore it is hoped that the fisheries concerned will go on to demand similar levels of competence from those that supply fish, provide fishery management advice, transport fish and carry out health examinations. This will have direct and indirect benefits to the health and welfare of fish stocked into still-water coarse fisheries. This will be achieved by awarding an accreditation mark at one of three levels: bronze, silver or gold.

The scheme is based upon a set of standards designed to protect and sustain the welfare of fish within fisheries. Achieving these standards largely centres on the IFM 'codes of practice' for the management of still-water fisheries and its biosecurity codes for still-water fisheries.

Fisheries that apply will be inspected by a 'competent professional' who has received scheme training. The status of fisheries that meet the standards will be recorded on a database that will be available publicly through the fisheries accreditation website (www.fisheriesaccreditation.co.uk). Those fisheries that do not make the grade initially will be given time to comply, by working towards the required standards, before being re-assessed. The accreditation lasts for two years and is renewable thereafter.

28.3 Code of conduct for Coarse Anglers

This document was produced by the National Angling Alliance, having been prepared by the Specialist Anglers' Alliance.

The Code was originally designed as a guide for specialist coarse anglers through the pitfalls of suitable conduct and the methods used in the sport of angling. It was originally published through the Specialist Anglers Conservation Group. With the formation of the National Angling Alliance, the Code was widened to include all aspects of coarse angling. The Code is a detailed document intended to offer guidance to individuals and groups such as angling clubs (it could form the whole or part of a club rulebook).

The code also includes a 'Newcomers Guide-Starting Angling' section at the back. This is in recognition of the detailed nature of the whole code.

The code is designed to complement the existing legislative framework and provides a commonsense guide to values and behaviour to which anglers should aspire. The Code is also designed to allow anglers to interact sympathetically with the environment and with the other conservationists who share water bodies. All fisheries in the UK should be members of the Angling Trust. Not only are they a unified body supporting angling but they have numerous further benefits for member fisheries.

28.4 Fisheries and conservation

Freshwater and wetlands are very important habitats for a wide variety of flora and fauna and are often protected by legislation. The public tends to recognise the value of freshwater in the environment, whether as rivers, still-waters or wetlands. They therefore have a very high perceived value.

As fisheries managers it is important to recognise the value of the habitat we manage, not only so that we can comply with any legislation but also because, by recognising the importance of the habitat, we recognise the value given to these aspects by our customers and the public.

28.4.1 Protected areas and conservation legislation

In the UK, protection of the landscape is achieved through many pieces of legislation and policy at local, UK, European and international levels. The UK has many national, European and international wildlife and landscape designations that provide protection to several sites of particular value. The management of these sites can involve legislation, voluntary management agreements with landowners, management by government agencies and non-government organisations and almost any combination of these.

The most important are areas designated by local government, such as local nature reserves, country parks and areas designated by national government which have a legal status, intended to control unsuitable developments that would change the inherent qualities of the area.

Sites of Special Scientific Interest

Sites of Special Scientific Interest (SSSIs) are intended to form a national network of sites, each representing a significant fragment of the last remnants of a wildlife resource in the UK. The statutory nature conservation agencies are responsible for SSSIs and for developing, with the owner, a list of potentially damaging operations. This list will hopefully ensure that the SSSI will be maintained in a suitable condition.

National Parks and Areas of Outstanding Natural Beauty

These were created by the National Parks (NPs) and Access to the Countryside Act, 1949. There are 11 National Parks in England and Wales, including wetlands like the Norfolk Broads, and over 41 Areas of Outstanding Natural Beauty (AONBs).

National Nature Reserves

National Nature Reserves (NNRs) are nationally important areas of habitat. There are around 200 National Nature Reserves in England.

Environmentally Sensitive Areas

Environmentally Sensitive Areas (ESAs) are particular areas of recognised importance from an ecological and landscape point of view, in which agricultural practices must be compatible with the requirements of conserving the natural habitat, as well as ensuring an adequate income for farmers. Administered by the UK Government's Department for Environment, Food and Rural Affairs (Defra), grants are available to landowners who attempt to preserve characteristic landscapes, by adopting or continuing traditional practices. The scheme is voluntary.

There are also areas or habitats designated by European Union or international legislation. These are described below

28.4.2 Conservation (Natural Habitats, etc.) Regulations, 1994

The regulations enact the European Union's Habitats Directive (92/94/EEC) in the UK. The Habitats Directive enables the maintenance of biodiversity within Member States, through the conservation of sites containing habitats and species of European importance. These sites make up the EU Natura 2000 network.

The regulations impose planning restrictions on Special Protection Areas (SPAs; designated for birds) and Special Areas of Conservation (SACs; detailed for habitat types and animal or plant species):

The network includes obligations to maintain and restore, and involves a strategy of 'integrating damage to biodiversity into liability mechanisms'. The Directives are implemented in the UK by the Conservation (Natural Habitats, etc.) Regulations, 1994.

28.4.3 Ramsar sites

The Ramsar Convention on Wetlands of International Importance, especially as Waterfowl Habitat, is an international, intergovernmental convention. The Convention requires signatories to designate and protect wetlands of international importance, to promote wetlands generally, and to foster the wise use of wetlands.

28.5 Wildlife and Countryside Act, 1981 (amended 1991)

The Act refers to the treatment and management of protected species listed as Schedule 1 (birds), 5 (mammals, reptiles, fish, invertebrates) and 8 (plants). It is an offence to intentionally kill, injure or take a scheduled species that is living wild at the time; to possess a scheduled species; to damage, destroy or obstruct access to the place of refuge used by the protected species.

28.6 Countryside and Rights of Way Act for England and Wales, 2000

The Countryside and Rights of Way Act 2000 (CRoW Act 2000) updates the laws on nature and wildlife conservation, and makes several changes to the provisions of the Wildlife and Countryside Act to protect Sites of Special Scientific Interest (SSSIs). It places a statutory duty on public bodies to conserve SSSIs and enhance their value, and enables Natural England and the Countryside Council for Wales with the power to impose management Schemes on SSSI owners. It also places a statutory duty on Government Departments and the National Assembly of Wales to regard biodiversity conservation fully and to promote conservation action by others. The Act strengthens the legal protection for threatened species and habitats in terms of killing, injuring, disturbing and destroying places of refuge. Further, the CRoW Act makes new provision for public access to the countryside, the 'right to roam', in England and Wales.

28.7 Wild Mammals Protection Act, 1996

The Act protects wild UK mammals from acts of wilful cruelty, with exceptions for lawful shooting, hunting, coursing or pest control activity. The Act is important for wetland mammals like otters and the water vole.

28.8 International legislation

The most important international legislation for UK wetland conservation is the Convention on Biological Diversity, the first legal framework for biodiversity conservation. The Convention on Biological Diversity is an international agreement that came out of the 1992 Rio Earth Summit (now ratified by 179 countries). Signatories are required to develop and implement national strategies on conservation of biodiversity and sustainable development.

The UK produced its national Biodiversity Action Plan in 1994 after consultation with around 300 organisations. The national Plan is divided into many Local Biodiversity Action Plans (LBAPs) with component Species Action Plans (SAPs) and Habitat Action Plans (HAPs). Normally each plan is managed by a committee, or a working party of interested parties from government and non-government organisations to industry and individuals.

28.8.1 European Union Directives

The Council Directives 79/409/EEC (April 1979) on the Conservation of Wild Birds and 92/43/EEC (May 1992) on the Conservation of Habitats and of Wild Fauna and Flora aim to maintain, or restore to favourable conservation status, species listed according to their conservation concern in a series of annexes.

28.8.2 European Union Water Framework Directive

This directive (2000/60/EC) requires member governments to produce management plans for every major river basin, aiming to meet strict ecological targets by 2015. The Water Framework Directive (WFD) highlights wetland restoration as a means of improving water quality and quantity. In the UK, the Directive will be implemented by a partnership of government departments and agencies (like the Environment Agency) and non-government organisations like the Wildfowl & Wetlands Trust.

28.8.3 Bern Convention on the Conservation of European Wildlife and Natural Habitats, 1979

The Convention aims to conserve wild flora and fauna and their habitats and to promote cooperation between countries. It also emphasises endangered and vulnerable species in a series of appendices. Article 2 requires parties to maintain populations at levels that match ecological, scientific and other requirements. Article 6 prohibits capture, killing, keeping, disturbance of, trade in live or dead specimens, and destruction of breeding or resting sites of species listed in its Appendix 11.

28.8.4 Bonn Convention on the Conservation of Migratory Species of Wild Animals (UK signed 1979)

The Convention requires the protection of listed endangered migratory species.

28.8.5 Convention on International Trade in Endangered Species (CITES)

The Convention prohibits or restricts trade in threatened species (or those likely to be threatened and affected by significant trade) that are listed in its appendices.

28.9 Organisations involved in wetland conservation

Wetland conservation involves a wide variety of organisations from government departments, government agencies, non-government organisations, the private sector and individuals with particular vested interests.

28.9.1 The European Union

The UK became a member of the European Economic Commission (EEC), a grouping of Western European countries for economic and political cooperation, in 1973. The EEC later became known as the European Commission (EC), which became the European Union (EU) in 1993, after the ratification of the Maastricht Treaty by the 12 Member States of the time. Several EU Directives impact on wetland management issues, such as Special Protection Areas and Special Areas of Conservation under the EU's Habitats Directive.

28.9.2 UK Government

Central government

Several UK Government departments impinge upon environmental issues. These are described below.

Defra

The Department for Environment, Food and Rural Affairs (Defra) is the main Government agency concerned with the environment. It aims to achieve sustainable development, including protecting and improving the rural, urban, marine and global environment; conserving and enhancing biodiversity; tackling social exclusion through the promotion of sustainable rural areas; improving enjoyment of the countryside for all; promoting sustainable farming and natural resource management; and to protect the public's interest in relation to environmental impacts and health.

Defra administers grants to farmers and landowners for agri-environment schemes that involve the enhancement of wetlands, woods, moors and other important habitats. Such schemes include Environmentally Sensitive Areas, Countryside Stewardship, Woodland Premium and Organic Farming.

Natural England

Natural England is a statutory agency of Defra formed in October 2006. It has four principal functions: (1) responsibility for conservation of nature in England in obeisance of the Wildlife and Countryside Act 1981; (2) enhancement of, access to and enjoyment of the natural environment; (3) the sustainable use of natural resources, and landscape management, including the overseeing of £0.5 billion in agri-environment funding; (4) provision of information to inform future government policy on the environment. Natural England coordinates with its two regional equivalents, the Countryside Council for Wales and Scottish Natural Heritage, through the Joint Nature Conservation Committee.

Natural England facilitates Local Nature Reserve projects, and it owns or controls all English National Nature Reserves (NNRs). It identifies and notifies all English Sites of Special Scientific Interest (SSSIs), Special Areas of Conservation (SACs), Special Protection Areas (SPAs) and Ramsar sites. It has powers under the Countryside and Rights of Way Act 2000 (CRoW) to serve Management Notices on owners, to compel them to undertake works to manage SSSIs appropriately for nature conservation. Natural England set up a Species Recovery Programme in 1991, to fund research into the needs of declining species. Natural England also aims to conserve rural landscapes as a national asset for leisure activity, using its responsibility for designating Areas of Outstanding National Beauty and English National Parks, which benefit from protection from development and some £23 million in government funding annually.

Environment Agency

The Environment Agency (EA) is the statutory body for protecting the environment in England and Wales. It is concerned with flood protection, pollution, industrial impacts on air, land and water, the cleaning of rivers and coastal areas, and contaminated land. The Scottish equivalent of the EA is the Scottish Environmental Protection Agency (SEPA).

Local government

At the local government level, the UK has a mixture of borough councils, county councils, metropolitan authorities, district councils and unitary authorities. All comprise officers and elected councillors. Local government is active in Biodiversity Action Plans, and in implementing (with partners) Local Agenda 21, which is an action plan for sustainability that originated at the 1992 Rio Earth Summit.

Several government agencies have statutory powers for UK wetland (and general) conservation.

Non-government organisations

These include many charities that own, lease and manage significant areas of important habitat. These include the Wildfowl & Wetlands Trust (WWT), the largest international wetland conservation charity based in the UK.

The National Trust, the Royal Society for the Protection of Birds (RSPB) and the Wildlife Trusts are also important organisations in this area.

Within the private (and public company) sector, many industries are concerned with the environment, including the major water companies. Many individuals also have a vested interest in UK wetland conservation. These include anglers, farmers, local people, members of hunting associations, providers of food, transport and accommodation, tourists, voters and you!

Appendix A: Example of Risk Assessment

Rabbits Fishing Lake, 25 February 2005

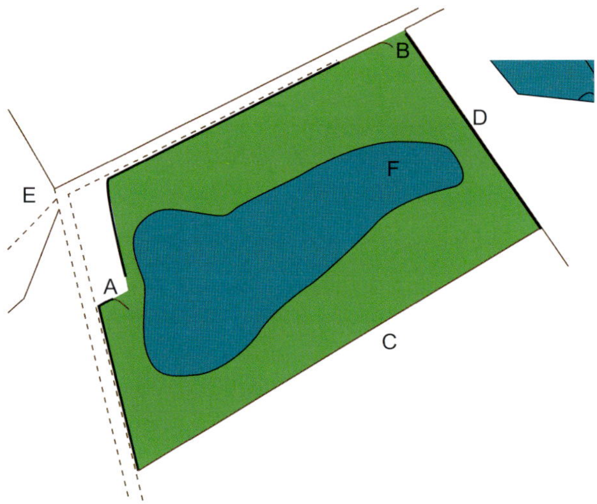

Figure A1 A plan diagram of Rabbits Fishing Lake.

Risk assessment (Fig. A1)

Points of access

Main pedestrian entry by car park. (A)

Gate in north fence gives vehicle access. (B)

There is possible vehicle access through south border fence. Simple wire sheep fence. (C)

Points of pedestrian access

As above plus one small possible access through east fence (D).

Well-used footpath passes just north of lake. Rarely do walkers go through fishery, as there is a good well signposted path. (E)

Likely forms of poaching, in order of likelihood

Theft of fish by authorised angler.	Throughout year
Seine netting in shallow east bay. (F)	Spring and summer
Fyke traps set on shallower north bank.	Spring and summer
Fixed lines.	Throughout year.

Selling fish

Largest carp 18–20 pounds. No fish tagged or readily identifiable.

Current fishery rules

- No barbed hooks.
- No peanuts or tiger nuts.
- No overnight fishing.
- No dogs.
- No fires.

Current bailiffing regime

Ticket collector visits fishery twice per day (roughly 10 a.m. and 2.30 p.m.) to collect day ticket money.

Fishery regularly passed on road by ticket collector and fishery manager 9–5.

Safe access

Several pegs in need of repair (particularly pegs 18, 26 and 32).

All pegs require surfacing with an anti-slip cover.

One of the two lifebuoys is missing.

Action plan

- Reduce access points to front car park.
- Reduce ability to operate a seine net in shallow areas.
- Reduce fish saleability.
- Display rules on signs in car park.
- Review ticket collection and fishery manager routines.
- Act on all safety aspects.

Preventative measure, undertaken 23 February – 2 August 2008

Access

- Gate into fishery from car park improved to make vehicle access possible.
- Gate on North fence removed and blocked with temporary concrete blocks and hedge planting. Concrete blocks to be removed when hedge has grown enough to prevent access.

- Ditch dug along the line of the double wire fence on south border. Ditch is designed to prevent vehicle access. Hedge planting undertaken along this fence, once established this would also prevent pedestrian access.

Poaching

- Lilies planted in wire cages alongside southern shallow bay. Addition of sunken logs in same area. These are marked to allow anglers to avoid the snags.
- Quotations sought for pit tagging the 40 largest fish within the lake. Review the use of visible implant tags as possible deterrent.
- Letter sent to all fisheries within 10-mile radius of rabbit fisheries. This proposes the establishment of a local Fishwatch scheme. Awaiting response.

Work Programme

- Ticket collector to vary routine.
- Fishery manager to make at least two walk rounds a week.
- Out of hours visits to be made at least twice a week.

Safe access issues

- Repairs undertaken on 18, 26 and 32.
- Anti-slip covers added to all pegs
- Lifebuoy replaced.

Appendix B: Example Model of a Disciplinary Procedure

1. The disciplinary procedure should be invoked to deal with any member who it is alleged has breached any club rule.
2. The disciplinary procedure should only be put into force once a written report or complaint has been received by the organisation's pre-appointed official, normally the Hon. Secretary. All complaints and/or reports of a breach of club rules must be in writing.
3. The written report/complaint should clearly identify the alleged offence and state the place, time and date of the breach. It must bear both the name and address of the person bringing the complaint (and membership number if appropriate), and bear the signature of that person. Club bailiffs need only include their name and membership number.
4. All reports/complaints should be notified to the Hon. Secretary within twenty-eight (28) days of the alleged offence/breach. Club bailiffs should send notification of any booking made to the Hon. Secretary as soon as is practicable and in any case within fourteen (14) days.
5. The Hon. Secretary will inform the member accused of committing any offence or breach of club rules, in writing, as soon as is practicable and in any case within twenty-eight (28) days of receipt of the report/complaint.
   The communication should be a letter warning as to the member's future conduct, or requesting the member attendance at a disciplinary hearing. The letter should be sent by recorded delivery.
6. The Hon. Secretary will convene the disciplinary sub-committee as soon as practicable. The member should be made aware of the allegation made against him. The member should be offered two dates for a hearing, both should, where practicable, be within twenty-eight (28) days of the committee meeting at which the Hon. Secretary first raised the matter. The letter will warn that if no response is received, the matter will be dealt with in their absence. The letter should also make it clear that a guilty plea may be accepted in writing. A copy of the disciplinary procedure should accompany the letter.

7. The Disciplinary Committee shall consist of the three primary Officers of the club: e.g. Chairman, Secretary and Treasurer. The Vice-Chairman will take the place of any officer in situations of absence or conflict of interest.
8. Should the defendant have pleaded guilty in writing, the sub-committee will follow the laid down procedure. The sub-committee Chairman may still summon witnesses, as he may deem appropriate.
9. Should the defendant not have made a plea or have protested their innocence, the sub-committee Chairman should ensure that the sub-committee have read all relevant statements as supplied by the Hon. Secretary.
10. The defendant should be asked into the hearing.
11. The bailiff or person originating the complaint (the plaintiff) will be called and questioned by the sub-committee, on points of clarification. The bailiff/plaintiff will then be invited to add any other material facts, as known to them. 'Hear say' evidence will not be accepted.
12. The bailiff/plaintiff will then be asked to retire.
13. The sub-committee will then call, individually, any witnesses for the complaint/breach. The witnesses will be questioned individually by the sub-committee and offered the chance to speak freely to add any further pertinent factual information.
14. They will then be asked to retire.
15. The sub-committee will read any statement of the defendant. The defendant will be questioned by the sub-committee, on points of clarification.
16. The Defendant will be asked if they wish to add any other material facts, as known to them. 'Hear say' evidence will not be accepted.
17. The defendant will be asked to retire.
18. The sub-committee will then call individually, any witnesses for the defence. The witnesses will be questioned individually by the sub-committee and offered the chance to speak freely to add any further pertinent factual information.
19. The witnesses will be asked to retire. Witness statements/evidence for both sides must be factual. The sub-committee will not be interested in hearing evidence of 'hear say' or 'character references'.
20. The Defendant will be asked to retire.
21. The Chairman having satisfied himself that all available material evidence and relevant testimony has been placed before the committee shall ask each individual member of the sub-committee to vote guilty or not guilty, there shall be no abstentions. A verdict will be decided by a simple majority. The Chairman has a vote, which must be exercised.
22. The defendant will be called into the hearing. They will be given the verdict of the sub-committee.

    Not Guilty. The defendant should be thanked for their time.

    Guilty. The defendant should be asked if they would like to make a statement in mitigation and present to the sub-committee any such written statements they have in mitigation or written character references.
23. The defendant will be asked to retire, while the Chairman and sub-committee consider any mitigation, and appropriate punishment.
24. The Hon. Secretary should at this time make the defendant's past disciplinary record available. It is acknowledged that in a small angling club, past disciplinary matters

may well be common knowledge. The principle that any past disciplinary record is set-aside until a decision has been reached as to the guilt or not of the Defendant is important.

25. The chairman should recommend his considered appropriate punishment, having taken into consideration all of the facts. He should then ask each individual member of the sub-committee for their endorsement or alternative suggested sentence.

In all cases of appropriate punishment the decision of the Chairman is final.

Appendix C: Example of a Fisheries Survey

Background

The still-water to be assessed is a 2-acre, man-made lake, dug direct into clay. The lake is a long, oval shape with two small bays and no islands. There is good marginal vegetation, with 75% bank coverage, but little or no submerged vegetation. Canopy shading from marginal trees is minimal; hence there is good marginal plant flora. The water was turbid and an algal bloom was evident on the day of the site visit. The lake has a natural population of roach, rudd and perch. It has also been stocked in the past (around 10 years previously) with carp and tench; no other stockings have been reported. Most anglers are pleasure or match anglers and the most common technique is pole fishing. There are a total of 27 pegs around the lake, and the manager estimates that there are 75 angler visits a week in the summer. There are between one and three matches a week on the fishery. The largest recorded carp is just over 10 kg (21 pounds) and carp dominate the catch records, particularly in recent years.

The anglers are complaining that the carp are getting too big to land on the pole and there 'are no other fish in the place'. There is also a problem with baits being pestered by small perch.

A fisheries assessment was commissioned to assess the population levels of the key species – roach, tench, carp and rudd – and to use these results to advise on a fisheries action plan to improve the angling performance of the fishery.

Sampling

A depletion survey using a 100-m seine net was selected as the method to obtain a population estimate of the carp, roach and rudd. Floating cages were used to hold the catch from the first sweep while the second sweep was undertaken (Figure C1). A large aerator was deployed to ensure that the fish in the cages did not suffer during their period of captivity.

The first sweep of the net covered most of the lake. The fish caught are listed in Table C1.

The numbers caught in the second sweep caught are given in Table C2.

All the fish were measured in centimetres. Obviously, with so many roach and perch in the nets, it was impractical to measure every fish, so these were grouped into size ranges and only those over 20 cm were measured individually. Scales were taken from a selection of the various sizes of the roach and the rudd.

Figure C1 Floating cage used to hold fish.

Table C1 Catch by species and number from the first sweep of the net

Species	Number	Size range (cm)
Carp	34	24–53
Roach	2153	9–31
Rudd	44	11–27
Crucian carp	1	12
Perch	1144	8–26

Table C2 Catch by species and number from the second sweep of the net

Species	Number	Size range (cm)
Carp	25	17–47
Roach	1252	11.5–30.5
Rudd	12	10.5–22
Crucian carp	0	
Perch	947	8–28

Results

Population estimate N and variance of N are calculated as:

$$N = C_1^2/(C_1 - C_2)$$

where

C_1 = number of fish caught in first sample;
C_2 = number of fish caught in second sample.

Carp

$N = 34^2/(34 - 25) = 1156/9 = 128$.
The population estimate is a total of 128 carp in the lake.

Roach

$N = 2153^2/(2153 - 1252) = 4\,635\,409/901 = 5145$.
The population estimate is a total of 5145 roach in the lake.

Perch

$N = 1144^2/(1144 - 947) = 1\,308\,736/197 = 6643$.
The population estimate is a total of 6643 perch in the lake.

Population estimates achieved by these methods can be inaccurate owing to various factors, and some common sense should be used when interpreting the results. They are, however, useful as a guide to the actual numbers of these species present in the lake. From the catch results it is likely that the estimates for the roach and perch are reasonably accurate. However, we would expect there to be far more carp than is estimated by these results. This expectation is partly because there are recognisable fish present in the lake and only a few of these were caught during the survey, and partly because of angler catch results. This inaccuracy results from the fact that some fish managed to evade the net and a few were even seen to jump the float line. On the first net the lead line was also seen to lift during the netting operation, which may have allowed some carp to escape.

Scale ageing and population profiles

For the main three species a graph has been drawn of the length of the fish against the number of fish in this length range. This is called a length–frequency histogram (Figure C2). Some of the fish were aged; the results are shown in Table C3.

Roach growth curve

The graph in Figure C3 shows the roach ages against a national standard growth curve for roach. This clearly shows that the roach are growing slowly compared with a national average.

Health

The fish were visually examined, during the sampling operation, by an experienced fisheries expert. There were large numbers of the fish louse, *Argulus* sp., on some of the carp and on a few of the roach, although no significant pathology was associated with this infection. A few of the roach had minor wounds and bacterial infections. A few of the carp had hook damage to the mouth. Most of the fish appeared to be in good overall condition.

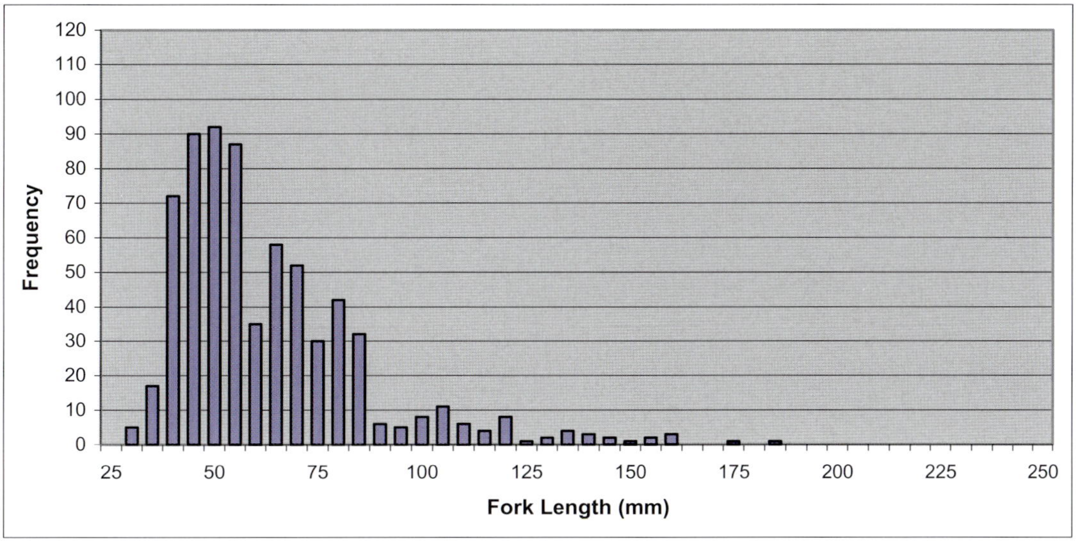

Figure C2 In this example the first peak in the graph probably relates to one age group (2 year olds) at about 50 mm then there is another peak at 70 mm, which could be 3 year old fish. So it is possible with some data to see the age groups of young fish up to three years. However, variability (plasticity) of growth means that this method of ageing is unreliable except for the very youngest age groups. The rest of the graph as expected shows a fairly steady decline in numbers, as the fish get older.

Results summary

Roach

The graph shows good populations of small roach between 9 and 16 cm (3.5–6 inches). Ageing of the fish shows these roach to be between 3 and 5 years in age. Smaller fish less than 9 cm are present but have escaped the net through the mesh. Larger fish are present in small numbers up to 31 cm and 9 years old.

Perch

The graph shows good populations of small perch between 9 and 16 cm (3.5–6 inches). Ageing of the fish shows these perch are mostly 3–4 years old. Smaller fish less than 9 cm are present but have escaped the net through the mesh. Only a very few fish over 16 cm are present.

Carp

Most of the carp were between 29 and 49 cm in length. From their appearance it is considered likely that these fish are part of the original stocking of carp. They have grown well and are in good condition. There are a few fish below 29 cm, which are likely to have been spawned in the lake since the original stocking. There are, however, very few of these fish present. Ageing of these smaller fish suggest that they were successfully spawned 2 or 3 years after the stocking of the larger fish. It appears that there has been no successful spawning since, as there are no smaller carp present.

Table C3 Length at age for fish from the sampling exercise

Species	Length (cm)	Age
Roach	9	3
Roach	9	3
Roach	9	3
Roach	10	3
Roach	10	3
Roach	10	4
Roach	13	4
Roach	13	4
Roach	13	4
Roach	16	5
Roach	16	5
Roach	16	5
Roach	18	7
Roach	19	6
Roach	19	7
Roach	23	9
Roach	27	9
Roach	30	9
Perch	10	3
Perch	11	3
Perch	12	4
Perch	14	4
Perch	14	6
Perch	14	5
Common carp	17	6
Common carp	21	8
Common carp	25	8

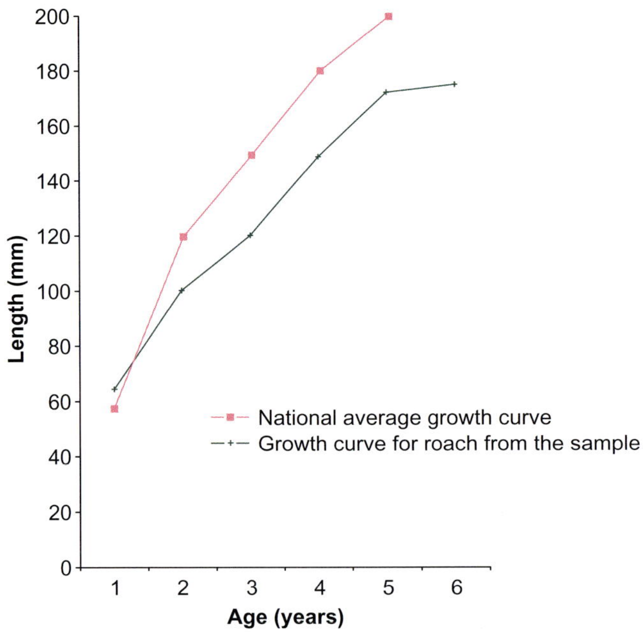

Figure C3 Growth curves of roach comparing the sample with the national average.

Other species

Rudd and crucian carp were caught during the survey, but only very few, and these species are unlikely to be making much of a contribution to the fishery. No tench were caught, although angling catches confirm they are present within the fishery, but not in great numbers. The sampling method used often means that tench are under-represented in the catch because, owing to their bottom-dwelling habit, they find it much easier to evade the net than other species. Fyke nets would be a much better way of getting an estimate of the population of tench.

Conclusions and recommendations

The fishery is dominated by a large number of large carp. Most of the carp caught were over 35 cm and over 3.5 kg (8 pounds) in weight. This is larger than the average angler on this water wishes to catch. The dominance of carp has led to the fishery becoming very turbid throughout the year; there is now very little in the way of submerged vegetation, as light can no longer penetrate far enough into the water. The carp will also be grazing directly on any submerged and some emergent weed, preventing colonisation.

Although there are plenty of roach in the lake, they are growing very slowly compared with an average growth curve for this species. There are also good quantities of perch within the lake but these are not growing beyond a certain point, staying quite small. The growth shown by both these species indicates that there is a general a lack of food and habitat available. Perch certainly will remain small if there is a lack of small fish prey species available to them.

The lack of rudd and crucian carp probably indicates a lack of suitable habitat for these species. The lack of tench in the netting operation probably indicates that this species finds it easier to evade a net than the other species.

Although the roach and perch appear to be spawning reasonably successfully, it is evident that there is little or no recruitment of carp or any of the other species. This is almost certainly due to the lack of habitat, which not only means that spawning is likely to be less successful, but also that the fry have no refuge from the large number of small perch, which will eat them before they reach any size. Furthermore, the water will not support good populations of the zooplankton that all fish fry require to develop.

We therefore recommend that the carp population is thinned out. This should take the form of a netting and electrofishing operation to catch as many of the large (above 37 cm, approximately 2.5 kg) as possible. These can be sold or transferred to one of the other specimen lakes. It is important that a health check is completed before this operation takes place.

A new addition of around 100 carp of 0.4–0.7 kg (1.0–1.5 pounds), can then be introduced to support anglers' catches in the short term. These should, if at all possible, be sourced from the same farm that supplied the original stock of carp. This farm is still in operation and produces good-quality fish and has a good biosecurity programme.

A replanting scheme should be established to encourage submerged and emergent plants. The two shallow bays offer good areas to attempt to establish these plants. They should be cut off with pre-planted coir rolls across most of the entrance. These will help to stop the larger carp gaining access to this area and grubbing up the young plants.

Once these areas are established, they will make good spawning and fry habitat for the fish, particularly the roach. Two 2 metre × 2 metre floating islands should be placed in the middle of the pond. Again, these will provide important spawning habitat and fry refuges once they are established. Establishing further emergent vegetation along the bank margins should be encouraged.

This planting should allow a much greater range of habitats, which should promote natural spawning and fry survival. The increase in numbers of small fish should give the perch an additional, larger prey source, allowing them to grow to a larger size. These larger perch should then help to reduce the numbers of small perch.

The additional planted areas will also provide a larger 'natural larder' for the fish in the lake, ensuring faster growth rates and larger fish.

Glossary

Abiotic: Non-living.
Acute: Temporarily severe, quick.
Aggregate: A loose mixture of various rocks and minerals.
Anthropogenic: Derived from human activity.
Autotrophic: An organism capable of utilising inorganic sources of carbon, nitrogen, etc. as starting materials for life.
Benthic: Pertaining to the bottom of a waterbody.
Bio-engineering: The use of biological material, usually plants, to solve engineering problems such as bank erosion.
Biomass: Biomass is the weight of any given group of organisms present in the system at any one time.
Bloom: Relating to the sudden, massive increase in amount of planktonic algae.
Chitinous: Composed of or containing chitin, a long-chain polysaccharide.
Chronic: Deep seated or long lasting.
Cold blooded (poikilothermic): Animals whose internal body temperature varies with that of the surrounding medium.
Coir: Stiff coarse fibre from the outer husk of a coconut.
Community: A well-defined group of plants or animals clearly distinguishable from other similar groups.
Disease: A disorder or want of health, an ailment.
Ecosystem: The community of different species interdependent on each other together with the non-living components.
Endemic: Prevalent to a particular area or region.
Epiphyton: A community of plants living attached to other plants.
Eukaryotic: Organisms with cells that possess a membrane bound nucleus with DNA organised into chromosomes. Also have many cellular functions sequestered into membrane-bound organelles such as mitochondria.
Eutrophication: The enrichment of bodies of freshwater with inorganic plant nutrients.
Evapotranspiration: Loss of water from the soil or water body by evaporation (transpiration) from the plants growing there.
Fluvial: Of, or pertaining to rivers.
Habitat: The locality or environment in which an organism lives.

Heterotrophic: Organisms that cannot synthesise their own food. They must obtain organic carbon and nitrogen from other animals or plants.

Holistic: An analysis of the whole rather than specific parts. Recognition that the interactions between all parts of the ecosystem are important.

Invertebrate: Animals without backbones.

Lease: A contract granting use or occupation of property during a specified period in exchange for a specified rent.

Macrophytes: Large aquatic plant.

mg/l: Milligrams per litre.

Pathogen: A disease-causing organism.

Phytoplankton: Photosynthetic algae that drifts free in the water column.

Prokaryotic: Organisms that lack a membrane-bound nucleus or other internal organelles such as mitochondria. The bacteria and cyanobacteria (blue-green algae).

Riparian: Frequenting, growing on or living on the banks of water bodies.

Spoil: The waste material removed during excavations.

Stress: A reaction to conditions that are outside normal parameters.

Submerged plants: Plants that are normally on dry land but may be temporarily underwater during flooding.

Submersed plants: Plants that naturally grow under water.

Systemic: Throughout the body. Involving the whole body.

Tanalised: Description of wood that has been impregnated with particular preservatives.

Temperate: The climate of temperate regions is exemplified by short cold winters and longer warmer summers.

Topography: The surface features of an area.

Trophic: Connected to nutrition and feeding.

Vector: Any organism that acts as a carrier for a pathogenic organism and transmits it to a susceptible host.

Zooplankton: Animals (usually small invertebrates) that live free in the water column.

Further Reading

Part I: The Ecology of Still-Water Fisheries

Alabaster, J.S. and Lloyd, R. (1982) *Water Quality Criteria for Freshwater Fish*. Butterworths, for the Food and Agriculture Organization of the United Nations, London and Boston.

Bone, Q., Marshall, N.B. and Blaxter, J.H.S. (Eds) (1999) *Biology of Fishes*. Stanley Thornes Publishers Ltd., Cheltenham.

Bronmark, C. and Hanson, L.-A. (2005) *The Biology of Lakes and Ponds* (2nd edn). Oxford University Press.

Brown, L. (Ed.) (1993) *Aquaculture for Veterinarians*. Pergamon Press, Oxford.

Burgis, M.J. and Morris, P. (2007) *The World of Lakes: Lakes of the World*. The Freshwater Biological Association.

Ellis, A.E. (1988) *Fish Vaccination*. Academic Press, London.

Fitter, R. and Manuel, R. (1994) *Collins Photo Guide: Lakes, Rivers, Streams and Ponds of Britain and North West Europe*. Harper Collins.

Fryer, G. (1982) *The Parasitic Copepoda and Branchiura of British Freshwater Fishes. A handbook and key*. Freshwater Biological Association Scientific Paper No.46.

Hoole, D., Burgess, P., Bucke, D, and Wellby, I. (2001) *Diseases of Carp and Other Cyprinid Fish*. Wiley-Blackwell, Oxford.

Jobling, M. (1995) *Environmental Biology of Fishes*. Fish and Fisheries Series, Chapman and Hall.

Kabata, Z. (1979) *Parasitic Copepoda of British Fishes*. The Ray Society, London.

Khalil, L.F., Jones, A. and Bray, R.A. (1994) *Keys to Cestode Parasites of Vertebrates*. CABI International, St Albans.

Macan, T.T. (1974) *Freshwater Ecology*. Longman Group.

Maitland P.S. (2004) *Keys to the Freshwater Fish of Britain and Ireland*. Scientific Publication 62, Freshwater Biological Association. 248 pp.

Maitland, P.S. and Linsell, K. (2006) *Guide to Freshwater Fish of Britain and Europe*. Phillips, UK.

Mason, C.F. (1996) *Biology of Freshwater Pollution*. Longman Group.

Moss, B. (1988) *Ecology of Freshwaters*. Wiley-Blackwell.

Noga, E.J. (1996) *Fish Disease: Diagnosis and treatment*. Mosby Year-Book, St Louis, USA.

Pike, A.W. and Lewis, J.W. (1994) *Parasitic Diseases of Fish*. Samara Publishing, Dyfed, Wales.

Schaperclaus, W. (1991) *Fish Diseases (Vols I–III)*. Amerind Publishing, Delhi, India.

Schlotfeldt, H.-J. and Alderman, D.J. (1995) What should I do? A practical guide for the fresh water fish farmer. *Bulletin of the European Association of Fish Pathologists*, **15** (4) (Supplement): 60 pp.

Williams, H. and Jones, A. (1994) *Parasitic Worms of Fish*. Taylor & Francis, London.

Wooton, R.J. (1995) *Ecology of Teleost Fishes*. Fish and Fisheries Series, Chapman and Hall.

Davies C., Shelley J., Harding P., Mclean I., Gardiner R. and Peirson G. (2004) *Freshwater Fishes In Britain: The species and their distribution*. Harley Books.

Part II: Management of Still-Water Fisheries

Amend, D.F., Croy T.R., Goven B.A., *et al.* (1982) Transportation of fish in closed systems: methods to control ammonia, carbon dioxide, pH, and bacterial growth. *Transactions of the American Fisheries Society*, **111** (5): 603–611.

Bennett, G.W. (1962) *Management of Artificial Lakes and Ponds*. Reinhold Publishing.

Berka, R. (1986) Transportation of live fish – a review. *EIFAC Technical Paper*, **48**. 52 pp.

Boyd, C.E. (1982) *Water Quality Management for Pond Fish Culture*. Elsevier.

Brooks, A. and Agate, E. (1997) *Waterways and Wetlands: A practical handbook*. BTCV Handbooks.

Cefas (2005) *Risk Identification and Assessment of Non-Native Freshwater Fishes: Concepts and perspectives on protocols in the UK*. Cefas, Lowestoft.

Copp, G. H., *et al.* (2005) To be or not to be, a non-native freshwater fish? *Journal of Applied Ichthyology*, **21** (4): 242–262.

Defra (2007) *The Invasive Non-Native Species Framework Strategy for Great Britain – Protecting our natural heritage from invasive species*. Defra.

English Nature (2005) *Audit of Non-Native Species in Britain*. English Nature.

Environment Agency (2005) *Guidance for the Control of Invasive Weeds in or Near Freshwater*. Environment Agency.

FAO (1994) *Handbook On Small-Scale Freshwater Fish Farming*. FAO Training Series 24. FAO, Rome.

Hawke, C.J. and Jose, P.V. (1996) *Reedbed Management for Commercial and Wildlife Interests*. RSPB.

Heywood, V. (Ed), (1995) *Global Biodiversity Assessment*. Cambridge University Press.

Holt, J. and Watten, B. (1988) Applications of pure oxygen in fish culture. *Aquaculture Engineering*, 7: 397–441.

Horvarth, L., Tamas, G. and Seagrave, C. (2002) *Carp and Pond Fish Culture*. Wiley-Blackwell.

Horváth, L., G. Tamás and I. Tölg, (1984) *Special Methods in Pond Fish Husbandry*. Halver Corporation, Seattle. 146 pp.

Institute of Fisheries Management. (1991) *The Creation and Management of Pond Fisheries*. Institute of Fisheries Management.

Lawrence, J.M. (1968) Aquatic weed control in fish ponds. In *World Symposium on Warm-Water Pond Fish Culture. FAO Fisheries Reports* **44** (5): 76–91.

Newman, J. (1994) *Control of algae with straw*. Information sheet 3, Aquatic Weeds Research Unit.

Piper, R.G., *et al.* (1982) *Fish Hatchery Management*. Washington, DC, US Department of the Interior, Fish and Wildlife Service, pp. 348–371.

Seagrave, C.P. (1988). *Aquatic Weed Control*. Fishing News Books.

Seagrave, C.P. (2001) *The Sparsholt Guide to the Management of Carp Fisheries*. Mitchellwing Publications.

Sheperd, C.J. and Bromage, N.R. (1992) *Intensive Fish Farming*. Blackwell Science.

Swingle, H.S. (1947) Experiments on pond fertilisation. *Alabama Polytechnic Institute Bulletin* **264**: pp. 36.

Templeton, R.G. (1996) *Freshwater Fisheries Management*. Fishing News Books.

UK Biodiversity Research Advisory Group (2005) *Strategy for Non-Native Species Research.*

Wheeler, A.C. (2000) Status of the crucian carp (*Carassius carassius* (L)) in the UK. *Fisheries Management and Ecology*, **7**: 315–322.

Woynarowich, E. and Horváth, L. (1980) The artificial propagation of warm-water finfishes – a manual for extension. *FAO Fisheries Technical Paper* **201**: 138–147.

Part III: Legal and Social Framework

Arlinghaus, R., Cowx, I.G., *et al.* (2008) EIFAC code of practice for recreational fisheries. EIFAC Occasional Paper 42.

Carty, P. and Payne, S. (1998) *Angling and the Law*. Merlin Unwin Books.

Hickley, P. and Tompkins, H. (1998) *Recreational Fisheries, Social, Economic and Management Aspects*. Wiley-Blackwell.

Index